Shipping and the Environment

Karin Andersson · Selma Brynolf
J. Fredrik Lindgren · Magda Wilewska-Bien
Editors

Shipping and the Environment

Improving Environmental Performance in Marine Transportation

Springer

Editors
Karin Andersson
Shipping and Marine Technology
Chalmers University of Technology
Gothenburg
Sweden

J. Fredrik Lindgren
Shipping and Marine Technology
Chalmers University of Technology
Gothenburg
Sweden

Selma Brynolf
Shipping and Marine Technology
Chalmers University of Technology
Gothenburg
Sweden

Magda Wilewska-Bien
Shipping and Marine Technology
Chalmers University of Technology
Gothenburg
Sweden

ISBN 978-3-662-49043-3 ISBN 978-3-662-49045-7 (eBook)
DOI 10.1007/978-3-662-49045-7

Library of Congress Control Number: 2015959585

Printed on acid-free paper

This Springer imprint is published by SpringerNature
The registered company is Springer-Verlag GmbH Berlin Heidelberg

Foreword

This book provides a timely and focused contribution to broader understanding of environmental impacts and pollution prevention measures of maritime transport. As a researcher and colleague in the interdisciplinary effort to help identify, characterise, and address compelling challenges with regard to maritime vessel operations, I thank the editors and commend the authors and contributors for their timely and carefully developed text.

Readers may think I was invited to contribute this foreword because of two decades work pioneering research in several areas that are now mature enough to merit chapters in modern maritime texts such as these. Perhaps that explanation works. However, this foreword may also be considered as the reflections of a sailor, a merchant marine engineering officer, whose original training included how to operate very few devices designed primarily for pollution control. One of them was an engine-room periscope that could view a light bulb through a boiler stack only if the visible smoke was minimised. "The smoke periscope is a simple arrangement of mirrors and a light bulb which shines across the uptakes, giving the operator an indication of the opacity of the combustion gasses. It is difficult to distinguish between white and black smoke with the periscope" (source: Massachusetts Merchant Marine Academy training manual, circa 1986–99). In fact, the ability to minimise smoke was also a means to achieving more complete combustion, thereby improving fuel consumption. In retrospect, my research as a science and technology policy analyst focused on twenty-first-century innovation in maritime and freight systems is bound to those few years operating the world's largest moving power plants aboard merchant ships at the start of my career.

Similarly, this text connects shipping and maritime operations with current scientific, policy, and technology knowledge about our natural environment. The book may appeal to the next generation of maritime professionals, some who may staff watch aboard a new fleet of ships designed for environmental stewardship as well as economic service under challenging and changing sea conditions. The chapters may inform you scientists working to understand changing ocean and coastal environments where the impacts of shipping are part of the ambient conditions they observe. The text may also serve as a launching point for policy makers and maritime business leaders looking to navigate global shipping towards cleaner seas, skies, and shorelines. Mutual understanding is needed among those who design and operate integrated systems aboard ships and those who care about the coupled natural–human systems in our world.

Students and professionals using this text may share at least one attribute: the motivation to act upon good information to achieve better understanding and improve performance. This text is designed to assist today's mariners, environmental scientists, and regulatory administrators in this regard. By connecting a brief historic overview of shipping and environment with some fundamental introduction to environmental impacts, the book introduces pollution prevention measures focused on energy efficiency, discharge and emission controls, and tools for better environmental management.

One thing is certainly different since my days operating ship power systems: It is no longer sufficient to view environmental stewardship through a periscope. Today's professionals will see a changing ocean system, affected by increasing human activity along coastlines and shipping lanes. Some of us will witness and others of us will invent new and better ship systems that safely deliver cargoes with better attention to environmental stewardship. And these innovations will partly depend upon policy signals that identify the needs for timely new achievements in ship performance, port operations, and the world supply chains. This text contributes to a better understanding of shipping and environment, and expands the horizons for twenty-first-century shipping.

James J. Corbett Ph.D.,
Professor of Marine Science and Policy
Former Merchant Marine Officer, and Graduate of the California Maritime Academy

Preface

How come we wrote a book? I guess this is what you ask yourself when a large manuscript is ready for print. I have seen colleagues write textbooks a number of times during my years as a university teacher. Each time I have concluded that book-writing is a very large and time-consuming challenge and I have promised myself that I will never do it. Still—now the book is obviously there, and in some way it has happened. One conclusion is that you should not try to write a book on your own—the combined work of a group is what drives the work forward, increases quality, and provides challenging discussions. This book is really a cooperative project that has grown more or less by itself, although I do not know if we all tell the same story of how it started.

The writing process was initiated by the need for a textbook to be used in courses at the department of Shipping and Marine Technology. Furthermore, we had a need to meet the demand of providing information and answering questions from shipping companies and authorities. Before starting the main work, we had the opportunity to perform a "verification project" where we made a survey of need in target groups among students as well as in the shipping industry.

A book on shipping and the environment will involve a large number of disciplines and competences. The diversity in research focus and expertise of the people working at the department of Shipping and Marine Technology at Chalmers and at the department of Law at Gothenburg University was a good starting condition. The authors come from many different scientific backgrounds; engineers of different disciplines, marine scientists as well as scientists working with legal research, and we have all learnt a lot from each other during the project. The efforts in writing texts as well as in reading and discussing other author's text are greatly acknowledged. Thanks to all my co-authors.

There are also a number of people who have been reading parts of the text and been providing specific expertise and input. Thank you all.

Special thanks to my co-editors, Selma Brynolf, Fredrik Lindgren, and Magda Wilewska-Bien, for their never-ending patience and ambition in making the manuscript consistent and correct and also in gently reminding the rest of us that it is time to deliver. You are the heroes of the book project.

Important prerequisites for the book have been the Lighthouse maritime competence centre and the Chalmers Area of Advance Transport. The Lighthouse funding for senior scientists and doctorate students as well as the contribution to funding of senior scientists from the Area of Advance has given us the possibility to work on the manuscript. In the "verification project", we got practical support and funding by Innovationskontor Väst (Chalmers Innovation Office).

So, finally, when summer is over and the autumn storms are approaching the Swedish west coast, the manuscript is ready for print. We all hope that it will turn out to be useful to the readers and contribute to make shipping at least a little more sustainable.

Gothenburg
September 2015

Karin Andersson

Acknowledgments

The authors would like to thank a number of professionals from both Chalmers University of Technology and other places who generously gave their time and provided comments on the draft chapters and draft sections of the book including Gabriela Argüello (University of Gothenburg), Göran Bark (Chalmers University of Technology), Rickard Bensow (Chalmers University of Technology), Josefin Borg (Chalmers University of Technology), Francesco Di Natale (University of Naples), Erik Fridell (IVL Swedish Environmental Research Institute), Maria Grahn (Chalmers University of Technology), Paul Gilbert (The University of Manchester), Linus Hammar (Swedish Agency for Marine and Water Management), Mathias Janssen (Chalmers University of Technology), Roger Karlsson (SSPA), Niclas Karlsson (Cleanship Scandinavia), Henrik Pahlm (Chalmers University of Technology), Erik Røsæg (University of Oslo), Aslak Suopanki (Wärtsilä), and Erik Ytreberg (Chalmers University of Technology). The contribution of Andreas Hanning (Chalmers University of Technology) to the initial discussion, reviewing parts of the book and performing the verification study on creating the educational material, is acknowledged. The authors thank Ida-Maja Hassellöv (Chalmers University of Technology), for contribution to the initial discussions, defining the scope of the book and providing comments on parts of the book.

The authors thank Manuel Frias Vega (HELCOM) for adjusting the map of the maritime traffic in the Baltic Sea.

The authors acknowledge Caroline Pamp (Chalmers University of Technology), Marje Berzins (Chalmers University of Technology) and Jonas Gilbert (Chalmers University of Technology) who provided assistance with legal aspects. Furthermore, gratitude for sharing advice and experiences regarding textbook writing goes to Madeleine Miller and Katarina Streiffert (University of Gothenburg).

Sincere gratitude goes to various organisations and institutions for giving the authors permission to print some graphical material in the book.

The authors are grateful to Innovationskontor Väst for financing the verification project and Bo Norrman (Innovationskontor Väst) for valuable discussions on utilisation of research.

The Lighthouse base funding for senior scientists as well as for doctorate students together with base support from the Chalmers Area of Advance Transport has given us the opportunity to work on the manuscript.

Contents

Part I Introduction

Editors and Contributors

About the Editors

Prof. Karin Andersson Ph.D. is a professor in Maritime Environment at the Department of Shipping and Marine Technology, Chalmers University of Technology since 2009. Her scientific background is in Chemical Engineering (M.Sc., KTH 1975), Nuclear Chemistry (Ph.D., Chalmers 1983), and archaeology. After seven years as a consultant and researcher in environmental systems analysis, she was one of the founders of the research group for Environmental Systems Analysis at Chalmers in 1990. This group has been developing master's courses in environment/sustainable development and was pioneer in research on Life Cycle Assessment (LCA) in Sweden. In the mid-1990's a national competence centre, CPM (now Swedish Life Cycle Center) with participation of more than 10 large Swedish companies, was started with the group as host. Karin was also a vice president at Chalmers 2004–2007. Present research interests focuses on decrease of environmental impact and resource use by the technical systems in shipping.

Selma Brynolf Ph.D. is a researcher at the Department of Shipping and Marine Technology and the Department of Energy and Environment at Chalmers University of Technology conducting research about the global energy system with focus on shipping and marine fuels. She received her master of science in Industrial Ecology in 2009. After her master of science, she started Ph.D. at the Department of Shipping and Marine Technology at Chalmers University of Technology. The research involved environmental assessment of present and potential future marine fuels and she earned her Ph.D. degree in 2014. She has published papers within the area of marine fuels, life cycle assessment, and global energy system modelling.

J. Fredrik Lindgren Ph.D. is a researcher at the Department of Shipping and Marine Technology at Chalmers University of Technology. He studied marine biology at the Department of Marine Ecology, University of Gothenburg and received his master of science in 2003, doing work related to the settlement of barnacle larvas and the development of non-toxic antifouling paints. After his master of science, he continued to work in this area, before starting his doctoral studies. There he studied ecotoxicological effects of small but frequent oil spills and factors that can influence the effects of the spills, earning his Ph.D. degree in 2015.

He has published papers within the areas of marine biofouling, risk assessment, and ecotoxicology of oil.

Magda Wilewska-Bien Lic. Tech. is a Ph.D. student at the Department of Shipping and Marine Technology at Chalmers University of Technology. She has a Lic. Tech. degree in Environmental Inorganic Chemistry in 2004 and M.Sc. in Applied Environmental Measurement Techniques. She maintains a professional interest in environmental science and in particular management of wastes.

Contributors

Maria Zetterdahl M.Sc. is a Ph.D. student at Chalmers University of Technology working with particle emissions from ship operations and investigating the impact of fuel type and abatement technologies on the emissions. The studies are related to the implementation of ECAs considering sulphur and NO_x. She has a master of science in Biology from Swedish University of Agricultural Sciences.

Francesco Baldi M.Sc. is a Ph.D. student at Chalmers University of Technology in Gothenburg, Sweden, working with modelling, analysis, and optimisation of ship energy systems. He will defend his thesis in May 2016. Before starting his doctoral studies, he obtained his Master's degree in Energy Engineering from the University of Bologna including experiences at the UPMC, Areva and Cemagref in France. His research interests focus on modelling and optimisation of energy systems.

Cecilia Gabrielii Senior Lecturer Ph.D. has a M.Sc. in Chemical Engineering and a Ph.D. in the area of refrigeration and heat pump technology, focusing on the replacement on ozone-depleting refrigerants. She works at Chalmers University of Technology as a lecturer mainly in the Marine Engineering Programme and as an assistant supervisor for Ph.D. projects aiming at increasing the efficiency of a ship's energy system.

Lena Granhag Ph.D. in Marine Ecology working with impact of shipping on the marine environment at Chalmers University of Technology. Her main interest is in biofouling on ship hulls and its mitigation. She also works with questions related to ballast water management and invasive species in marine environments.

Hannes Johnson M.Sc. is a Ph.D. student interested in organising policy aspects of energy efficiency in the maritime field. He has managed action research projects on the implementation of energy management systems in shipping companies. His Ph.D. thesis is due early 2016.

Hanna Landquist M.Sc. is a Ph.D. student in environmental risk assessment at Chalmers University of Technology with a background in civil engineering. Her research is focused on risk analysis and decision support for mitigation of potentially polluting shipwrecks.

Philip Linné LL.M. is a Ph.D. student in Environmental Law at the Department of Law, School of Business, Economics and Law, University of Gothenburg. His Ph.D. thesis examines the international regulation of air emissions from ships in the form of sulphur oxides.

Mathias Magnusson Ph.D. has a M.Sc. in Naval Architecture and a Ph.D. in the area of NO_x abatement technologies, with special focus on catalytic exhaust aftertreatment systems for marine diesel engines. After finishing his Ph.D., he started to work at the Advanced Technology and Research division at Volvo Trucks, with specific focus on exhaust aftertreatment system.

K. Martin Eriksson Ph.D. has a Ph.D. in Environmental Sciences from University of Gothenburg. His primary research areas are Ecotoxicology and Microbial Ecology. Martin's research has focused on the effects of contaminants on marine organisms and communities, and on the structure and function of fouling communities. Currently, Martin is working on environmental effects of contaminants released from shipping and holds a researcher position at Chalmers University of Technology.

Kent Salo Senior Lecturer has a Ph.D. in Natural Science, specialising in Chemistry with a focus on physical properties and processes of secondary organic aerosols. He works as a senior lecturer and a researcher at Chalmers university of Technology on the topic of the environmental impact from shipping with a focus on emissions to the atmosphere.

Erik Svensson Ph.D. has a background in environmental science and has since 2008 specialised in issues of international environmental cooperation and policy. His analytical work encompasses different fields of social sciences and natural science and his Ph.D. thesis focused on different perspectives to understand decisions by the International Maritime Organization (IMO).

About this Book

This book focuses on the interaction between shipping and the natural environment and how shipping can strive to become more sustainable. The book is designed in such way that the reader can either read the chapters in order to get a broad and complete picture or concentrate on individual chapters that are written as independent parts with clear links in-between. This book is divided in four parts: Introduction, Environmental Impacts, Pollution Prevention Measures, and Outlook.

The first part *Introduction* introduces ships, shipping, and environmental impacts of ship operations in Chap. 1. This is followed in Chap. 2 where the reader gets a background on the Earth's major systems and how they function. Environmental issues connected to human activities are also included. This chapter is especially intended for readers without deep knowledge in environmental science. The last introductory chapter covers regulation of pollution from ships and is intended primarily at other groups than legal professionals.

The second part *Environmental Impacts* guides the reader in the emissions and discharges of ships and the connected environmental impacts. Chapter 4 describes the discharges to the sea while Chap. 5 describes the emissions to air. Anthropogenic noise is described in Chap. 6, and Chap. 7 deals with issues connected to shipping infrastructure, marine spatial planning, and shipwrecks.

The third part *Pollution Prevention Measures* gives the reader a detailed overview on ways to minimise the environmental issues connected to shipping. The section starts with a chapter introduces how to work with environmental management which is followed by a chapter about methods and tools for environmental assessment (Chap. 9). Chapter 10 guides the reader in how to reduce environmental impacts by increasing energy efficiency or changing fuels. The last chapter in this part starts from the emissions and discharges described in Part II and presents solutions to reduce or hinder these.

The fourth part *Outlook* includes the last chapter in the book which summarises and concludes the book and discusses ways forward to improve the environmental performance of shipping.

Part I
Introduction

Shipping and the Environment

1

Karin Andersson, Francesco Baldi, Selma Brynolf, J. Fredrik Lindgren, Lena Granhag and Erik Svensson

Abstract

Humans have always had a close relationship with the aquatic environment, including the early use of the sea for food harvesting and communication. Today, the sea is an important component of the transportation system, with large amounts of cargo and passengers. This chapter provides a short introduction to ships and shipping, focussing primarily on commercial ships; nonetheless, many of the emissions, impacts and measures discussed throughout this book are common to other sectors, such as leisure, research and fishing. This chapter also introduces the environmental impacts related to ship operations. Ship transportation has increased tremendously since the industrial revolution, which has resulted in increased emissions due to shipping and increased stresses on the environment. However, this trend is not only related to shipping. Currently, there are several

K. Andersson (✉) · F. Baldi · S. Brynolf · J.F. Lindgren · L. Granhag
Shipping and Marine Technology, Chalmers University of Technology, Gothenburg, Sweden
e-mail: karin.andersson@chalmers.se

F. Baldi
e-mail: francesco.baldi@chalmers.se

S. Brynolf
e-mail: selma.brynolf@chalmers.se

J.F. Lindgren
e-mail: f.lindgren@chalmers.se

L. Granhag
e-mail: lena.granhag@chalmers.se

E. Svensson
Gothenburg, Sweden
e-mail: e-svensson@outlook.com

K. Andersson et al. (eds.), *Shipping and the Environment*,
DOI 10.1007/978-3-662-49045-7_1

warning signs that we are not taking care of the Earth and its ecosystem in a sustainable manner, that the Earth's ecosystems are degrading and that natural capital is being exploited, e.g., by the burning of fossil fuels. The marine industry is a component of our society; similar to all industry sectors, it contributes to unsustainable patterns in our society. Although the marine industry is a contributor to these problems, it can also be part of the solution, yet several challenges must be addressed. Sustainability and related concepts, such as ecosystem services, planetary boundaries and resilience thinking, could be used as guidance in addressing these challenges.

Humans have always had a close relationship with the aquatic environment. Indeed, a scientific discussion debates whether the first humans evolved in a dry land environment, on the savannah, or in shallow water environments (as the "water man" or "aquatic ape") [1]. With respect to environmental awareness, the sea has come into focus relatively late compared with other natural areas. Independent of this observation, the sea has served as an important transportation route and a source of food and recreation throughout history. In a world where more than 70 % of the surface is covered by oceans, our interaction with and dependence on the sea in numerous aspects is obvious.

1.1 Man and the Sea

Early use of the sea consisted of food harvesting and communication. Increased trade and the population growth in Europe launched global explorations and led to "discovery" of the New World and new communication routes to Asia. A romantic view of nature in the 19th century evoked a new interest in the sea in art and in recreation. Much of this interest was related to the sea in terms of something to look at and enjoy. In the mid-20th century, science made discoveries below the surface of the sea possible, and public interest in diving and learning about the sea's ecosystem developed. These developments also led to additional observations of adverse effects on the environment and an increased awareness of the marine environment [2]. The increased use of oil for propulsion and increased global transportation of oil during the 20th century led to discharges of oil that were highly visible and produced demands for safer and less damaging shipping. The formation of what later became the International Maritime Organization (IMO) was a result of the need to enact international regulations for safety at sea and to prevent accidental oil discharges.

The fact that most of the water volume around us is not visible has rendered the events occurring below the sea's surface both frightening and unknown. There are many myths about strange beings, large sea monsters, mermaids, and "bottomless" lakes. An illustration from 1572 in the maritime atlas "Carta Marina" [3] by the Swedish cartographer Olaus Magnus that portrays creatures believed to be living in the sea gives an impression of the myths of the time (Fig. 1.1). In addition, the

Fig. 1.1 Map of the North Sea and the Baltic sea 1572. From: "Carta marina, opus Olai Magni Gotti Lincoensis, ex typis Antonii Lafreri Sequani", Rome 1572, National Library of Sweden, KoB, Kartor, 1ab

belief that the sea is greatly resilient and is a suitable place for discharging waste has survived for a long time. Organised dumping of waste at sea was a common practice around the world, including Europe, until the 1970s. An instructional clip from Swedish Television produced in the early 1960s and aimed at leisure boat owners provides the advice to attach a stone to on-board waste parcels before sinking them into the sea. The amount of waste ammunition and chemicals dumped into the sea after World War II is still a potential problem for fishing and installations at sea. Even today, large amounts of waste end up in the oceans on purpose or from uncontrolled sources. Recently, problems related to "micro-plastics" and "ghost fishing nets" have garnered more attention (see also Sect. 4.5).

In shipping and in other maritime activities, the sea remains close to working people, and the relationship between activities and environmental impacts is sometimes highly visible, see Fig. 1.2, although some shipping impacts occur far from the source of emissions or activity.

This book focuses on the interaction between shipping and the natural environment and discusses how the use of the oceans by humans is affecting the environment as well as and how these activities can be made more sustainable.

Fig. 1.2 Examples of observations of impacts on and emissions into the sea. *On the left*, algae belts have formed due to excess nutrients in the Baltic Sea; *on the right* an exhaust plume from a ship on the horizon. *Photo* Karin Andersson

1.2 Ships and Shipping

To discuss the relation among shipping, the environment, and sustainable development, it is useful to define certain terminology. The main focus of shipping is on commercial ships, although many of the emissions, impacts and measures are common to other sectors, such as leisure, research, and fishing. The different regulations also require definitions of a ship. Different definitions exist, although according to the glossary of the US Navy, a "boat" usually refers to small vessels that are often open, whereas "ships" are vessels of considerable size that are intended for deep-water navigation [4]. A ship can also be defined by its size: vessels longer than 12 m and wider than 4 m are referred to as ships, and smaller vessels are known as boats [5].

Common ship types can be identified according to their type of use:

- *Container ships*: These vessels carry most of the world's manufactured goods and products in standardised containers that also can be transported by rail and truck. These ships are usually scheduled liner services.
- *Bulk carriers*: These vessels transport unpacked cargo in large volumes. The cargo might be grain (e.g., wheat, oats, and maize), products such as concrete, or raw materials (e.g., iron ore, limestone and coal).
- *Tankers*: The vessels transport liquids, such as crude oil, chemicals and petroleum products.
- *Ferries*: Ferries usually perform short journeys that carry mixtures of passengers, cars and commercial vehicles. Most of these ships are RoRo (roll on–roll off) ferries, in which vehicles can drive straight on and off. Ferries that combine passengers and RoRo transport are often referred to as RoPax.
- *Cruise ships*: Cruise ships have different sizes, and several thousand passengers and crew are common on these vessels. These ships combine transport with the role of 'floating hotels'.

Many other types of ships operate regionally or locally. One size limit is up to 500 passenger vessels, which includes road ferries and public transport/shuttle ferries. Vessels might be intended for special purposes, such as pilot boats, fishing vessels, icebreakers and military vessels. Different ships are also adapted for transport on inland waterways in areas with rivers and canals. No universally applicable definitions of ship types exist, although IMO provides a list of ship types mentioned in various IMO instruments (see Chap. 3) [6].

1.2.1 The Infrastructure: Fairways, Canals and Ports

The infrastructure for ships covers a rather large proportion of the open sea, where there is no need for the specific construction of infrastructure and it is available for free. However, the need exists for connections to land. Close to land, fairways occur, which are commonly marked, as do ports for ships to enter. Inland infrastructure also exists in the form of canals and sluices.

In terms of environmental impacts, the fraction of shipping that occurs close to land accounts for an important contribution. Ports are often located in or close to large cites, and the sea traffic in and out of ports can affect the population in the area due to emissions to the air and water and from noise and waves.

In contrast to other shipping activities, the location and construction of ports and fairways are often regulated and assessed from an environmental perspective as a component of land-based activities in environmental impact assessments (EIAs, see Chap. 9) or in planning processes. The activities at sea are regulated in international, regional and national frameworks (see Chap. 3).

1.2.2 Marine Spatial Planning

Spatial planning is an essential process for managing land in several areas of the world. The use of spatial planning was triggered by the industrial revolution, when coal became an important raw material. People began to aggregate in areas close to a site of excavation, and villages soon became overcrowded with a subsequent lack of water or contaminated water supplies because no infrastructure was available to accommodate the rapidly increasing population. The need for and advantages of proper spatial planning soon became obvious. Today, terrestrial land-use planning and management is standard. This process of future spatial development and planning has not been implemented in marine areas (with a few exceptions). This lack of implementation does not mean that the ocean is fully unregulated or unmanaged; for example, shipping lanes, military zones, and marine protected sites exist, although only within individual economic zones. However, few frameworks have integrated the regulation and management of all activities occurring within an area. Marine spatial planning (MSP) is a tool that can be applied to avoid the problems that can arise when multiple activities occur simultaneously within a marine area [7]. The large development of wind power and fish, shellfish or biomaterial production at sea in coastal areas creates a challenge in terms of competition for space with shipping; a MSP document that covers these regions is important to avoid future conflicts. MSP is further discussed in Chap. 7. Within the European Union, legislation on a common framework for maritime spatial planning was adopted in 2014 [8].

1.2.3 What Types of Cargo Are Transported by Ships, and Where Is the Cargo Transported?

Shipping is a means of transport by different types of vessels, as mentioned previously, and the variety of cargo types and passengers is quite large. There are nearly 50,000 registered ships with over 1000 gross tons dead weight (GT DW), including offshore drilling and offshore production units [9] (military vessels and fishing vessels are not included in the statistics). If also smaller ships are included, over 70,000 ships with more than 400 GT DW are registered [10]. The majority of large ships transport goods of different types that are either packed in containers and tanks or handled as bulk goods in cargo holds. Common tank cargo includes petroleum products and chemicals. Bulk transport is used for various solids, such as grain, minerals or ores.

The use of container transport for various goods is a growing sector; more than 6000 large ships are used in international trade [11]. By value of transported goods, this segment is the largest, with over 50 % of the value of goods transported by sea [12]. In terms of fuel consumption, container and oil/gas transport are the largest categories, which indicates that oil and gas transport is a large sector in terms of volume, although it constitutes only approximately 20 % of the economic value.

The large cargo trade routes are located between continents, primarily North America to Europe and to South East Asia.

Possible future routes that could reduce transport distances while increasing environmental impacts are located in the Arctic region. Arctic shipping raises specific environmental issues, as discussed in Box 1.1.

In the passenger sector, the cruise ship industry is a growing sector, accounting for nearly 300 cruise ships and a total annual passenger capacity of 21 million in 2014 [13]. Together with RoPax ships (passenger/car ferries), this sector constitutes the largest sector involved in passenger transport in terms of fuel consumption.

However, sea transport includes local transport and a wide range of other applications; those mentioned above are important examples.

Box 1.1 The Arctic

The Arctic region has in recent years gained much attention due to its warmer climate and decreasing ice cover. The conditions for shipping routes in the Arctic area are changing, and as global energy reserves are declining, natural resources such as oil and gas in the Arctic are being explored.

There are today several definitions of the Arctic area: the area north of the Arctic Circle; the area north of the isotherm, with a mean temperature of 10 ° C in July; or the area north of the tree-line. The Arctic coastal states with maritime jurisdictional claims are Canada, Denmark (Greenland), Iceland, Norway (Jan Mayen, Svalbard) the Russian Federation and the United States (Table 1.1, [14]) (see Chap. 3 for definitions). Together with Sweden and Finland, the Arctic coastal states and six representatives from the Arctic indigenous communities comprise the Arctic Council, a forum formed in 1996 that works towards the responsible development of the region.

The ice cover in the Arctic, which reaches its each year maximum in March and minimum in September, has in recent years decreased (recorded since 1970, [15]). There are different modes of transport in the Arctic. In the trans-Arctic transport, ships use either the Northern Sea Route/North East Passage or in some cases, the Northwest Passage for routes across the Arctic. In destinational transport, a ship goes to one Arctic destination; in intra-Arctic transport, ships are in route between destinations within the Arctic. One of the major driving forces for increased trans-Arctic transport is the shorter routes for shipping. As opposed to going through the Suez Canal, the distance can be shortened by 40 % by taking the Northern Sea route from Rotterdam in Holland to Yokohama in Japan, leading to savings in both time and fuel consumption.

Table 1.1 Arctic coastal state maritime jurisdictional zone claims. Data from the Arctic Council [14]

Arctic states	Territorial sea		200 NM zones		
	3 NM	12 NM	EEZ	Extended fisheries protection	Fisheries protection
Canada		x	x		
Denmark (Greenland)	x		x		
Iceland		x	x		
Norway		x	x		
-Jan Mayen		x		x	
-Svalbard		x			x
Russian Federation		x	x		
United States		x	x		

1.3 Sustainability and Shipping

The human population has increased by more than a factor of 10 since the industrial revolution, and it is expected to continue to increase to approximately 9 billion people by 2050. The standard of living for most people on Earth has improved during this period due to technical and social innovations, economic growth and international collaboration and trade. However, approximately one billion people still live in poverty. Several signs have also emerged that humans are not taking care of the Earth and its ecosystem in a sustainable manner. For example, we are consuming and producing an increasing number of products, leading to large energy and material requirements. We are degrading the Earth's ecosystems and exploiting the Earth's natural capital, e.g., by burning fossil fuel, which emits carbon dioxide to the atmosphere. Additional information on human impacts and environmental issues can be found in Sect. 2.7. The gaps between the richest and the poorest people on Earth are increasing. The marine industry is a component of our society. Similar to all industry sectors, it contributes to unsustainable patterns in our society. Although this industry is a contributor, it can also act as a component of the solution.

The following question is important: "What is sustainability?" Sustainability is a mainstream concept that is often used as an equivalent of all that is good and desirable in society. Concepts such as sustainable development, resilience thinking, socio-ecological principles and planetary boundaries can be helpful to place the Earth on a sustainable track. These concepts are introduced in the following sections and summarised in Box 1.2.

1.3.1 Sustainability and Sustainable Development

Sustainable development is a global goal that gained international attention due to the Report of the World Commission on Environment and Development in 1987, which is also known as the Brundtland Report [16]. This concept is related to a series of normative ideas that include protecting the environment, promoting human welfare (especially the urgent development needs of the poor), concern for the well-being of future generations, and public participation in environment and development decision-making [17]. However, sustainable development and sustainability are terms that lack consensus and suffer from a variety of different and vague definitions [18, 19]. Key questions for a relevant definition are provided below [19]:

- What is intended for sustainability?
  - Nature (earth, biodiversity and ecosystems)
  - Life support (ecosystem services, resources and environment)
  - Community (cultures, groups and places)
- What is intended for development?
  - People (child survival, life expectancy, education, equity, and equal opportunity)
  - Economy (wealth, productive sectors and consumption)

  - Society (institutions, social capital, states, and regions)
- For how long?
  - For example, 25 years or forever

The most common international definition of sustainable development is "*development that meets the needs of the present without compromising the ability of future generations to meet their own needs*", which was presented in the 1987 Brundtland Report [16]. Four primary characteristics of sustainable development also have been derived from the Brundtland Report: (1) safeguarding long-term ecological sustainability, (2) satisfying basic human needs, (3) promoting intra-generational equity, and (4) promoting inter-generational equity [20]. Several secondary characteristics are also important for sustainable development, e.g., preserving nature's intrinsic value, endorsing long-term effects, promoting public participation, and satisfying aspirations for an improved quality of life [21].

The importance of *safeguarding long-term ecological sustainability* is expressed in the Brundtland report, e.g., through such statements as, "At a minimum, sustainable development must not endanger the natural systems that support life on Earth: the atmosphere, the waters, the soils, and the living beings" [16] and "There is still time to save species and their ecosystems. It is an indispensable prerequisite for sustainable development" [16]. This characteristic has its origin in ecology and represents the conditions that must be present for the world's ecosystems to sustain themselves over long periods of time.

Satisfying basic human needs is at the core of the Brundtland definition of sustainable development. What are the basic human needs? The Brundtland Report mentions food, water, sanitation, clothing, shelter, energy and jobs as essential needs [16]. Other than these basic needs, aspirations for an improved quality of life can also be met as long as they do not endanger long-term ecological sustainability.

Promoting intra-generational equity and inter-generational equity is also at the core of the Brundtland definition, which emphasises the needs of the current and future generations. All previous and future generations share the Earth; therefore, each generation must pass on the Earth and its natural resources to the next generation in at least as good of a condition as they received them (inter-generational equity). It is argued in the Brundtland Report that social equity between generations "must logically be extended to equity within each generation" [16]. The allocation of resources among all members of a single generation should also be guided by fairness. The Brundtland Report also notes that "a world in which poverty and inequity are endemic will always be prone to ecological and other crises" [16].

Sustainable development is commonly represented as three pillars: *economic*, *social* and *environmental*. Another method used to visualise sustainable development uses the concept of carrying capacity, which represents how both economy and society are constrained by environmental limits (Fig. 1.3).

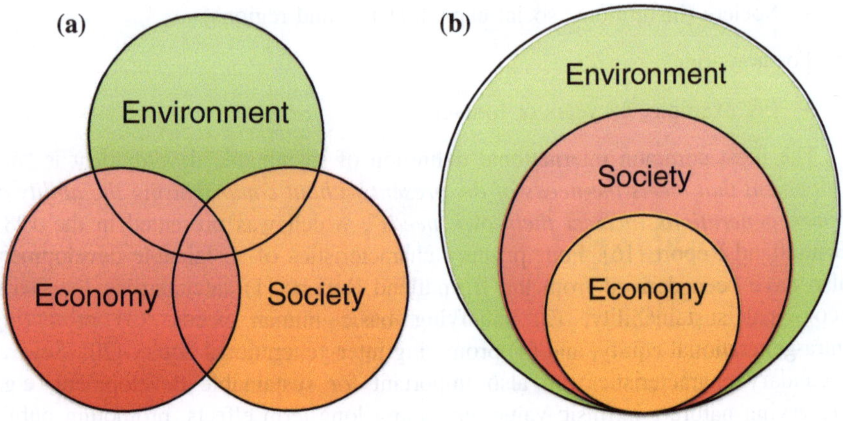

One of the most common views on sustainable development

Another perspective – carrying capacity
The environment sets the outer limits. It is not sustainable development if it is not ecologically sustainable

Fig. 1.3 Visualisations of sustainable development showing the three pillars of sustainable development as overlapping circles (**a**) and the environment, as the outer limit for society and the economy, representing the carrying capacity (**b**) of the Earth's systems

1.3.2 What Is an Environmental Concern?

The extent of human activities is increasing with the population, and changes in the natural environment and in the use of resources are also increasing. Humans are a natural component of the environment; thus, human activities should be a natural component of the environment. However, there is an increased concern over the negative effects of human activities in public debate and legislation.

In contrast to other species, humans have developed aspects that differ in nature, such as the large-scale use of tools, the creation of products ("artefacts"), and the development of a culture with an exchange of information and thoughts, including ethical and religious aspects. These developments have not only made the potential impact on the natural environment from human activities large but have also enabled the emergence of views on acceptability in terms of environmental impacts and resource use. The discussion of negative impacts on the environment is not new, although it has been increasing since World War II, with large public debates and legislations appearing in many places around the world. The early discussions on environmental issues were primarily related to the land environment, freshwater systems, such as lakes and rivers, and threats to human health. For a long time, the ocean was viewed as a place for dumping waste and an environment that was resilient to human impact. The only concern was over oil spills, which could impact sea birds and contaminate beaches. The potential impact "under the surface" was less known or observed.

Before discussing environmental impacts, it is necessary to define what is meant by environmental concerns and priorities. When the global population was lower, nature could easily recover from the effects of human activities, and the visible effects from use of non-renewable resources were small. An increased population means that the use of natural and renewable resources has become increasingly intense, and nature might face difficulties in recovering. Additionally, non-renewable resources have become depleted. Even in a situation with few inhabitants, the effects of the use of nature can be observed. What we consider as "nature" today is to a small extent untouched by humans, although it is in fact a product of culture. We typically do not regard farmed land or planted forests as environmental problems; instead, these regions are viewed as natural components of our environment. This viewpoint means that traces of human activities in nature are not viewed as an environmental problem. When will they become a problem?

Changes in the environment from the state untouched by humans are not sufficient to constitute a concern or problem; thus, concerns must stem from another source. That source is a general agreement in society that an effect or change is unwanted. At that point, it becomes an environmental problem. A sociologist would say that environmental problems are a "social construction", meaning that nature does not have an opinion on this topic and that the values come from society. This

Fig. 1.4 Different views on nature may cause conflicts. *Photo* Karin Andersson

philosophy implies that the process for agreement on what constitutes an environmental concern is long and involves stakeholders with different perspectives and priorities. Certain priorities stem from necessities, such as prioritising the growth of food and preventing disease before protecting nature, whereas others might originate from the prioritisation of either economic or natural preservation interests. These differences can cause conflicts, as illustrated in Fig. 1.4.

Additionally, the process of observing a change in nature, identifying the change as a problem that should be handled, and enacting different measures (such as regulations) can be extensive and involve many areas of society, including international bodies. One common observation is that the public debate and opinion process might be much faster than the reaction from society in the form of regulations. This aspect can force changes in technology or behavioural changes before they are requested via regulations.

1.3.3 Ecosystem Services

Human society depends on the Earth and its ecosystems. In many ways, the services obtained directly or indirectly by humans from ecosystems are known as ecosystem services. Examples of ecosystem services include the maintenance of hydrological cycles, cleaning of air and water, biological diversity, seafood, and recreational services [22]. A diverse and healthy environment is important to human well-being, and protecting the ecosystems and the services they provide is necessary for life. Three different types of ecosystem services are generally acknowledged: (1) provisioning (which includes food and bio-energy), (2) regulating (which includes the regulation of waste and pollution or of the physical environment), and (3) cultural services (which can be symbolic or intellectual/experience-based) [23].

In addition to providing waterways for the shipping industry, the ocean provides other functions or ecosystem services. The ocean serves as a source of food and natural resources, participates in the ecological and geochemical cycling of elements, and also provides a place for recreation. It is estimated that approximately 10 % of human protein intake comes from the sea. Minerals, such as salt (NaCl), are harvested from the sea, and life in the oceans also provides a source of oxygen to the atmosphere. The surface layer of the oceans absorbs approximately half of the heat radiating from the Sun to the Earth and distributes it around the world, thus playing a major role in determining the climate on our continents. Recreation and tourism have developed into a multibillion Euro and fast-growing industry [24].

The concept of ecosystem services can be used as a tool to solve environmental issues because these services demonstrate the importance of the ocean to the general public and provide arguments in policymaking debates [25]. Comparing the values of coastal and open ocean areas is difficult due to the large range of various ecosystems in the two areas. However, the most biologically productive zones and most of the world's fisheries are located in coastal areas, where human impacts are greatest in general [26, 27]. In addition, many important shipping routes pass through coastal areas.

1.3.4 Planetary Boundaries

The Earth's environment has been relatively stable for the last 10,000 years, making it possible for humans to develop, thrive, settle, and invent agriculture and industrialisation. This stable state, which is known as the Holocene, might be threatened due to impacts from human actions. The Earth is severely affected by the activities of humans; some scientists have claimed that we have entered the Anthropocene, a human-dominated geological epoch [28]. To avoid a shift from this environmentally stable period in Earth's history, a research group has developed the concept of *planetary boundaries* [29, 30]. Planetary boundaries are boundaries that we cannot cross if we want to sustain the Earth in its current stable state. The boundaries are human-determined values of the control variables set at a "safe" distance from a dangerous level or threshold [29]. A threshold or a tipping point is a point in a system that will cause the system to react in an abrupt and non-linear manner if crossed and will most likely result in irreversible changes. The planetary boundaries and selected other environmental issues of special importance for ship operations are described in Sect. 2.7.

1.3.5 Resilience Thinking

Resilience thinking addresses the dynamics and development of complex systems, such as ecosystems and societies. Resilience can be simply described as the long-term capacity of a system, e.g., an ocean, port or economy, to address change and continue to develop [31]. A resilient system will maintain the same "identity" or essentially the same functions, structure and feedback during periods of change. The term resilience was originally used to understand the ability of ecosystems to persist during perturbation, although it is now used as a wider concept for all types of complex systems [32]. Ensuring resilient systems can contribute to a more sustainable Earth.

Box 1.2 Some concepts related to sustainability

Biosphere: The biosphere contains all of Earth's living things, including all microorganisms, plants, and animals, together with the dead organic matter produced by them (see also Chap. 2).

Carrying capacity: The number of people that can be supported by the earth; describes that economy and society are constrained by environmental limits (see Sect. 1.3.1).

Circular economy: An economy built on the concept of reuse, recycling and refurbishment of materials and products.

Decoupling: In environmental context, the ability of an economy to grow without corresponding increases in environmental pressure.

Ecosystem: A community of organisms interacting with each other and with their physical environment.

Ecosystem services: The benefits that people obtain from ecosystems, e.g., provisioning of clean water, decomposition of wastes and fisheries (see Sect. 1.3.3).

Planetary boundaries: Planetary boundaries are boundaries that we cannot cross if we want to sustain the Earth in its current stable state. The concept as developed by a group of researchers in 2009 (see Sect. 1.3.4 and Sect. 2.7).

Rebound effect: The effect of increased consumption when prices are decreasing due to energy efficiency measures. For example, a more energy-efficient car might be used more frequently because the cost of fuel per kilometre driven is lower. Therefore, the increased consumption might offset the energy savings that could otherwise be achieved. This concept is important to consider if using energy efficiency as a strategy to reduce environmental impact.

Resilience: Resilience can be simply described as the long-term capacity of a system, e.g., an ocean, port or economy, to address change and continue to develop (see Sect. 1.3.5).

Sustainable development: "Development that meets the needs of the present without compromising the ability of future generations to meet their own needs" Defined in the "Brundtland report" [16] (see Sect. 1.3.1).

1.4 Ships and Their Environmental Impacts

Before discussing the various impacts that ships have on the environment, it is important to provide a short background on ships and how they interact with the maritime environment. The main environmental impacts associated with shipping are shown in Fig. 1.5.

The following chapters describe the different types of environmental impacts from ships. Chap. 4 focuses on *discharges to the sea* and their impacts on the marine environment, and Chap. 5 addresses *emissions to the air* and their impacts on the atmosphere. However, for a correct understanding, it is important to discuss introductory information on how ships are built and the reasons for the presence of components that generate pollution.

A ship is a vessel for use at sea that has a hull and can be steered, e.g., by a rudder. A ship's mission can vary substantially depending on the ship, such as the transportation of passengers or goods through international waters, servicing of other vessels, exploiting the sea in the form of fishing, or building underwater pipelines. The different systems on a ship must be able to perform the functions that are necessary to fulfil its mission.

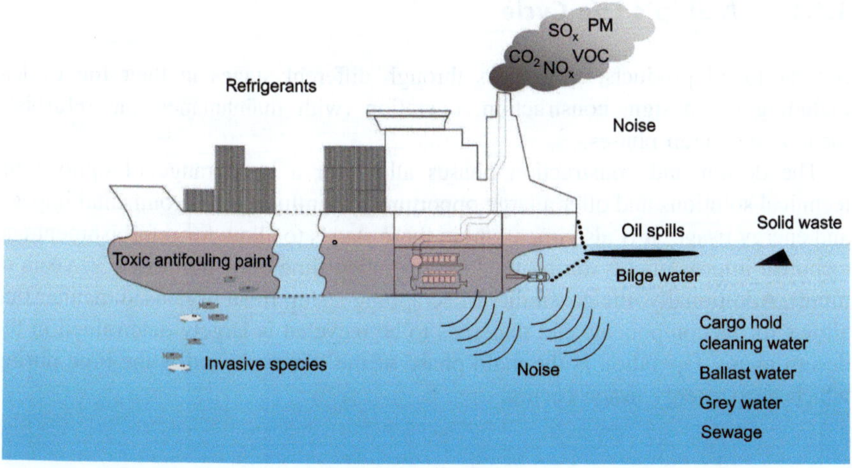

Fig. 1.5 Environmental impacts of marine transportation during the use of a vessel

Even if absolute generality can rarely be obtained, it is reasonable to state that every ship must be able to *provide mobility*. This basic function is provided by the propulsion system and by several additional systems related to the functions of steering, navigation, and anchoring. The propulsion system and its main components are described in Sect. 1.4.3.

Because a ship is assumed to perform a specific mission, this capacity will often require specific *operational functions* related to the specific mission. This functionality involves the need for given systems that are largely dependent on the mission, i.e., container cranes for small container ships, cargo pumps for tankers, equipment for handling fishing gear for trawlers, and kitchens and sanitary systems for cruise ships. Even if the large variety of possible operational functions does not allow us to specifically address each and every of them, these functions are often associated with a large consumption of heat or electric power and, sometimes, with the handling of hazardous material.

Every ship has several on-board operations that must be directly performed by humans. Therefore, the crew must be provided with *hotel facilities* that fulfil the basic needs of accommodation, food, and services. Additional details on this topic are provided in Sect. 1.4.4.

Finally, several *general support functions* must be performed, such as providing electric and hydraulic supplies, fuelling and lubrication, and heating and cooling. Selected details on the machinery involved in these functions are given in Sect. 1.4.5.

1.4.1 A Ship's Life Cycle

Similar to all products, ships pass through different stages in their life cycles, including the design, construction, operation (with maintenance and refurbishments), and scrap phases.

The design and construction phases allow for a large range of options for technical solutions and offer a large opportunity to influence environmental impacts and energy usage. It is also important in these stages to allow for refurbishment and technical improvements during the long operation time of a ship (often 30 years or more). Additionally, the possibility of scrapping a ship in an organised manner that allows for its components and materials to be recycled is largely determined in the design phase. Operation is the main phase of the life cycle and is the time during which most energy usage occurs.

1.4.2 The Hull and Ship Structure

The main function of a ship is to safely carry its cargo, crew and passengers. Therefore, a ship must be a safe and trustworthy vehicle that can handle various sea states and reduce damage in the case of accidents.

The main structure of a ship is the hull, which provides a carrying platform and protection against the environment. The hull must be resistant to loads of different types and intensity over its entire lifetime and to possible collisions. These requirements translate into different solutions and technologies depending on the ship type and its trade.

Ships sailing on international routes, particularly in the North Atlantic Ocean, tend to be specially loaded as a consequence of frequent heavy seas, and they must meet higher requirements in terms of hull structure resistance compared with ships sailing in inland waterways or in less harsh seas.

The possible consequences of accidents also influence the selection of hull structure, shape and materials. The extensive consequences of accidents with large oil tankers (e.g., the *Exxon Valdez*, the *Amoco Cadiz* and the *Prestige*) led to stricter regulations for these vessels, including requirements for double-hull construction to limit the consequences of such accidents (see Sect. 3.2). However, ferries and passenger ships have also experienced several accidents (e.g., the *Estonia* and the *Costa Concordia*) that have led to increased standards in safety requirements.

A stronger and thicker hull comes at a cost. Additional material is required for construction, which impacts both the investment cost and the life cycle demand of materials. In addition, a heavier hull requires a higher lightweight (i.e., the weight of the ship's structure alone), which results in reduced cargo carrying capacity (weight) for a given ship size and shape. This trade-off can pose a challenge for tankers and bulk carriers, for which weight is the limiting factor.

Modern ship hulls are almost always constructed from steel. Lighter materials, such as aluminium and composite materials, are currently being investigated and have been used in highly specific applications. The choice of materials used in ship manufacturing has an impact on the emissions associated with shipping (see Sect. 7.4).

1.4.3 The Propulsion System

Several methods are available to generate the thrust required for a ship to move through water. However, nearly all of the world's commercial fleet is currently based on the concept of converting the chemical energy contained in fuel to mechanical energy, which in turn is converted into ship thrust. Box 1.3 depicts the historical changes in marine propulsion.

1.4.3.1 Ship Resistance

Ship movement through water generates a resistance from the water. This resistance depends primarily on a ship's speed (a standard approximation correlates the propulsive power requirement with the third power of a ship's speed) and on a resistance coefficient, which in turn depends on the hull (e.g., the shape, state, and wetted surface). However, a ship operates in the natural environment, which can lead to the attachment and subsequent growth of various marine organisms on the surface of the hull. These organisms can significantly enhance the hull drag, increasing the power needed by the engine to propel the ship (see Sect. 4.3). It has been estimated that fuel consumption increases by 6 % for every 0.1 mm increase of hull roughness due to fouling [33]. To reduce this phenomenon, so-called "antifouling" treatments are often used to hinder marine growth. Antifouling paints are applied to hulls to prevent the growth of fouling organisms, such as barnacles, mussels, bryozoans and algae. Antifouling systems are required when unwanted biological growth occurs, and the need to protect ship hulls from fouling is as old as the use of ships [34]. However, the release of biocides from antifouling into the water can result in a harmful impact on the marine environment (see Sect. 4.3).

1.4.3.2 Propulsors

Several different types of propulsors are used on ships. The *screw propeller* is the most commonly used and generates thrust through its rotation in water, thus converting the mechanical power delivered from the engine into the thrust required to overcome the ship's resistance and maintain the required speed. Propellers are often highly loaded, and this loading can generate the typical phenomenon of cavitation. This event, in addition to damaging the propeller surface, also generates intense noise, which is a source of disturbance to marine life (see Chap. 6).

1.4.3.3 Transmission Components

Mechanical power produced by the prime mover is subsequently transferred to the propeller by the propeller shaft, and the thrust bearing. The thrust shaft transmits the thrust generated by the propeller to the hull.

Because the propeller is located outside of the hull, the need exists for a sealing system that prevents sea water from entering the ship and discharge of the bearing lubrication oil to the sea. Even if the stern tube fulfils this purpose, small discharges of lubricant to the sea are common. The presence of lubricating oils in different areas of a ship can lead to oil leakage, which is collected in the bilge. This bilge water must be treated before it is released into the ocean (see Sect. 4.1).

1.4.3.4 The Prime Mover

Although several different technologies (mostly diesel engines, gas turbines, and steam turbines) are used as prime movers for ships, all of these options are based on the conversion of the chemical energy contained in the fuel to thermal energy via a combustion process and to mechanical energy via a thermodynamic cycle.

Combustion is the process that generates the largest amount of emissions to the air from ships (see Chap. 5). The exhaust emissions from internal combustion engines depend on the combustion process, the fuel used and the engine. The main compounds that are emitted include carbon dioxide (CO_2), carbon monoxide (CO), nitrogen oxides (NO_x), hydrocarbons (HCs),[1] sulphur dioxide (SO_2) and particulate matter (PM). The emissions of exhaust gases and particles from ocean-going ships contribute to the environmental and health impacts caused by shipping, especially in coastal communities because nearly 70 % of the exhaust emissions from ships occur within 400 km of land [36].

The reaction of carbon with oxygen (which generates the largest portion of energy during the combustion process using modern fuels) generates CO_2, which is one of the main anthropogenic contributors to the greenhouse effect from fossil fuels (see Sect. 5.1). The sulphur contained in the fuel reacts with oxygen to form sulphur oxides (Sect. 5.2), which are precursors to the formation of secondary pollutants (Sect. 5.4.2). Modern marine fuels have much higher sulphur contents than road fuels. Sulphur oxides contribute to the formation of acid rain and impact human health [37]. The high temperatures reached during the combustion process make it possible for nitrogen (which comprises nearly 80 % of combustion air) to react with oxygen, thus generating nitrogen oxides (Sect. 5.3). Incomplete combustion and ashes lead to the formation of particulate matter (Sect. 5.4). Secondary particles are formed in the atmosphere, e.g., from SO_2 and NO_x emissions, and create sulphate and nitrate aerosols via coagulation and condensation of vapours (see Sects. 5.3.2 and 5.4.2). The main concerns related to particle emissions are health effects [37],[2] although particles also contribute to climate change due to both direct effects on the radiative balance and indirect effects via increased cloud

[1]Hydrocarbons are compounds consisting of hydrogen and carbon; the term *volatile organic carbons* (VOCs) is also used. VOCs are formally defined as organic compounds with boiling points between 50 and 260 °C [35]. It is also common to separate VOCs into methane and non-methane volatile organic compounds (NMVOCs).

[2]The smallest particles are considered to be most harmful to humans [38].

formation [36, 39]. Emissions of HCs are a consequence of the incomplete combustion of fuel and consist of unburned and partially oxidised HCs. Unburned lubrication oil from cylinder lubrication is also a major contributor to HC emissions from two-stroke engines [40]. HCs act as precursors to photochemical ozone, and certain HCs are toxic, e.g., benzene and polycyclic aromatic hydrocarbons (PAHs) [37]. In addition, the HC methane is a strong greenhouse gas.

1.4.3.5 Oil Spills

Accidental oil spills from tanker vessels have decreased since the 1970s, although numerous small spills still occur in ecologically sensitive locations [41]. The release of oil to the environment from shipping originates from the transportation of fuels in tanker vessels and from fuel used for propulsion. The portion that originates from fuel used for propulsion is affected by the choice of fuel for marine transportation. Only approximately 7 % of oil spills originated from non-tank vessels during the period 1990–1999 [42]. Operational oil pollution also originates from various sources, such as bilge water and propeller shaft bearings (see also Sect. 4.1).

Box 1.3 Historical development of marine propulsion

Marine propulsion has changed over the course of history (Fig. 1.6). Human power (oars) and wind power were initially used, followed by steam engines and steam turbines fuelled with coal at the beginning of the nineteenth century. Early steam ships used masts and sails because the engines were generally treated as auxiliaries for assisting the sails [43], and the full transition from sail to steam power spanned more than 50 years [10]. The steam engine changed marine transport in the sense that marine transportation was no longer dependent on the wind.

Most steam engines were replaced with marine engines fuelled by diesel and residual oil. Between the shift from steam engines to internal combustion engines (ICEs), a fuel shift also occurred from coal to oil that made this transition possible. During World War I, warships were built with oil-fired boilers or were converted from coal to oil. This shift increased the steam boiler output and/or reduced the storage requirements, thereby increasing the power output of warships [43]. Furthermore, oil-powered steam engines required smaller crews and provided a greater operational range and the possibility of easier refuelling at sea [44]. The first diesel-powered ship went into service in 1912 and was followed by a transition to diesel engines from steam over the next 50 years except for the most powerful ships [10]. Steam turbines are still used in most LNG carriers, which use the boil-off gas as fuel.[3] However, over the past decade, other types of propulsion systems have been considered for LNG carriers, involving various configurations of diesel engines, electric drives and gas turbines [45]. Steam turbines are less efficient

[3]The boil-off gas is the vapour created due to the ambient heat input (while maintaining constant pressure in the storage of cryogenics).

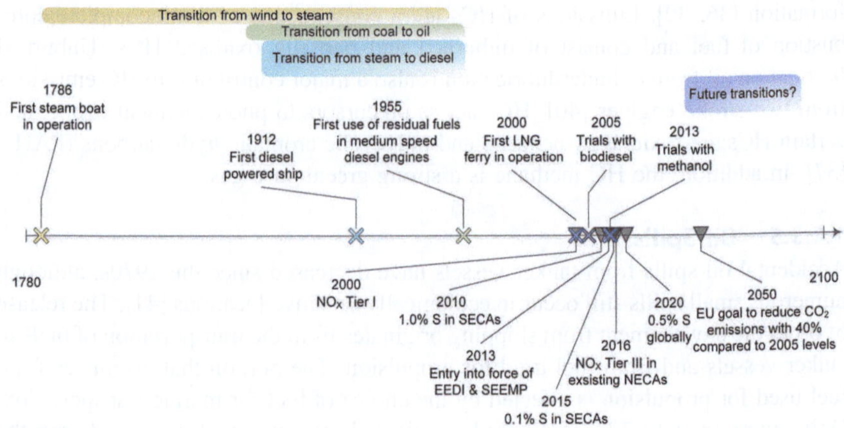

Fig. 1.6 Timeline for transition of marine fuels from 1780 to 2100; selected events in history and environmental regulations are depicted

than diesel engines, and the trend in new LNG carriers points towards propulsion by diesel or dual-fuel engines [46, 47].

Currently, residual fuel or heavy fuel oil, HFO[4] is used in most marine engines. In 2007, nearly 350 million tonnes of fuel were consumed by shipping, and approximately 250 million tonnes of these fuels were residual fuels [48]. The use of HFOs in marine engines followed John Lamb's experiments in the early 1950s, where it was first applied to slow-speed diesel engines and came into general use in medium-speed diesel engines in the 1960s [49].

Although HFO is now the dominant shipping fuel, several *alternative fuels*[5] have been proposed and tested, including liquefied natural gas (LNG), biodiesel, methanol and glycerol. LNG is used in Norway and is promoted as a future ship fuel. Biodiesel has been tested for marine propulsion by a large container shipping company [50] and by the US Navy [51]. This fuel also has been promoted as a suitable fuel for marine propulsion [52, 53]. The potential use of methanol as a marine fuel gained attention in a Swedish research and development project [54]. As a result of this work, a large ferry operator in northern Europe plans to convert several of their vessels to methanol propulsion [55]. A Japanese ocean shipping company signed a contract in December 2013 to build and charter three methanol carriers equipped with flex-fuel engines that can run on methanol, fuel oil or gas oil [56]. Glycerol is under evaluation in the GLEAMS project in the UK [57], and the use of

[4]The terms residual fuels and HFO are used interchangeably in this book.

[5]The term *alternative fuels* is used in this book to describe fuels that are alternatives to HFO in shipping. The term is occasionally used elsewhere as a synonym for renewable fuels.

synthetic fuels produced from carbon dioxide and hydrogen has also been suggested [58]. Thus, many alternatives exist for selection and evaluation. These fuels vary in terms of the energy carrier of the fuel, the primary energy sources used to produce the fuel, and the type of prime mover used to convert the energy carrier to work. A set of criteria for future marine fuels is used to compare the fuels and their various characteristics; the marine fuels are subsequently evaluated using these criteria

1.4.4 Hotel Facilities

A ship can sail for several days without reaching land; thus, it must be able to provide accommodation, food, and services for the people on-board. Cruise ships that can carry several thousand passengers and a crew of several hundred will specifically require a large capacity, and the waste and wastewater that is generated must be handled either on-board or in port (see Sect. 11.4).

A ship is generally designed for operation in many different environmental settings, i.e., from the hot and humid tropical seas to the cold conditions of the North Sea and Arctic areas.[6] For this reason, in addition to considerations related to space and comfort, accommodation of the crew on-board involves fulfilling substantial needs for heating or cooling depending on the period of the year and location. Therefore, heat, ventilation, and air conditioning (HVAC) systems are installed on-board. In particular, the systems related to cooling involve the use of cooling media that can have strong impacts on the environment (Sect. 5.6). Provisions required to feed the crew (between 10 and 40 people on cargo ships and up to one thousand crew members and up to several thousand passengers on cruise ships) for several days must be stored on-board in safe conditions, which requires extensive cooling for the provisions at low temperatures (−20 °C). In addition, accommodations for the crew require several systems to handle various waste products and sewage (Sects. 4.2, 11.4 and 11.5).

1.4.5 Auxiliary Systems

Many technical auxiliary systems are installed on-board to perform supporting functions, such as providing heating, cooling, and lubrication. Together with the propulsion train, these systems require the handling of fuel and lubricating oil, and spills result in the formation of bilge water, which is a mix of water, oil and several different chemicals. Bilge water must be treated on-board and is a source of oil released into the sea (Sects. 4.1.1 and 11.1.3).

[6] Arctic navigation is closely associated with a highly specific design because navigating through ice and extreme weather conditions requires specific measures and equipment.

Tanks are used to store fuel, engine oil, and fresh water. Ballast water is needed to ensure vessel stability during operation without cargo and to balance the weight when the cargo is not evenly distributed. In port, the ballast water might be pumped into specially designed tanks to compensate for changes in the weight distribution as cargo is removed and subsequently released when cargo is loaded. It is estimated that at any given time, approximately 10,000 different species are transported between geographic regions in the ballast tanks alone [59]. Although many alien species become integrated components of the background flora and fauna, others are invasive and will eventually take over and dominate the native flora and fauna. This may have associated economic impacts such as a decrease in economic production by fisheries, aquaculture, tourism and marine infrastructure. Human health can also be affected. For example, the Asian strain of the bacterium responsible for cholera was probably introduced into Latin America via the discharge of ballast water [60]. The environmental impacts related to ballast water are further discussed in Sect. 4.4.

1.5 Sustainability Challenges for the Maritime Industry

Many problems remain to be formulated and solved before the shipping industry can be deemed sustainable [61], e.g., the combustion of fuel in ship engines impacts global warming, acidification, eutrophication and human health (Chap. 5); invasive species spread via ship ballast water (Chap. 4); the scrapping of old ships on beaches causes heavy metal contamination (Chap. 7) [62]; and seafarer working conditions vary depending on the flag state [63]. Another question that arises is related to how shipping can contribute to sustainable development.

IMO has developed the concept of a Sustainable Maritime Transportation System for the "safe, secure, efficient and reliable transport of goods across the world, while minimising pollution, maximising energy efficiency and ensuring resource conservation" (p. 9) [64]. This concept is divided into the following areas with specific goals and actions for each area:

- Safety culture and environmental stewardship
- Education and training in maritime professions and support for seafarers
- Energy efficiency and ship-port interfaces
- Energy supply for ships
- Maritime traffic support and advisory systems
- Maritime security
- Technical cooperation
- New technology and innovation
- Finance, liability and insurance mechanisms
- Ocean governance [64].

In addition, independent organisations as well as different consortia in industry and academia have discussed strategies on sustainable shipping [65, 66].

The future challenges and opportunities for shipping to contribute to a sustainable future are further discussed in Chap. 12.

We hope that this book can contribute to a more sustainable shipping sector by providing knowledge, information, and selected methods and tools related to the environmental issues associated with shipping.

References

1. Kuliukas, A. V. & Morgan, E., *Aquatic Scenarios in the Thinking on human Evolution: What are they and how do they Compare?* in *Was Man More Aquatic in the Past? Fifty Years After Alister Hardy - Waterside Hypotheses of Human Evolution*. 2011. p. 106-119.
2. Rozwadowski, H. M., *Fathoming the Ocean: The Discovery and Exploration of the Deep Sea*. 2008, Harward University Press.
3. Olai Magni Gotti Linkopensis, *Carta marina, opus Olai Magni Gotti Lincopensis, ex typis Antonii Lafreri Sequani. Rome*. 1572 National Library of Sweden, KoB, Kartor 1 ab. kolorerat kopparstick.
4. US Department of Transportation. *Glossary of shipping terms*. 2008, U S Department of Transportation.
5. Swedish Transport Administration. 2015, https://www.transportstyrelsen.se/sv/Sjofart/Fartyg/, last accessed on 31 August 2015.
6. IMO. *Safety regulations for different types of ships*. 2015: London, http://www.imo.org/en/OurWork/Safety/Regulations/Pages/Default.aspx, last accessed on 31 August 2015.
7. Douvere, F., *The importance of marine spatial planning in advancing ecosystem-based sea use management.* Marine Policy, 2008. **32**(5): p. 762–771.
8. European Union. *Maritime Spatial Planning*. 2014, http://ec.europa.eu/maritimeaffairs/policy/maritime_spatial_planning/index_en.htm, last accessed on 31 August 2015.
9. UNCTAD. *UNCTAD stat*. 2015, http://unctadstat.unctad.org/EN, last accessed on August 31 2015.
10. Stopford, M., *Maritime economics*. 2008, London: Routledge.
11. Alphaliner. *Operated fleets as per 11 August 2015*. 2015, http://www.alphaliner.com/top100/, last accessed on August 31 2015.
12. UNCTAD, *Review of maritime transport*. Vol. UNCTAD/RMT/2014 United Nations Publications. Sales no. E.14.II.D.5. 2014, New York and Geneva.
13. News, C. *Cruise Industry News Annual Reoprt 2014–2015*. 2014.
14. AMSA. *Arctic Marine Shipping Assessment 2009 Report, 2nd ed.* 2009, Arctic Council.
15. Polyakov, I. V. *et al.*, *Arctic Ocean Warming Contributes to Reduced Polar Ice Cap.* Journal of Physical Oceanography, 2010. **40**(12): p. 2743–2756.
16. Brundtland, G. H. *Report of the World Commission on Environment and Development: Our Common Future*. 1987, UN: New York.
17. Meadowcroft, J., *Who is in Charge here? Governance for Sustainable Development in a Complex World.* Journal of Environmental Policy & Planning, 2007. **9**(3–4): p. 299-314.
18. Bell, S. & Morse, S., *Sustainability indicators: measuring the immeasurable?* 2nd ed. 2008, London: Earthscan.
19. Kates, R. W., Parris, T. M. & Leiserowitz, A. A., *What is sustainable development? Goals, indicators, values, and practice.* Environment: Science and Policy for Sustainable Development, 2005. **47**(3): p. 8–21.
20. Holden, E., Linnerud, K. & Banister, D., *Sustainable development: Our Common Future revisited.* Global Environmental Change, 2014. **26**(0): p. 130–139.
21. Holden, E. & Linnerud, K., *The sustainable development area: satisfying basic needs and safeguarding ecological sustainability.* Sustainable Development, 2007. **15**(3): p. 174–187.
22. Hooper, D. *et al.*, *Effects of biodiversity on ecosystem functioning: a consensus of current knowledge.* Ecological Monographs, 2005. **75**(1): p. 3–35.
23. Swedish Environmental Protection Agency. *Sammanställd information om Ekosystemtjänster (Compiled Information on Ecosystem Services) (In Swedish)*. 2012: Stockholm.

24. Schrag, D. P. *The Habitable Planet - A systems approach to environmental science*. 2014, Annenberg Learner.
25. Engel, S. & Schaefer, M., *Ecosystem services — a useful concept for addressing water challenges?* Current Opinion in Environmental Sustainability, 2013. **5**(6): p. 696–707.
26. Beaudoin, Y. & Pendleton, L. *Why value the ocean?* 2012, UNEP/Grid-Arendal.
27. Guinotte, J. M. & Fabry, V. J., *Ocean Acidification and Its Potential Effects on Marine Ecosystems.* Annals of the New York Academy of Sciences, 2008. **1134**(1): p. 320–342.
28. Crutzen, P. J., *Geology of mankind.* Nature, 2002. **415**(6867): p. 23–23.
29. Rockstrom, J. *et al.*, *A safe operating space for humanity.* Nature, 2009. **461**(7263): p. 472–475.
30. Steffen, W. *et al.*, *Planetary boundaries: Guiding human development on a changing planet.* Science, 2015. **347**(6223).
31. Moberg, F. & Simonsen, S. H. *What is resilience? - An introduction to social-ecological research*. Stockholm Resilience Centre: Stockholm.
32. Folke, C. *et al.*, *Resilience Thinking: Integrating Resilience, Adaptability and Transformability.* Ecology and Society, 2010. **15**.
33. Voulvoulis, N., Scrimshaw, M. D. & Lester, J. N., *Alternative Antifouling Biocides.* Applied organometal chemistry, 1999. **13**: p. 135–143.
34. Almeida, E., Diamantino, T. C. & de Sousa, O., *Marine paints: The particular case of antifouling paints.* Progress in Organic Coatings, 2007. **59**(1): p. 2–20.
35. Baird, C. & Cann, M., *Environmental chemistry*. 2008, New York: W.H. Freeman & Co.
36. Eyring, V. *et al.*, *Transport impacts on atmosphere and climate: Shipping.* Atmospheric Environment, 2010. **44**(37): p. 4735–4771.
37. Harrison, R. M., *Pollution: causes, effects and control.* 2001, Cambridge: Royal Society of Chemistry.
38. Pope, C. A. *et al.*, *Lung Cancer, Cardiopulmonary Mortality, and Long-term Exposure to Fine Particulate Air Pollution.* JAMA: The Journal of the American Medical Association, 2002. **287**(9): p. 1132–1141.
39. Lauer, A., Eyring, V., Hendricks, J., Jockel, P. & Lohmann, U., *Global model simulations of the impact of ocean-going ships on aerosols, clouds, and the radiation budget.* Atmospheric Chemistry and Physics, 2007. **7**(19): p. 5061–5079.
40. Henningsen, S., *Chapter 14 - Air Pollution from Large Two-Stroke Diesel Engines and Technologies to Control It*, in *Handbook of Air Pollution From Internal Combustion Engines*, Sher Eran, Editor. 1998, Academic Press: San Diego. p. 477–534.
41. Burgherr, P., *In-depth analysis of accidental oil spills from tankers in the context of global spill trends from all sources.* Journal of Hazardous Materials, 2007. **140**(1–2): p. 245-256.
42. National Research Council, *Oil in the sea III: inputs, fates, and effects*, ed. National Research Council. 2003, Washington, D.C: National Academy Press.
43. Clark, G. H., *Industrial and marine fuels reference book.* 1988, London: Butterworths.
44. Corbett, J. J., *Marine Transportation and Energy Use*, in *Encyclopedia of Energy*, J. Cleveland Cutler, Editor. 2004, Elsevier: New York. p. 745–758.
45. MAN Diesel & Turbo. *Propulsion trends in LNG Carriers*. 2007: Copenhagen, Denmark.
46. Wiggins, E., *COGAS propulsion for LNG ships.* Journal of Marine Science and Application, 2011. **10**(2): p. 175–183.
47. Chang, D. *et al.*, *A study on availability and safety of new propulsion systems for LNG carriers.* Reliability Engineering & System Safety, 2008. **93**(12): p. 1877–1885.
48. Buhaug, Ø. *et al. Second IMO GHG Study 2009*. 2009, International Maritime Organization: London, UK.
49. Tree, D. *Future fuels*. in *Marine Propulsion and Future Fuels*. 1979. London: IPC Industrial Press Ltd.
50. Gallagher, T. L. *Maersk to Run Biodiesel Test*. Journal of Commerce Online, 2010.
51. Bruckner-Menchelli, N. *US Navy targets 2016 for green fuel use*. Sustainable shipping, 2011.

52. Mihic, S., Golusin, M. & Mihajlovic, M., *Policy and promotion of sustainable inland waterway transport in Europe - Danube River.* Renewable and Sustainable Energy Reviews, 2011. **15**(4): p. 1801–1809.
53. Lin, C.-Y. & Huang, T.-H., *Cost–benefit evaluation of using biodiesel as an alternative fuel for fishing boats in Taiwan.* Marine Policy, 2012. **36**(1): p. 103–107.
54. Fagerlund, P. & Ramne, B. *WP 9 Summary Report - EffShip Project Summary and Conclusions.* 2013: Gothenburg.
55. Einemo, U. *Stena highlights 'fourth ECA compliance option'.* Sustainable Shipping, 2013.
56. Mitsui OSK Lines. *MOL, WFS Sign Deal to Build, Charter Methanol Carrier - World's First Dual-Fuel Engine Uses Methanol and Heavy Oil, Press release.* 2013, http://www.mol.co.jp/en/pr/2013/13085.html, last accessed on September 16 2015.
57. Macqueen, J. *Moves afoot to establish glycerine as a marine fuel.* Sustainable Shipping, 2014.
58. Vergara, J., McKesson, C. & Walczak, M., *Sustainable energy for the marine sector.* Energy Policy, 2012. **49**(0): p. 333–345.
59. Bax, N., Williamson, A., Aguero, M., Gonzalez, E. & Geeves, W., *Marine invasive alien species: a threat to global biodiversity.* Marine Policy, 2003. **27**(4): p. 313–323.
60. Hensher, D. A. & Button, K. J., *Handbook of transport and the environment.* Vol. 4. 2003, Amsterdam: Pergamon.
61. Cabezas-Basurko, O., Mesbahi, E. & Moloney, S. R., *Methodology for sustainability analysis of ships.* Ships and Offshore Structures, 2008. **3**(1): p. 1–11.
62. Reddy, M. S., Basha, S., Sravan Kumar, V. G., Joshi, H. V. & Ramachandraiah, G., *Distribution, enrichment and accumulation of heavy metals in coastal sediments of Alang–Sosiya ship scrapping yard, India.* Marine Pollution Bulletin, 2004. **48**(11–12): p. 1055–1059.
63. Michel, C. D. & Ward, A. S., *Abamdonment of Seafarers - Solving the Problem*, in *Proceedings of the Marine Safety & Security Council, the Coast Guard Journal of Safety at Sea.* 2009, U.S. Coast Guard: Washington, DC. p. 75–78.
64. IMO. *World Maritime Day - A Concept of a Sustainabel Maritime Transport System.* 2013, International Maritime Organization: London.
65. European Commission. *Sustainable Transport.* 2013, http://ec.europa.eu/transport/themes/sustainable/news/2013-09-25-essf-call-for-applications_en.htm, last accessed on August 31 2015.
66. *Sustainalbe Shipping Initiative.* 2015, http://ssi2040.org, last accessed on August 31 2015.

The Natural Environment and Human Impacts

2

J. Fredrik Lindgren, Kent Salo, Selma Brynolf, Karin Andersson, Erik Svensson, Maria Zetterdahl, Lena Granhag and Mathias Magnusson

Abstract

To comprehend the implications of the various environmental issues that man is inducing on the Earth (with a focus on the shipping industry), an understanding of the Earth's major systems is necessary. The natural environment, which consists of air, water, land and living organisms, is a dynamic system in which material and energy are exchanged within and between the individual components. The system is divided into four spheres (atmosphere, hydrosphere,

J.F. Lindgren (✉) · K. Salo · S. Brynolf · K. Andersson · M. Zetterdahl · L. Granhag
Shipping and Marine Technology, Chalmers University of Technology, Gothenburg, Sweden
e-mail: f.lindgren@chalmers.se

K. Salo
e-mail: kent.salo@chalmers.se

S. Brynolf
e-mail: selma.brynolf@chalmers.se

K. Andersson
e-mail: karin.andersson@chalmers.se

M. Zetterdahl
e-mail: maria.zetterdahl@chalmers.se

L. Granhag
e-mail: lena.granhag@chalmers.se

E. Svensson
Researcher, Gothenburg, Sweden
e-mail: e-svensson@outlook.com

M. Magnusson
Volvo Group Trucks Technology, AB Volvo, Gothenburg, Sweden
e-mail: mathias.magnusson.2@volvo.com

K. Andersson et al. (eds.), *Shipping and the Environment*,
DOI 10.1007/978-3-662-49045-7_2

geosphere, and biosphere), and fluxes of energy and material are exchanged amongst these spheres. The spheres also largely govern the fate of various environmental problems originating from the shipping industry. Therefore, background information related to these spheres is provided, and their major properties and implications are explained. Regarding the atmosphere, radiation and energy budgets are explained in conjunction with the weather and climate. Concerning the hydrosphere, oceanography is introduced together with marine ecology. Addressing the geosphere, the elements in the Earth's crust and mineral commodities are discussed. Regarding the biosphere, energy is transferred through food chains; the differences between life in water and life on land are examined. Energy flows through and is stored in these spheres; this stored energy is essential to the natural environment and human society. The different primary energy sources are described and divided into non-renewable and renewable sources. Finally, an introduction to human impacts on the natural environment and to major environmental issues is provided.

Keywords
Hydrosphere · Atmosphere · Geosphere · Biosphere · Biogeochemical cycles · Primary energy sources · Environmental issues

The natural environment, which consists of air, water, land and living organisms, is a dynamic system in which material and energy are exchanged within and between the individual components. Solar energy flows through these systems, warms the atmosphere, evaporates and recycles water, generates wind and supports plant growth. Matter is transported both within and between the spheres in various cycles (the cycles are further discussed in Sects. 2.3–2.6). Therefore, emissions and discharges entering the system will be distributed among the different components, sometimes becoming diluted or accumulating temporarily or permanently (Fig. 2.1). Chemical transformations will also change the properties of the released substances over time, often moving back and forth among gaseous, liquid and solid states. Thus, the fate of a substance entering the natural environment from a ship or marine structure can be complex, and environmental impacts can arise far from the emission source. Shipping and potential environmental impacts that are partially or solely connected to the industry ultimately affect different parts of the Earth's systems. Emissions and discharges from ships are released into both the atmosphere and the sea, where they are unlikely to remain. The various systems are related in many ways, and hazardous chemicals can travel between them. Therefore, this chapter will introduce and describe these various systems.

The Earth can be divided into four major systems: the atmosphere (air), the hydrosphere (water), the geosphere (land) and the biosphere (living organisms).[1]

[1]There are different ways to divide the Earth and its system and slightly different names are sometimes used than the ones used here. As an example, the Biosphere sometimes also includes all air, water and land on the planet.

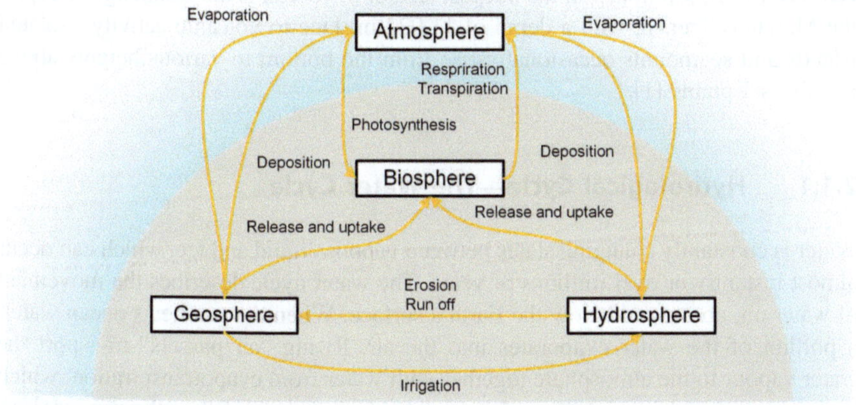

Fig. 2.1 Illustration of the relationships between the atmosphere, hydrosphere, geosphere and biosphere through the exchange of energy, water, biomass and chemicals. Modified from Environmental Chemistry

These systems are strongly connected through fluxes of energy, chemicals, biomass, nutrients, and water (Fig. 2.1). The atmosphere is the thin layer of gases that cover the Earth's surface. Among other things, it moderates the Earth's temperature and absorbs energy and ultraviolet radiation from the sun. The hydrosphere contains the Earth's water. Approximately 70 % of the surface of the globe is covered with water, and approximately 10 % of the land is covered by ice. Approximately 1 % of the mass of the hydrosphere is fresh water, 2 % is ice, and the remaining portion is oceanic water. Water is recycled though the hydrosphere via evaporation, condensation and run-off. The lithosphere is the uppermost portion of the geosphere, containing the solid rocky crust that covers the entire planet. The lithosphere is the source of raw materials, e.g., crude oil, iron and copper. The biosphere is composed of all living organisms, including plants, animals, and single-celled organisms.

2.1 The Hydrosphere

Because 97 % of all water on the planet is saltwater, the focus of this section will be on the marine environment and the challenges that marine organisms are confronted with. Furthermore, most shipping activities occur in marine systems in the ocean. The marine ecosystem is the largest aquatic ecosystem on Earth and is mainly divided into areas with different depths. Continental shelf areas are the shallowest, gently sloping down 100–200 m. At the edge of the shelf, an abrupt steepening of the bottom forms the continental slopes, with depths down to 3–5 km. The abyssal plains cover the sea bottom with monotonously flat, sediment-covered areas. These enormous plains are sometimes cut by deep, narrow troughs called trenches, which can reach 7–10 km in depth; the deepest area of the ocean is the Challenger Deep in the Marianas Trench, with a depth of 11,022 m. Due to volcanic activity, isolated islands and seamounts occasionally rise from the bottom to various heights above the abyssal plains [1].

2.1.1 Hydrological Cycle—The Water Cycle

Water is constantly changing states between vapour, liquid and ice, which can occur almost instantly or over millions of years. The water cycle describes the movement of water on, above and below the Earth's surface. When the sun heats ocean water, a portion of the water evaporates into the air. Rising "air parcels" transport the water vapour to the atmosphere together with water from evapotranspiration, which is water transpired from plants and evaporated from the soil. When the vapour rises, cooler air temperatures cause it to condense into clouds. Air currents transport the clouds around the Earth, and cloud particles collide, grow and eventually fall out of the sky as precipitation. A portion of the precipitation falls as snow and ice, forming ice caps and glaciers that can store the frozen water for thousands of years. In warmer climate zones, the ice pack thaws and melts in the summer, and the melted water flows over the ground as snowmelt. However, most precipitation falls from the clouds back into the oceans or onto land, where it flows over the ground as surface run-off due to gravity. Some of this water enters rivers that end up in the ocean. Alternatively, this water can accumulate with groundwater seepage and become stored as freshwater in lakes. However, not all surface run-off flows into rivers. Much of it soaks into the ground as infiltration, where it is stored in subsurface rock formations called aquifers, which can store massive amounts of water for long periods of time. Some of the infiltration remains near the surface and, through groundwater discharge, can seep into surface water bodies, i.e., rivers, lakes and oceans. Even more groundwater is absorbed by plant roots and eventually becomes evapotranspirated from the leaves. When groundwater re-enters the oceans or the atmosphere through evapotranspiration, the hydrological cycle starts over.

2.1.2 Chemical and Physical Properties of Water

The term "water" refers to numerous different things and uses. The water molecule, H_2O, is the common denominator and the prerequisite for all life on earth. However, for most applications, the role of water is to act as a solvent and carrier for major and minor components and a medium in which aquatic species live. The water that is a necessity for human consumption is not pure H_2O; its usability is dependent on the dissolved salt contents. Life in the oceans is adapted to a higher salinity than most terrestrial organisms can use for drinking.

The water molecule is a relatively small chemical compound, and if the logical relation between molecule size and physical state were true, it would be present as a gas at ambient temperature (20–26 °C) and pressure. However, water is a liquid at this temperature because of the charge distribution within the molecule, which yields stronger forces between the molecules than would be expected (Fig. 2.2) [2]. The same characteristic also results in a high freezing point and good properties as a solvent. In addition, water has a high heat capacity compared to many other fluids, providing a large storage capacity for heat and a large capability to regulate temperature on the Earth.

One important property of water is the potential to dissociate into a hydroxide ion, OH^-, and a hydrogen ion, H^+. Pure water has a hydrogen concentration of 10^{-7} M (moles/l). The hydrogen concentration of a solution is often expressed in terms of pH. pH is a measure of the acidity of a solution, defined as $-(\log [H^+])$, where $[H^+]$ is the concentration of hydrogen ions. A "neutral" solution has pH 7 (or $[H^+] = 10^{-7}$ M), as with pure water. A lower pH value indicates a surplus of hydrogen ions, i.e., the solution is acidic. Solutions that have a higher value are considered to be basic/alkaline. Because the scale is logarithmic, a solution with a pH of 6 is ten times more acidic than a solution with a pH of 7 [3, 4]. The properties of dissolved species in the water determine the resulting pH value. Strong acids, such as sulphuric and nitrous (from dissolved sulphur and nitrogen oxides), strongly affect pH, whereas weak acids, such as carbonic acid (from dissolved carbon dioxide), help establish equilibria, therein reducing the acidifying effects.

2.1.2.1 Water as a Solvent

Natural water contains dissolved inorganic species (salts) as well as soluble organic compounds originating from the degradation of living material, solid suspended organic compounds and living organisms. The dissolved matter exhibits some variability, although the salts are generally enriched in water, beginning with rainwater, which dissolves some gases from the atmosphere. More salts are added as water flows through soil and minerals in the ground. In this process, groundwater is formed; water also percolates (gradually filtering through a porous surface or substance) into lakes and rivers. When it reaches the sea, water is usually still drinkable. In estuaries, water is mixed with seawater, and the salinity increases.

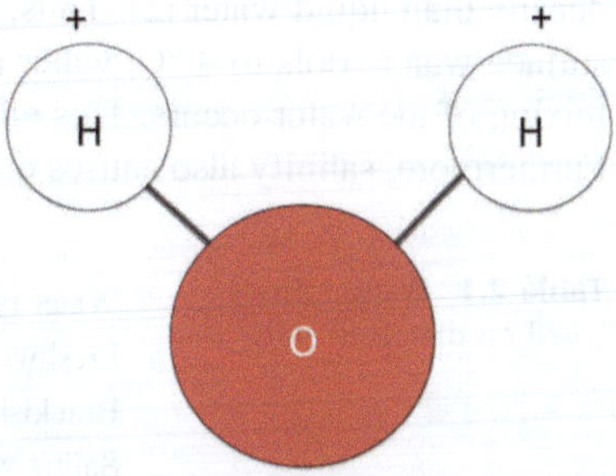

Fig. 2.2 Water molecule, consisting of one oxygen atom and two hydrogen atoms

Ocean water salinity is generally measured in practical salinity units (PSU) and is more saline than lake and river water. In some areas, the mixing of salt and freshwater occurs over large distances, resulting in brackish water.

2.1.2.2 Water Composition

Water composition, resulting from the interactions with the geological material in the hydrologic cycle, will thus vary over a large range. Many inorganic ions are added due to these interactions. The constituents that vary the most in different bodies of water are sodium and chloride. The oceans have a relatively constant composition, whereas the variation in salt content is large in lakes, rivers, estuaries, other mixed water bodies, and some very saline lakes (Table 2.1). For seawater, the total salinity may vary between 34 and 37 Practical Salinity Units (PSUs), although the relative proportions are quite constant for most major ions [5]. Salinity is today measured in PSUs, which are based on the electrical conductivity of seawater. It has been previously measured in grams dissolved salts per litre of seawater, and the measurements have the same meaning: 35 PSU is 35 g dissolved salt in a litre of seawater. However, the measurement using conductivity is more precise, and therefore, this method is currently used. Salinity determines a number of properties of water.

The carbonate system is also important for determining the composition of water. Carbonate speciation and the solubility of carbon dioxide depend on pH. An increase in the carbon dioxide concentration in the air will lead to an increase in the amount of carbonate in the water, which may cause a decrease in pH. This process is discussed further in Sect. 2.7.5. *Alkalinity*, which is the sum of the contents of several other ions, determines the buffering capacity of water, which is the ability to counteract a change in pH. The chemical processes and the importance of buffering are discussed in further detail in Chap. 5. In addition, organic dissolved components contribute to the properties of water. Organic compounds originate primarily from the degradation of plants and animals, usually called humic substances. These substances are large molecules with different functional groups, often carboxylic or hydroxyl, that cause the compounds to act as weak acids.

2.1.2.3 Density and Stratification

Among the properties distinguishing water from other liquids are that its maximum density occurs at a temperature of +4 °C and that solid water (ice) has a lower density than liquid water [2]. Thus, ice covers the surface of water. Moreover, when surface water cools to 4 °C, water at greater depths remains warmer, and a sudden mixing of the water occurs. This effect occurs in Arctic and Antarctic ocean waters. Furthermore, salinity also causes differences in density, which can affect the mixing

Table 2.1 Water salinity based on dissolved salts

Water type	PSU
Freshwater	>0.5
Brackish water	0.5–30
Saline water	30–50

of water masses in estuaries and other brackish water. When heavier seawater enters an area with outflowing fresh or less saline water, the heavy water sinks to the sea floor, providing oxygen-rich water to benthic areas with oxygen depletion, for example, in the Baltic Sea (Box 2.1).

2.1.2.4 Properties of Oceanic Water Affecting Organisms

Approximately 71 % of the Earth's surface is covered by saltwater, which is a special environment for organisms compared to limnic (non-seawater) or terrestrial life. The main components that distinguish saltwater from other aquatic systems are the presence of dissolved salts in higher concentrations, e.g., chloride, sodium, sulphur (sulphate), magnesium, calcium and potassium. Furthermore, saltwater contains inorganic nutrients, such as nitrate $\left(NO_3^-\right)$, phosphate $\left(PO_4^{3-}\right)$ and trace elements, including iron, manganese, cobalt and copper that limit growth in marine ecosystems. The salinity in the open ocean ranges between 34 and 37 PSU (Practical Salinity Units), although the average is 35 PSU. The pH typically varies from 7.5 to 8.4. Gaseous oxygen and carbon dioxide are also dissolved in saltwater. The solubility of oxygen is a function of temperature (akin to other gases), in which lower water temperatures yield higher oxygen solubility. Carbon dioxide exists in a relationship with carbonic acid (H_2CO_3) and the bicarbonate ion $\left(HCO_3^-\right)$, in which more CO_2 dissolved in the ocean yields more carbonic acid, more bicarbonate ions and more hydrogen ions (H^+), which slowly lead to the acidification of seawater through lowering the pH [6].

Box 2.1 The Baltic Sea

The Baltic Sea is a sensitive sea area with brackish water, i.e., a mixture of salt water from the North Sea and freshwater from rivers, and has been declared a particularly sensitive sea area (PSSA) according to IMO. In the Baltic Sea, aquatic life is a unique composition of marine and freshwater species that have adapted to brackish water. The Baltic Sea is subject to dense maritime traffic concentrated along the main shipping routes (Fig. 2.3). Approximately 15 % of all goods worldwide are transported within this relatively small sea area [7]. There is also ferry traffic, both national and international, between the nine countries with coastlines on the Baltic Sea and seasonally international cruise ships are operating in the area.

The anthropogenic effects on the Baltic Sea originate from a large land-area of river runoff. This runoff area is several times larger than the sea area. Both river run-off and air deposition contribute to high nitrogen (N) and phosphorous (P) loadings into the Baltic Sea. With very limited water exchange due to the narrow, shallow waters of Öresund and the Belt Sea, the excess nutrients remain in the sea and cause problems for the marine ecosystems. As a result of this substantial excess of nutrients, phytoplankton (small marine microscopic plants) can grow unrestrained and form blooms. When the mass of plankton dies and sinks to the bottom, the degradation of

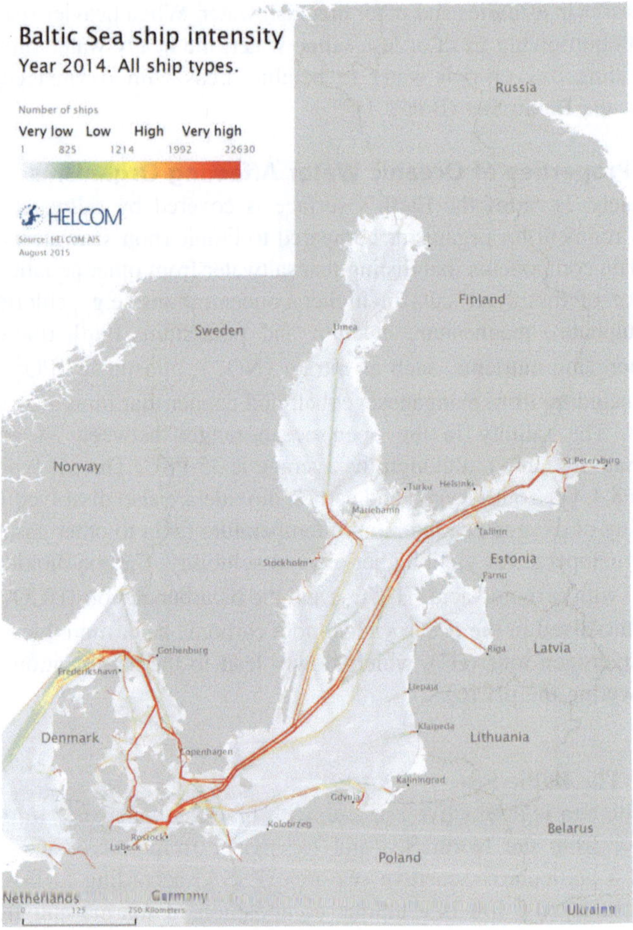

Fig. 2.3 Baltic Sea with ship traffic, 2014 (Printed with permission from HELCOM AIS)

the material consumes substantial amounts of oxygen, which can lead to oxygen depletion near the sea floor. The small animals living on the bottom cannot survive, and there will be lack of food for larger animals, such as fish, which are forced to vacate the oxygen-depleted areas.

The nitrogen inputs from shipping come in the form of nitrogen oxides (NO_x) and sewage. Nitrogen in the form of nitrate and phosphorous in the form of phosphate reach the sea via sewage. Even if the nutrient inputs from sewage are small compared with the total nutrient inputs, sewage can have large local effects if released directly into the water, especially during summer, when temperatures and productivity in the sea are high.

2.1.3 Oceanography

Oceanography covers a wide range of related topics, including ocean currents, waves, physical properties within the ocean, fluxes of various chemical substances and geophysical fluid dynamics, e.g., marine organisms and ecosystem dynamics.

The upper water masses in the ocean are constantly moving due to winds that create waves and currents. Waves vary in size from small ripples to storm waves of up to 30 m. Deep water is not affected by winds, although the movement of this water originates from density changes in the surface waters. Ocean surface currents do not flow in the same direction as the wind. Instead, they flow in circular gyres, moving clockwise in the Northern Hemisphere and counter clockwise in the Southern Hemisphere. These large gyres are the result of the *Coriolis Effect*—the Earth's rotation around its axis displaces water to the right in the Northern Hemisphere and to the left in the Southern Hemisphere. The surface currents affect the water layers beneath them and set the deeper layers of water in motion with, however, decreasing amounts of energy and reduced speeds at deeper depths in the water column. Simultaneously, due to the Coriolis effect, each affected water layer is deflected with respect to the layer directly above it, resulting in a net water movement that is perpendicular (90°) to the wind direction. The result of the different current directions and speeds from the surface and downwards is called the *Ekman spiral* (Fig. 2.4).

In some areas and under certain conditions, wind-driven currents can produce upwelling. As a result, nutrient-rich water (rich in inorganic nitrogen compounds, phosphate and silicate) from the deep ocean is driven to the surface in eastern water basins by southerly winds and by the Coriolis Effect, which pushes the surface water offshore. In equatorial waters, two currents flowing towards the west are deflected by the Coriolis Effect, one deflected right and northward and the other left and southward. This surface water is then replaced by upwelling subsurface water.

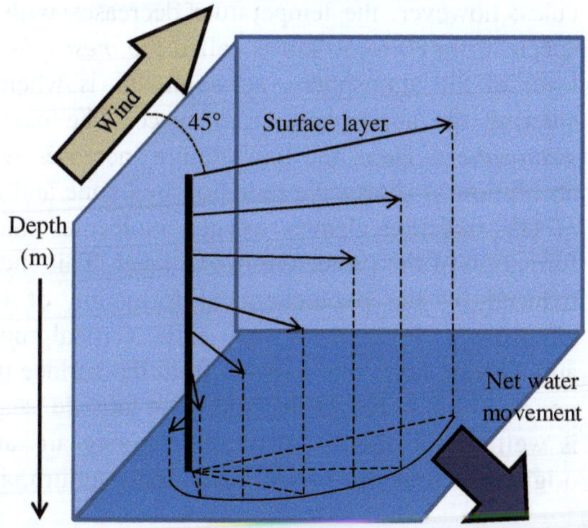

Fig. 2.4 Schematic of the Ekman spiral, in which the movement of the surface water layer is 45° from the wind direction and the net water movement is perpendicular to the wind direction

When warm, oxygen-rich, saline water is transported to the poles by surface currents, e.g., the Gulf Stream, the water is cooled, which increases its density. This enormous amount of water then sinks to the bottom of the sea, thus providing the deep abyssal plains with oxygenated water. The water slowly flows south to Antarctica. Eventually, these currents warm and slowly rise to the surface, restarting this large cycle of water currents. These currents, which form the *ocean conveyor belt*, move very slowly, requiring hundreds of years to move through the ocean basins [1, 8].

2.2 The Atmosphere

When viewing the Earth from space, something clearly distinguishes it from all other known planets. The most important differences between the Earth and our neighbour planets in the solar system are the presence of water and an atmosphere. The atmosphere is a thin layer of gas surrounding the Earth. If our planet were an apple, the atmosphere would be thinner than the apple peel. The atmosphere not only supplies us with oxygen but also protects us from cosmic radiation and ultraviolet (UV) radiation from the sun. By acting like the glass in a greenhouse, the atmosphere captures outgoing longwave radiation, keeping us warm in the cold of outer space.

2.2.1 The Structure and the Composition of the Atmosphere

The outermost layer of the atmosphere, which borders outer space, is called the *thermosphere*. This layer extends upward from a height of 80 km above the Earth. In the outer layer of the thermosphere, the temperature can exceed hundreds of degrees Celsius due to intense solar radiation and the presence of very few molecules; however, the temperature decreases with altitude. The layer of the atmosphere from 80 to 50 km is called the *mesosphere*. The mesosphere is a very thin layer of the atmosphere, although this is where most meteorites burn up when entering the atmosphere from space. The next layer of the atmosphere is the *stratosphere*. Here, the temperature increases with increasing altitude due to the absorption of shortwave radiation by ozone and oxygen (Fig. 2.5). The absorption of UV radiation cleaves oxygen molecules that subsequently participate in the formation of the protective ozone layer. This increase in temperature with altitude induced by the photochemical formation of ozone creates a stable, stratified atmospheric layer, with very little vertical movement. The *troposphere* is the atmospheric layer that extends from the surface of the Earth up to the stratosphere, i.e., 9–17 km in height depending on latitude, season and weather. The troposphere is well mixed due to convection (raising air) and winds. The name troposphere originates from the Greek word for overturn, i.e., trope (τροπή). *Convection* is

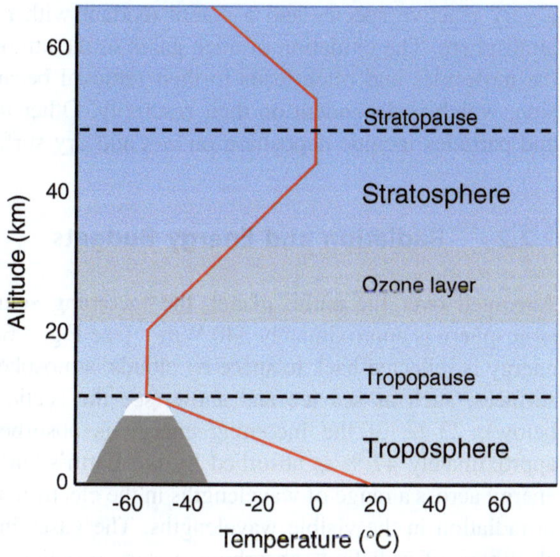

Fig. 2.5 The various layers in the lower parts of the atmosphere and temperature as a function of altitude

generated via energetic shortwave radiation from the sun heating the Earth's surface, which subsequently emits longwave radiation and heats the moist air closest to the ground. This heating causes air masses to rise and transport water vapour and aerosol particles higher into the troposphere. When a rising "air parcel" expands due to the reduced pressure at higher altitudes, the air cools and may become supersaturated with respect to water vapour. This allows the water to condense on aerosol particles and participate in the formation of clouds. When water condenses to form clouds, latent heat is released, causing even more convection. Thus, the troposphere is characterised by complex weather systems that propagate globally (Fig. 2.5).

The troposphere contains 90 % of the mass of the atmosphere. The main components of dry air are nitrogen (78.08 %), oxygen (20.90 %) and argon (0.93 %). The water content in the troposphere varies from nearly zero to 4 % depending on altitude, latitude and temperature. In addition, the troposphere contains many trace gases, typically in concentrations of parts per billion (ppb) or parts per million (ppm). The most important examples of these trace gases are hydrogen (H_2), carbon dioxide (CO_2), nitrous oxides (NO_x), sulphur dioxide (SO_2), ammonia (NH_3), methane (CH_4), and volatile organic compounds (VOCs). These gaseous compounds can originate from both natural and anthropogenic processes, and their tropospheric lifetimes vary from only a few seconds to thousands of years due to their different chemical stabilities. An important sink for chemical compounds in the atmosphere involves transformation into other compounds through chemical reactions, often involving oxidation by the hydroxyl radical (OH radical). These radicals are formed when O_3 is decomposed by UV radiation, forming reactive oxygen radicals that react with gaseous H_2O to form OH radicals. The OH radical is a very reactive species and a potent oxidant with a short lifetime (~1 s) in the atmosphere. The oxidation of trace gases in the atmosphere alters the properties of the molecules and often leads to their removal because of gas-to-particle conversion, which is dependent on their reactivity. Other important sinks for both gases and particles include deposition on wet and dry surfaces [8].

2.2.2 Radiation and Energy Budgets

Averaged over the entire planet, the incoming solar radiation at the top of the atmosphere is approximately 340 Wm^{-2} (see Fig. 2.6). Approximately 29 % of this energy is reflected back to space by clouds, atmospheric particles, or bright ground surfaces, such as sea ice and snow (see the section regarding the albedo effect below); 23 % of the incoming energy is absorbed by the atmosphere, while approximately 47 % is absorbed by the Earth's surface [9, 10]. The sun radiates energy across a range of wavelengths in the electromagnetic spectrum, with its peak in radiation in the visible wavelengths. The gases in the atmosphere are selective absorbers of radiation, absorbing certain wavelengths better than others. Water vapour, ozone and oxygen absorb direct solar energy; however, this occurs high in the atmosphere, and only non-visible wavelengths are absorbed. Visible

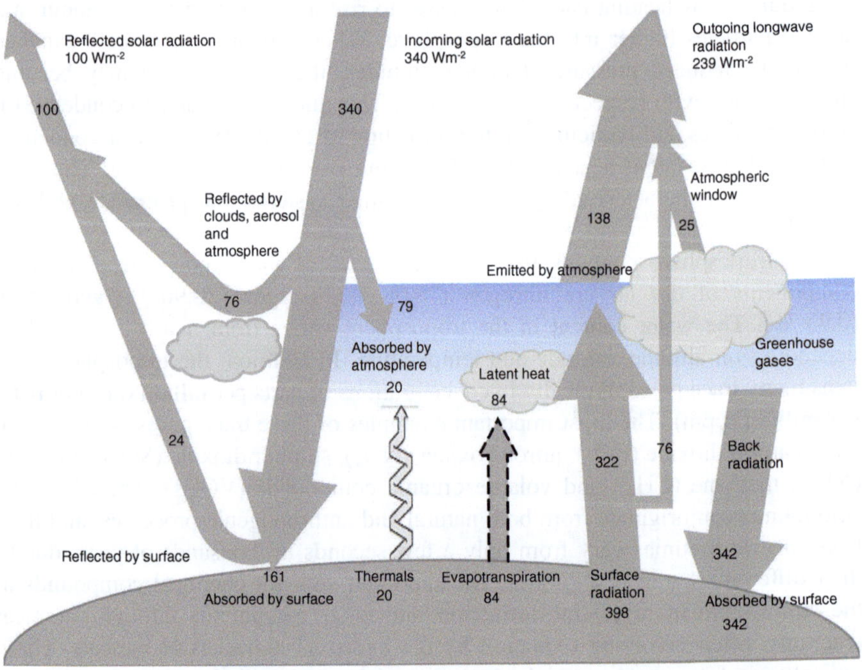

Fig. 2.6 The Earth's annual and global mean energy balances, expressed as watts per square metre of the Earth's spherical surface (Wm^{-2}) [10]

wavelengths constitute approximately 43 % of the energy radiated from the sun, and this radiation is transmitted to the Earth's surface without contributing to the heating of the atmosphere. In contrast, water vapour and carbon dioxide are good at absorbing longwave radiation emitted from the Earth, which heats the lower atmospheric layer, i.e., the troposphere. Thus, the atmosphere is heated from the ground up and not the other way around. When the absorbed energy heats the air, it in turn emits energy to space and back to the Earth. The energy emitted back to the Earth's surface causes the Earth to warm, resulting in more energy radiating from the surface (Fig. 2.6). This relationship, which is called the *greenhouse effect*, maintains the Earth's temperature at a reasonable temperature (average 33 °C [59 °F]) to sustain life. When the flow of incoming solar energy is balanced by an equal flow of energy to space, the Earth is in radiative equilibrium, and the global temperature is relatively stable. Anything that increases or decreases the amount of incoming or outgoing energy, i.e., *the energy budget*, disturbs the Earth's radiative equilibrium, and the global temperature can rise or fall in response. In 1896, CO_2 levels were already recognised as a source of fluctuating temperatures on Earth [11]. Increasing CO_2 concentrations in the atmosphere, mostly due to anthropogenic sources, leads to greater absorption of the energy emitted from the Earth, thus resulting in a higher average temperature of the atmosphere and in turn more energy being emitted back to Earth. This anthropogenic phenomenon has been termed *global warming* (see Sect. 2.7.4) [8].

2.2.2.1 Greenhouse Effect

Carbon dioxide (CO_2) only accounts for 0.037 % of the gases that constitute dry, clean air. However, CO_2 is an effective absorber of longwave radiation emitted from the Earth's surface and is transparent to incoming shortwave radiation from the sun. A portion of the radiation leaving the Earth's crust is absorbed by atmospheric CO_2 and reemitted. Some of this radiation is radiated back to the surface, causing the air near the ground to become warmer than it would have been without CO_2. Hence, combined with water vapour, CO_2 is largely responsible for the greenhouse effect, and an increase in CO_2 concentrations in the atmosphere leads to altered temperatures in the lower atmosphere. In addition to CO_2 and water vapour, many other greenhouse gases exist, e.g., ground-level ozone and methane, some of which are further described in Sect. 2.7.4 [8].

2.2.2.2 Albedo Effect

The portion of the radiation emitted from the sun that is reflected by a surface is called its *albedo*. This varies both spatially and temporally depending on the cloud cover, particles in the air and angle of the sun. Thick clouds and fresh snow have high albedos, reflecting approximately 70–90 % of the incident radiation, whereas dark soil has a low albedo, reflecting approximately 10–25 %. The total albedo of the Earth is approximately 29 %, with clouds being mostly responsible for this reflection [10]. In addition to the increasing CO_2 levels in the atmosphere causing the Earth's average temperature to increase, the snow and ice around the poles are experiencing greater melting during summer. This effect is caused by the albedo effect; smaller quantities of ice and snow reflect less radiation back into space, causing additional melting of the snow and ice cover [8, 12].

2.2.2.3 Ozone Layer

Ozone is an important component of the atmosphere. Its molecular form is three oxygen atoms bonded together (O_3). Very little ozone exists in the atmosphere, and the majority is concentrated in the stratosphere, forming the ozone layer and fulfilling a crucial task of absorbing harmful ultraviolet (UV) radiation emitted from the sun. Without the filtering of UV radiation, and if that radiation were to reach the Earth undiminished, our planet would not be inhabitable to many of the organisms living on it today [8].

2.2.3 Weather and Climate

Winds are generated when unequal heating of the Earth's surface leads to pressure differences. Global winds are generated because the tropics receive more solar radiation than the poles. Consequently, winds blow to the north and south, balancing the major surface temperature difference between the regions. However, the Coriolis Effect (see Sect. 2.1.3) changes the airflow, directing it (in the Northern Hemisphere) to the right and generating westerly winds. These generated macro-scale wind systems can extend around the globe and remain unchanged for weeks. However, the weather systems are not continuous around the Earth because the surface is not uniform. In the Northern Hemisphere, which has a higher proportion of land compared to ocean, these systems are often replaced by high- and low-pressure weather systems. In addition, seasonal variations contribute to the fluxes in global weather systems through temperature changes.

Where the atmosphere and water are in contact, energy is transferred through friction. As drag occurs on the water surface from blowing winds, the surface layer of the water begins to move. Thus, winds are the primary drivers of surface ocean currents. Currents can also be strengthened by winds; for example, the Gulf Stream is strengthened by the prevailing westerly winds. The Gulf Stream is also a good example of ocean currents playing a major role in the Earth's heat balance. In general, ocean currents transfer warm water from the equatorial regions to the colder polar regions. Ocean currents represent a quarter of this transport, and winds represent the remaining three quarters.

Warm parcels of air rise from the Earth's surface, condense and produce clouds. Water vapour in these clouds becomes saturated and condenses, forming cloud droplets. These droplets fall to the ground as *precipitation*. The amount of annual precipitation that falls in various regions is governed by the global wind and pressure systems. In general, regions influenced by high-pressure systems and divergent winds experience dry conditions. In contrast, regions with low pressure and converging winds receive large amounts of precipitation. This results in patterns of abundant precipitation around the equator, ample precipitation in sub-polar areas, and dry areas and seasonal precipitation patterns near the sub-tropical areas. The distribution of large landmasses complicates the precipitation patterns because less precipitation occurs in the middle of landmasses. Mountain barriers are another factor that can alter precipitation patterns [8].

Climate is an aggregate of weather over time, including not only the average temperatures and precipitation but also the variability in weather, including extreme weather events, e.g., El Niño (see Box 2.2), hurricanes and storms. In addition, climate is not based solely on the atmosphere but also on the exchange of energy and moisture through the four spheres: the biosphere, atmosphere, hydrosphere (also including ice and snow on the Earth's surface) and geosphere. Consequently, climate changes do not only affect one component of the climate system. Instead, when each part is affected, the other parts react. Reconstructions of past climates have demonstrated that the climate is naturally variable over both short (decades) and long (millions of years) temporal scales. However, research on human impacts has shown that the Earth's climate is changing due to human activities [8].

Box 2.2 Weather Phenomenon—El Niño

Normally, the cold Peruvian current flows along the coast of Ecuador and Peru, creating an upwelling of cold, nutrient-rich water that results in immense primary production. This primary production serves as the primary food source for many fish species, including economically valuable species such as anchovies. At the end of the year, this current is replaced by a warm current that flows southward. This episode typically lasts for a couple of weeks before being replaced by the cold Peruvian current. However, at irregular intervals (3–7 years), this episode along with the warm counter-current is unusually strong. This phenomenon has been termed El Niño and leads to abnormal weather patterns in Ecuador and Peru, with abnormal amounts of precipitation. This abnormality also affects other areas of the world, causing droughts in, e.g., Australia, Indonesia and the Philippines, leading to crop and property damage and human suffering. The state of California (USA) has also been affected by large storms due to El Niño [8, 9].

2.3 The Geosphere

The geosphere is the part of the Earth's system that includes its inner core, rock, minerals and the processes that shape the surface. Similar to the atmosphere, the portion of the geosphere used by humans is thin. The geosphere can be divided into three zones: the core, mantle and crust. The crust is the upper layer of the Earth, representing approximately 0.5 % of the radius of the Earth. The thickness of the crust varies from approximately 35 km in continental regions to approximately

5–10 km under the oceans. Continuous reformation of the crust occurs, in which high points are worn down by weathering and erosion, while low areas are filled with debris. Erosion involves the transport of soil particles by wind and water flows. Weathering includes the chemical breakdown of minerals into individual ions that are soluble in water [13].

The crust and the upper layer of the mantle are known as the *lithosphere.* The lithosphere contains several important resources that are used in society, for example, to produce mobile phones, computers and ships. The lithosphere contains both non-renewable (e.g., aluminium, iron, copper, oil, and natural gas) and renewable (e.g., soil) resources. A renewable resource is a resource that can be replenished reasonably quickly (from hours to a hundred years). The quantity of non-renewable resources is essentially fixed in the Earth's crust, although geological processes can renew such resources over time scales of millions to billions of years [13].

The crust consists of rocks that can be classified into three basic categories: igneous, sedimentary and metamorphic. Igneous rocks are formed from magma. The lava that flows out of a volcano and hardens is a well-known type of igneous rock, although most igneous rocks are formed well below ground where magma slowly cools and crystallises. Igneous rocks are often covered by sedimentary rocks. Sedimentary rocks form in one or more of the following ways: from the fragments produced during weathering and erosion, as a result of dissolved material precipitating from water, or as a consequence of biological activity. Metamorphic rocks are formed by the alteration of existing rocks by extreme heat, extreme pressure and/or the presence of hot gases and liquids. Rocks are not static objects; they are created, destroyed, and recycled, which is known as the rock cycle and is another example of how the Earth is continually being transformed [13].

Approximately 99 % of the mass of the Earth's crust is composed of eight elements: oxygen (47 %), silicon (29 %), aluminium (8 %), iron (4 %), calcium, sodium, magnesium, and potassium [13]. The remaining 1 % contains approximately 90 different elements. Some minerals are abundant and can be found in most areas around the world, whereas others are concentrated in relatively few places. Because minerals are used in various industrial processes, the reserves of these commodities are diminishing and variable (Table 2.2).

Table 2.2 Estimated worldwide reserves and life expectancy of selected mineral commodities [13]

Mineral commodity	1999 reserves (million tonnes)	Life expectancy with no growth in primary production (years)
Aluminium	25,000	202
Copper	340	28
Iron	74,000,000	132
Lead	64	21
Nickel	46	41
Silver	0.28	17
Zinc	190	25

Another important component of the lithosphere is soil, which is generated in the interphase between the crust and the atmosphere and can be defined as a region of the crust that has been altered by biological activity. Soil is very important from an environmental perspective because it supports nearly all plant life. Most soil contains inorganic matter, organic matter, soil water and soil atmosphere. Soil properties change over time; different soil types are formed depending on climate, vegetation and parent material. Soil is an open system that continuously changes due to weathering and biological activity. Weathering is the physical and chemical breakdown of rocks exposed to the surface of the Earth. Weathering releases elements that have no gaseous form, freeing them for plant uptake and contributing to the formation of secondary minerals [13].

2.4 The Biosphere

The biosphere contains all of Earth's living things, including all microorganisms, plants, and animals. The biosphere is the global ecological system that integrates all living things and their relationships, including their interactions with the elements of the other spheres (i.e., atmosphere, hydrosphere, and geosphere). Moreover, this sphere extends into the atmosphere, where birds and insects can be found, and down to the bottom of the ocean floor. The biosphere is generally believed to have been formed 3.5 billion years ago and can be divided into separate *biomes*, each consisting of living things that have adapted to a specific climate, for example, deserts, forests, grasslands, and tundra. The biomes are separated by latitude. A tropical biome is vastly different from the biome in Antarctica.

2.4.1 Primary Production and Food Chains

Life on the Earth depends greatly on the harvest of solar energy via the process of *photosynthesis*. During photosynthesis, light energy is absorbed by chlorophyll molecules, which convert carbon dioxide and water into carbohydrates and oxygen. Organisms that have this ability (*primary producers*), e.g., plants, algae and cyanobacteria, then store the converted solar energy in living tissue. This energy can then flow into herbivores, predators, parasites, decomposers and all other life forms. A small amount of the energy stored in plants, i.e., between 5 and 25 %, passes into herbivores, and a similar percentage is subsequently passed to carnivores from herbivores. The result is a pyramid of energy; most of the energy is concentrated in the lowest part of the food chain, i.e., in photosynthetic organisms. Hence, the top of the food chain contains the least energy. A large portion of the energy is not directly transferred into the photosynthetic organisms—herbivore—predator food chain; rather, it is passed to the detritus food chain. Bacteria and fungi are all consumers of detritus, which is dead organic material. Eventually, these consumers are consumed by other organisms.

The total productivity of the biosphere depends on the rate at which plants convert solar energy into chemical energy, which is approximately 1 %, and at what rates other parts of the ecosystem convert that chemical energy into biomass. Therefore, human-induced changes in the productivity of the biosphere have the largest consequences in areas where the productivity is greatest, such as tropical forests and estuaries [1].

2.4.2 Living in Sea Water—Implications for Marine Organisms

One major implication for organisms living in seawater is the importance of regulating the salt levels in their bodies. Charged ions, such as Na^+ and Cl^-, which constitute the major salt components in salt water, are unable to freely pass through biological membranes. *Diffusion* is the tendency for all matter to be equally concentrated in the environment. Due to diffusion, water will travel through biological membranes from regions of low salt concentrations to regions of high salt concentrations; this process is called osmosis. Therefore, organisms living in high-saline environments must be able to prevent the loss of water from their bodies. In contrast, organisms in low-saline environments (e.g., lakes and rivers) must prevent water from entering their bodies. Organisms that live in both environments, such as salmon or eels, must have the ability to regulate the saline content of their bodies in both ways [1].

Marine birds, marine mammals and some species of fish have the ability to regulate their body heat. Other marine organisms are *ectothermic*; their body heat varies with that of the surrounding water. This variability has led to the latitudinal distribution of species around the globe, dividing the marine ecosystem into four different zones: polar, cold temperate, warm temperate (sub-tropic) and tropical areas, with transition zones in between. Beneath the warm surface waters, the temperature of the water begins to decrease. Over a narrow depth range of 50–300 m, the temperature declines rapidly. A zone like this, of rapid temperature decrease is called a *thermocline*. Below the thermocline, the temperature continues to decrease, although at a much slower rate. The thermocline water layer hinders the mixing of water between the upper and lower water layers, preventing the depleted nutrients in the upper water layer from being resupplied from the deep, nutrient-rich water layers [1].

Light only penetrates to depths of 200–400 m in water, the so-called *photic zone*, which limits the activity of primary producers. These organisms utilise sunlight to produce energy. Thus, they only dwell in the upper water layers. Moreover, this also implies that there are numerous marine communities existing without a photoautotrophic component. Specific abiotic factors (non-living chemical and physical parts of the environment) differ between marine ecosystems and their terrestrial counterparts. The density of liquid water is approximately 800 times greater than that of air, and the viscosity of water is approximately 60 times greater than that of air. The higher density and gravity of seawater provides buoyancy. As a result, marine organisms invest less of their biomass (biological material derived from living organisms) in structural material, such as skeletons, to resist the force of gravity. In addition, marine organisms struggle less against gravity for movement, resulting in reduced energy consumption. Another consequence of the difference in density is that organic material and organisms in the water can float, allowing an entire ecosystem, i.e., plankton, to exist. *Plankton* is an organism living in the water column that is unable to propel itself against a current, e.g., phytoplankton, bacteria, copepods, and jellyfish. The group primarily consists of numerous small organisms that commonly dwell in the sunlit upper portion of the water column, acquiring energy through photosynthetic processes. As a result of the evolvement of plankton, another group of organisms the filter feeders, unique to aquatic environments, were able to evolve. *Filter feeders*, e.g., barnacles and mussels, filter organisms and organic material from seawater to sustain themselves and are often stationary and attached to surfaces. This sessile form of living has led to the development of motile larval forms that are used to establish and maintain benthic communities. Larvae are small and morphologically different from their adult stages following development from fertilised eggs. There are three different larval strategies. The chosen strategy often depends on environmental conditions. The substrate that the larva eventually selects to settle and metamorphose on is not randomly chosen. They have the ability to examine, to various degrees, possible substrates for different cues, e.g., chemical factors, light, salinity and surface texture, that the larvae find favourable. They also respond to the presence or absence of adults of their own species. Many larvae of sessile organisms find human structures, such as ship hulls, favourable, which causes large economic challenges for the shipping industry [1].

Food chains in the ocean are generally longer and more complex compared to their terrestrial counterparts because filter feeders feed on several different trophic levels and because juveniles of species might feed on different prey compared with adults. Moreover, adults may switch trophic prey to meet their needs. Autotrophs are generally single-cell organisms belonging to various groups of algae, which differs from large autotrophs dominating on land (trees). The dominant herbivore is also microscopic; various forms of copepods, i.e., a small crustacean, are the most numerous herbivores. The majority of the large animals in the ocean are carnivores, which also differs from the terrestrial environment, where the large animals are mostly herbivores [1].

The different interconnected communities of the ocean are typically divided into plankton, oceanic nekton, deep sea communities, shallow-water subtidal benthic associations, intertidal areas and estuaries and salt marshes [1]. These communities are often further divided into sub-communities. This classification reflects the large variation in communities and species diversity that can be found in the ocean. This variation is frequently forgotten because the water surface appears identical to humans from wherever we observe it from above. When evaluating the environmental impacts of pollutants in the ocean, you have to be aware of the fact that the recorded effects on one type of community probably cannot be extrapolated to other types of communities. Some areas are more sensitive to anthropogenic disturbance and have been classified by IMO as particularly sensitive sea areas (PSSAs) (Box 2.3).

Box 2.3 Particularly Sensitive Sea Areas (PSSAs)
Particularly sensitive sea areas are classified by IMO for ecological, socioeconomic or scientific reasons as being particularly sensitive and possibly vulnerable to damage from maritime activities. Actions taken within these areas can face stricter environmental regulations or routing systems to avoid particular areas. IMO has classified fourteen areas as PSSAs: the Great Barrier Reef (Australia) encompassing the south-west part of the Coral Sea, the Sabana-Camagüey Archipelago (Cuba), Malpelo Island (Colombia), the sea around the Florida Keys (USA), The Wadden Sea (Denmark, Germany, and the Netherlands), the Paracas National Reserve (Peru), Western European waters, the extension of the existing Great Barrier Reef PSSA to include the Torres Strait, the Canary Islands (Spain), the Galapagos Archipelago (Ecuador), the Baltic Sea area (Denmark, Estonia, Finland, Germany, Latvia, Lithuania, Poland and Sweden), the Papahānaumokuākea Marine National Monument (USA), the Strait of Bonifacio (France and Italy), and the Saba Bank (Caribbean area of the Kingdom of the Netherlands) [14].

2.5 Biogeochemical Cycles

Biogeochemical cycles describe how essential elements of living matter are a part of and travel through natural circulation pathways in the ecosystem. These elements can flow from non-living to living components and back again. For example, in the water cycle, water is stored in the atmosphere, released as precipitation, taken up and excreted by living organisms, and finally evaporated from the surface and once again stored in the atmosphere. Biogeochemical cycles, including chemical, biological and geological processes, have immensely important functions in ecosystems because essential components are continuously recycled [15]. Examples of three essential biogeochemical cycles are the sulphur cycle, the nitrogen cycle and the carbon cycle.

2.5.1 The Sulphur Cycle

The sulphur cycle describes how sulphur moves to and from living organisms and minerals (Fig. 2.7). Sulphur is an essential element in living organisms and is included in many biochemical processes. It is a component of all proteins, vitamins, and many enzymes. The majority of sulphur is stored in seawater and in sedimentary rocks. The burning of coal and fossil fuels in recent centuries has provided an input of sulphur into the atmosphere in a different chemical form, i.e., as sulphur dioxide, which acts as an air pollutant [15].

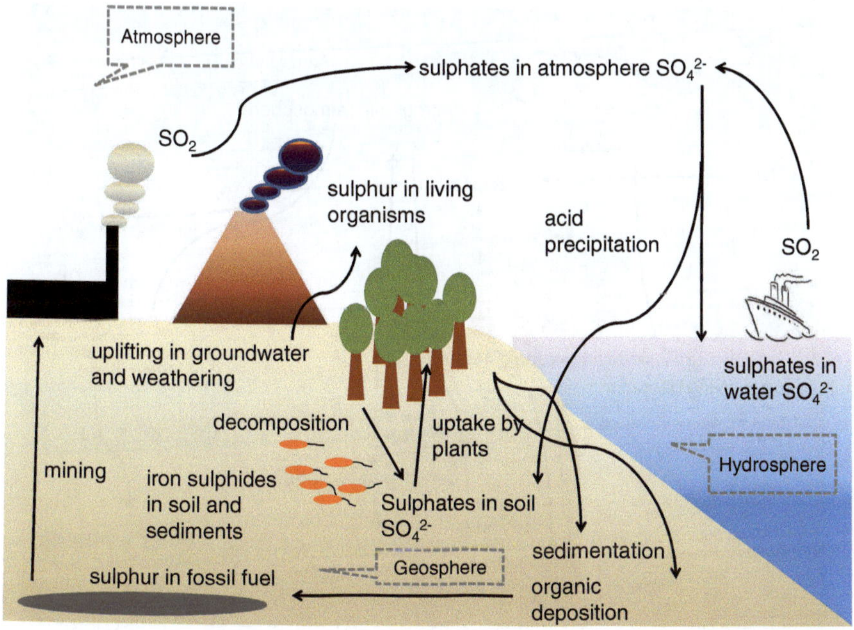

Fig. 2.7 Illustration of the sulphur cycle and how sulphur moves into and out of living organisms and minerals

2.5.2 The Nitrogen Cycle

Nitrogen affects numerous processes in ecosystems, including primary production and decomposition and is a component in biomolecules, such as proteins, DNA and chlorophyll. The largest pool of nitrogen is in the atmosphere, which is 78 % nitrogen. However, nitrogen is not available to living organisms in this form, causing nitrogen to be a limiting resource in ecosystems. Only when dinitrogen is converted into ammonia $\left(NH_3\left[NH_4^+\right]\right)$ is it accessible to primary producers, such as plants and bacteria [16]. In the nitrogen cycle, nitrogen is transformed between its various chemical forms, i.e., NH_4^+, NO_2^-, NO_3^- and N_2. This can be accomplished through both physical and biological forms, where the biological form dominates. On land, nitrogen-containing substances move from animal droppings, plants or fertilisers through denitrifying microorganisms in soil back to plants, water or the air [15]. In marine areas, organic material containing nitrogen in the form of ammonium is deposited on the sea floor and oxidised to nitrate in oxic sediments. Nitrate is subsequently used in denitrification processes (nitrate reduction, which ultimately produces molecular nitrogen [N_2]) in the anoxic part of the sediment, where the created nitrogen is once again recycled into the atmosphere [17, 18]. Ammonium can also be oxidised to nitrogen in anoxic sediments using nitrite via the anammox process (nitrite and ammonium are converted directly into N_2 gas) (Fig. 2.8) [19]. The excess loading of nitrogen in the ground, water and air

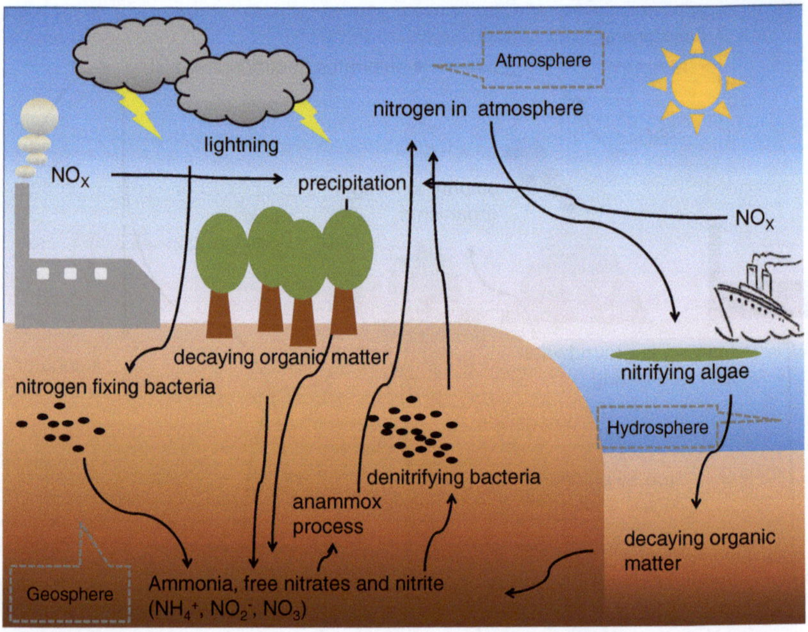

Fig. 2.8 Illustration of the nitrogen cycle and how nitrogen is transferred between its various chemical forms

from agricultural and industrial processes is currently a major environmental issue because natural processes are unable to capture all of the nitrogen.

2.5.3 The Carbon Cycle

Carbon is the fundamental building block for all living things and can either be used for structural integrity or as energy. Carbon is also a component of various minerals. Carbon is present in the atmosphere mainly as carbon dioxide (CO_2). Animals and plants exhale CO_2 as carbon compounds are used, therein releasing energy. Photosynthesising organisms later assimilate CO_2 from the air during photosynthesis to produce new carbon products, mainly carbohydrates (glucose, $C_6H_{12}O_6$). The carbon cycle is usually divided into four major carbon storage compartments between which carbon is exchanged: the atmosphere, biosphere, ocean and sediments. This exchange occurs through various biological, chemical, geological and physical processes. CO_2 is continuously exchanged between the atmosphere and the oceans. Because of the burning of carbon-containing fossil fuels, the CO_2 levels in the atmosphere have increased since the beginning of the industrial revolution in 1760 (Fig. 2.9) [10, 15].

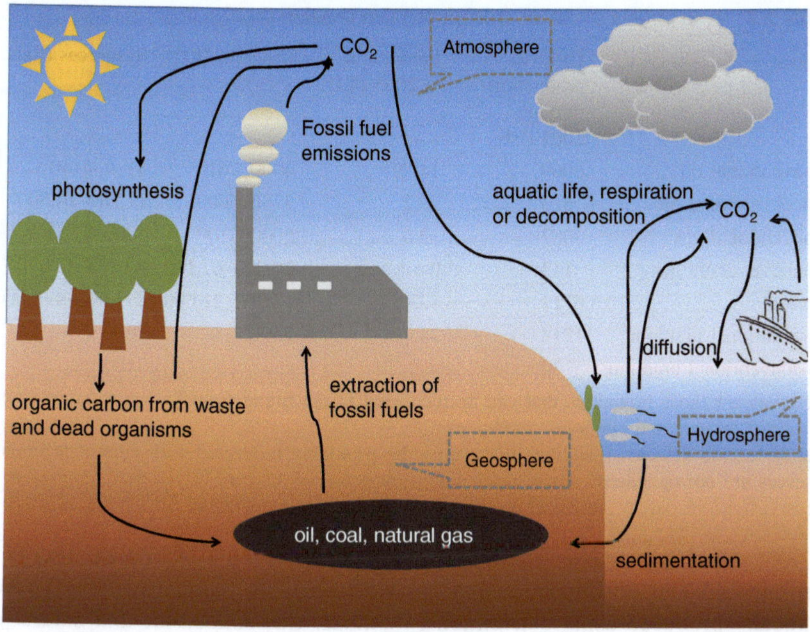

Fig. 2.9 Illustration of the carbon cycle and how the various forms of carbon are cycled through living organisms and the Earth's crust

2.6 Energy Sources

The Earth's largest energy source is the sun, which has an annual energy input of 3,900,000 EJ (exajoule, 10^{18}). This amount is so large that the average radiation striking the Earth's surface in one hour is approximately equal to all the energy consumed by all humans in one year. Energy is also stored in the different spheres. The energy from the sun and stored in the Earth's spheres can be harvested and used by society.

2.6.1 Fossil Energy Sources

The energy sources currently used in society are mostly non-renewable and are extracted from the Earth. Oil, natural gas, coal and nuclear energy represented approximately 80 % of the global primary energy usage in 2010 [20]. The supplies of oil and gas are expected to be seriously diminished by the middle of the twenty-first century. This picture is further complicated by the substantial regional disparities in the distribution of fossil fuel reserves; Table 2.3 presents the reserves, resources and distribution of fossil energy sources, including fossil fuels (oil, gas, and coal) and uranium. The table distinguishes between *conventional* and *unconventional resources*. However, no clear boundary exists between what is called

Table 2.3 Reserves and resources of fossil fuel and uranium [21]

	Historic production through 2005 (EJ)	Production in 2005 (EJ)	Reserves (EJ)[a]	Resources (EJ)[b]
Conventional oil	6069	147.9	4900–7610	4170–6150
Unconventional oil	513	20.2	3750–5600	11,280–14,800
Conventional gas	3087	89.8	5000–7100	7200–8900
Unconventional gas	113	9.6	20,200–67,100	40,200–121,900
Coal	6712	123.8	17,300–21,000	291,000–435,000
Conventional uranium[c]	1218	24.7	2400	7400
Unconventional uranium[c]	34	n.a.		7100

[a]Reserves are those quantities that can be recovered in the future from known reservoirs under existing economic and operating conditions
[b]Resources are detected quantities that cannot be profitably recovered with current technology (reserves are not included)
[c]Reserves and resources are based on once-through fuel cycle operation

conventional and unconventional. Conventional oil is mobile and can be economically produced using conventional methods. Gas condensate and natural gas liquids are also usually included. Unconventional oil includes high-viscosity oil and shale oil. Deep offshore oil and Arctic oil are also occasionally considered unconventional oil because of the difficulties related to deep-sea drilling and Arctic production. Currently, the total amount of oil extracted from the Earth's crust is approximately equal to the current remaining conventional oil reserves [21]. The production of conventional oil will most likely peak soon and begin to decline shortly thereafter. Massive development of unconventional oil occurrences would be necessary to shift the peak of oil production to 2050 or later [21]; the peak concept is discussed further in Box 2.5. Typical crude oil consists of hundreds of combustible hydrocarbons in addition to small amounts of sulphur, oxygen, nitrogen, metals and salts, although the amounts vary widely [22]. Crude oil is generally categorised as sweet (<0.5 % sulphur) or sour (>0.5 % sulphur) depending on the sulphur content and light or heavy depending on its density. Crude oil is the raw material for the fuels that are currently used in the shipping industry. The different fuel fractions derived from crude oil are described in Box 2.4.

Box 2.4 Crude Oil Refining

Crude oil refining consists of three main processes: *separation*, *conversion* and *purification*. Many different types of refineries exist, and the included processes and complexity vary; the simplest refineries use only separation and purification. After the removal of contaminants, crude oil is separated into its components in a series of distillation towers, with the bottom product from each tower feeding the next. The feed is heated from the bottom, forming vapours and liquids depending on the density of the components. The liquids remain at the bottom and the vapours rise. As the vapours rise, the temperature decreases, and the components condense. Products from the distillation tower range from gases at the top to very heavy, viscous liquids at the bottom.

The proportions of the streams from the distillation towers do not match consumer demand; in general, too little petrol and too much heavy oil are produced. Therefore, the conversion of heavier hydrocarbons (compounds with long carbon chains) to lighter hydrocarbons (compounds with shorter carbon chains) is needed. This conversion is completed by breaking the carbon chains in the molecule, which is performed by *fluidised catalytic crackers (FCCs)*, *cokers* and *hydrocrackers*. In addition to breaking the chains, it may be necessary to change the form of the chains or to join chains together; these processes are performed by catalytic reformers and alkylation units. The FCC unit uses catalysts and heat to convert gas oil into a mix of liquefied petroleum gas, petrol and diesel and is often an important conversion unit in refineries. Some refineries also use a delayed coker to convert the heaviest fractions from the distillation column, i.e., fractions that are too heavy and have too many contaminants to be converted in the FCC unit. The delayed coker uses high temperatures to break the chains, producing coker gas oil and petroleum coke. Hydrocrackers can also be used to complement the other units. Hydrocrackers use both heat and catalysts, and the reactions occur with high concentrations of hydrogen to produce products with low sulphur contents. The alkylation unit combines two butane molecules into a longer chain, whereas catalytic reforming increases the octane number of petrol by changing the length of the hydrocarbon chains, generating hydrogen for the hydrotreaters.

The last step in a refinery is purification, which primarily consists of sulphur removal. This process is completed by hydrotreating, in which unfinished products are exposed to hydrogen under heat and high pressure in the presence of catalysts, producing hydrogen sulphide and desulphurised products. Sulphur recovery converts hydrogen sulphide to elemental sulphur and water. In Fig. 2.10, the typical products generated from a barrel of crude oil are shown.

Many refinery processes can be used for the production of marine fuels. Gas oils are light, whereas heavy gas oil fractions and blends from straight-run and cracked origins have a boiling range between 200 and 350 ° C. They are predominantly used as automotive diesel fuels and as domestic heating fuels, although *marine gas oil* (*MGO*) is also produced from this fraction [23]. *Heavy fuel oil* (*HFO*) consists of various mixtures of residual oils from the distilling and conversion processes in the refinery. These products are used as marine bunker fuels, in power stations and in industrial furnaces. Moreover, HFOs can be blended with gas oils to adjust the density, viscosity and sulphur content [23].

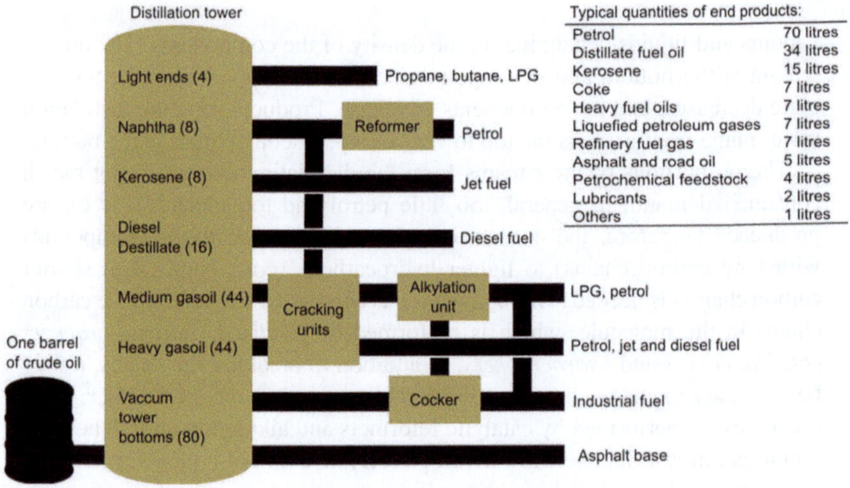

Fig. 2.10 Typical products produced from a barrel (159 l) of crude oil [24]

The remaining natural gas resources are abundant compared with oil, and estimates of these resources have risen steadily over recent decades [21]. Natural gas consists of a mixture of gases in which methane (CH_4) is the main component. Gas reservoirs are often associated with oil reservoirs. Associated gas is recovered together with oil and is separated above ground. The associated gas may be recovered, reinjected or flared.[2] The choice of method depends on the location, field size, geology and the amount of gas in the reservoir. Approximately 17 % of all gas that is recovered in association with oil is currently flared [21].

Box 2.5 The Peak Debate

How much fossil energy does the Earth's crust hold? This question has troubled many researchers and analysts since the beginning of the twenty-first century, and opinions differ, particularly between researchers from different disciplines. The focus has primarily been on the peak in conventional oil production, although peak gas, peak coal and peak uranium have also received attention.

The arguments regarding peak oil are based on the fact that large oil discoveries ceased occurring in the mid-1960s, which was followed by a decline in discoveries of new reserves. Extraction of more oil than is offset by new reserves will eventually result in a peak in oil production at approximately the time when half of the oil reserves have been used. A large number of estimates of the world's available conventional oil have been developed;

[2]To flare means to burn the fuel directly into the atmosphere, which releases the combustion products into the atmosphere.

the majority of these estimates are in the range of 12,600–16,700 EJ. It is assumed that conventional oil production will peak in the foreseeable future, most likely before 2040, with a peak production rate of approximately 4 Gt (≈200 EJ) per year [21]. When considering both conventional and unconventional oil production, it is expected that maximum production will be characterised by a fluctuating plateau instead of a peak.

Arguments against a peak include the assertion that human creativity will remain ahead of resource depletion, which has occurred in the past. Technological innovation will continue to unlock new reserves that have not yet been identified or understood or are not economically extractable with existing technology and market conditions. Furthermore, scarcity will result in higher prices, which will increase available reserves and decrease the demand. There may be a level at which renewable resources become more economical, thereby leaving a substantial portion of untapped oil in the ground.

The separation between conventional and unconventional gas is even more blurred than that of oil. Conventional gas is extracted using standard extraction technologies, although the technologies considered to be standard change over time. Shale gas, coal-bed methane, tight gas, deep gas, water-dissolved gas and gas hydrate are all considered to be unconventional gas. The recent development of shale gas in the United States has received a great deal of attention, and the United States is now estimated to become a net exporter of natural gas instead of a net importer by 2021 [25]. Shipping fuels based on natural gas may be produced from shale gas in the future. The development of shale gas is discussed further in Box 2.6.

Coal reserves remain substantial and are expected to last for more than 100 years. Portions of global coal deposits are located in remote areas or in areas with harsh conditions; bringing this coal to market will be challenging. The productivity of coal mining has increased significantly during recent years, and coal is the lowest-cost fossil energy source [21]. Uranium is another non-renewable resource that can be used to fuel nuclear reactors in shipping or to produce electrofuels. Fissile materials are abundant and are not expected to limit any future expansion of nuclear power [21].

Box 2.6 Shale Gas Production
Shale gas is gas trapped in the pore spaces of sedimentary rock, in vertical fractures in the rock, or adsorbed onto mineral grains and organic materials. Recent growth in the production of shale gas is the result of new technology that creates extensive artificial fractures around horizontal well bores [21]. The development has been concentrated in North America, although shale gas resources can be found in other parts of the world [26].

With increased exploration and production, questions arise regarding the nature of shale gas development, its potential environmental impacts and the ability of the current regulatory structure to address this development. The environmental impacts associated with shale gas development occur at both global and local levels. These impacts generate environmental concerns involving water issues, greenhouse gas emissions, induced earthquakes and human health [25].

Hydraulic fracturing includes high-pressure injection of large quantities of water and chemicals to split rock apart and release natural gas; the protection of water resources in areas of hydraulic fracturing can be challenging. The amount of water used can be reduced by recycling the used water, although this is challenging due to contaminants. Another problem related to water use is the potential for water contamination because several of the chemicals used are toxic and carcinogenic. In addition, methane contamination of groundwater is also a risk [25].

The life cycle greenhouse gas emissions associated with the use of shale gas are still debated, particularly those associated with methane leaks during well completion and extraction. There are authors who argue that the use of shale gas is better for the climate than coal and oil, whereas others have stated that shale gas may be worse than coal [25].

2.6.2 Renewable Energy Sources

The annual flow of solar energy is greater than all fossil and uranium reserves and resources that are presently known. The energy from the sun can be used directly after conversion to heat and electricity or be used after its natural conversion to flowing water, wind, waves and biomass. The annual flows of renewable energy and their technical potentials are shown in Table 2.4. Biomass was the primary source of energy for humans before the nineteenth century and remains the primary source in the least-developed countries.[3] Biomass is biological material that is derived from agricultural crops, forest products, aquatic plants, crop residue, animal

[3]The traditional use of biomass is associated with the inefficient use of animal dung, wood, charcoal and crop residues for domestic cooking and heating.

Table 2.4 Renewable energy flows, potentials and utilisations [21]

	Utilisation in 2005 (EJ)	Technical potential (EJ/year)	Annual flows (EJ/year)
Biomass, municipal solid waste, etc.	46.3	160–270	2200
Geothermal	2.3	810–1545	1500
Hydro	11.7	50–60	200
Solar	0.5	62,000–280,000	3,900,000
Wind	1.3	1250–2250	110,000
Ocean[a]	–	3240–10,500	1,000,000

The data are expressed as energy inputs. Considerable energy losses occur during conversion to useful energy carriers, such as electricity, heat or fuels; these losses depend on the specific technology

[a]Ocean energy refers to the kinetic energy carried by waves, tides and currents and the potential energy stored in ocean salinity and temperature differences

manure and waste. The annual flows of biomass are considerably larger than current global energy consumption, although harvesting large factions of the available biomass would result in severe adverse impacts on biodiversity, resilience and the Earth's ecosystems. Therefore, the following question must be addressed: how much utilisation is ecologically sustainable and socioeconomically desirable? The global technical potential has been estimated to range between 160 and 270 EJ per year [21]. Of this potential, dedicated energy crops, crop residues, manure, municipal solid waste, and forestry represent 44–133, 49, 39, 11 and 19–35 EJ, respectively. This estimated global potential is lower than previous estimates and is primarily based on lower expectations for growing dedicated energy crops [21].[4] However, these types of estimates are highly uncertain. These potentials can be compared with the world's primary bioenergy production of approximately 55 EJ 2012 [27] and the world's biofuel production of approximately 11 EJ in 2011 [28].

There is a potential for the shipping industry to using wind, solar and wave power directly or electrofuels produced from renewable sources. Electrofuels are discussed in Box 10.9.

2.7 Human Impacts and Environmental Issues

This chapter discusses the basic functions of the Earth's systems. These systems have been relatively stable for the last 10,000 years [29], although increasing evidence exists that human activities are threatening the functions of the Earth's systems and their stability. Human activities that have environmental impacts concern the alteration of the natural environment. Fundamentally, environmental

[4]Estimates in the range of 100 EJ per year to greater than 400 EJ per year in 2050 have also been suggested [28]. The future availability of land and the yields of energy crop production are very uncertain and a major source of the wide range in the estimates.

impacts involve either the *depletion of resources* (consumption) or the *pollution of sinks* (production of waste). Depletion occurs when accelerated cycling and flow of matter and energy occur faster than natural processes are able to renew them. Conversely, pollution occurs when the accelerated cycling and flow of matter are discharged into the environment faster than they can be purified and/or broken down.

Due to increased human population, the impacts on the natural environment have also increased, causing deforestation, overfishing and loss of biodiversity. The burning of fossil fuels began during the industrial age and has resulted in atmospheric deposition of the by-products of burning fossil fuels. Over time, the release of carbon dioxide from this burning has become a serious problem. The increased CO_2 levels in the atmosphere have caused climate change and the connected issue of ocean acidification. Increased intensity and industrial forms of agriculture to feed the growing population has yielded changes in land use and contributed to the eutrophication of aquatic systems. The impacts of deforestation, overfishing and loss of biodiversity have dramatically increased over time. Furthermore, the manufacturing of chemical products has introduced chemical pollution problems via the direct impacts of toxic compounds on organisms and through sub-lethal effects or interactions with other compounds in secondary reactions, for example, chlorofluorocarbons (CFCs), which have been shown to cause ozone depletion.

Several environmental issues exist, and various ways of grouping these issues together have been developed. One group of researchers has suggested nine environmental issues to be the most critical: (1) stratospheric ozone depletion, (2) loss of biodiversity, (3) chemical pollution and the release of novel entities, (4) climate change, (5) ocean acidification, (6) freshwater consumption and the global hydrological cycle, (7) land system change, (8) alteration of biogeochemical flows and (9) atmospheric aerosol loading [29]. We use this division of environmental issues in a slightly modified way below to describe the environmental issues related to the shipping industry. The ninth category is modified to include all types of air pollution because shipping is a large contributor to air pollution. This research group has developed planetary boundaries for seven of the nine environmental issued mentioned above [29, 30]. The planetary boundaries identify levels of man-made disruptions of the Earth's systems below which the risk of destabilising the Earth's systems are likely to remain low. Four of these environmental issues, namely, (1) climate change, (2) land system change, (3) loss of biosphere integrity and (4) alteration of biogeochemical flows to the biosphere and oceans, are suggested to already have surpassed the planetary boundary. The latter two are considered to be the most severe [30]. Here follows an overview regarding the status of the nine environmental issues.

2.7.1 Stratospheric Ozone Depletion

Stratospheric ozone depletion is an anthropogenic problem originating from the use of chemicals called CFCs, which are stable, odourless, nontoxic, noncorrosive and cheap to produce. Known areas of usage are as coolants for refrigeration and air-conditioning, cleaning solvents in electric products and propellants for aerosol sprays. In 1974, scientist raised alarms that CFCs were likely reducing the ozone concentration in the stratosphere. The chlorine atoms in CFCs react with ozone and remove it through a series of subsequent reactions. As the ozone layer thins, more UV radiation reaches the Earth's surface; the increased radiation can induce skin cancer in humans and have detrimental effects on crops, plants and microscopic marine plants (phytoplankton). It has also been shown that the depletion of ozone affects precipitation trends, foremost in the Southern Hemisphere [31]. Ozone depletion has occurred globally, although the ozone layer dramatically decreased over Antarctica during September and October. By 2001, the ozone hole covered an area larger than North America. Furthermore, another ozone hole was detected over the North Pole several years later, although this ozone hole was not as large. Based on the 1987 Montreal Protocol, more than 40 countries agreed to reduce CFC emissions by 50 % until the year 2000. In 1990 a new agreement was reached to completely phase out CFCs in the early twenty-first century. Subsequent to this phase out, CFC levels have declined, and the ozone concentration in the stratosphere has once again increased, although at slower rates than expected [8, 32]. The introduction of new types of refrigerants has contributed to solving the problem with ozone depletion, although several of the new chemicals have large greenhouse gas potentials (see Sect. 5.7).

2.7.2 Loss of Biodiversity

Substantial loss of biodiversity on Earth has occurred during the past 50 years. Humans have caused this loss by rapid and extensive ecosystem changes in an attempt to meet a growing demand for food, freshwater, timber, fibre and fuel. The changes to ecosystems have helped increase human well-being and economic development, although they have also degraded many ecosystem services. Ecosystem services can be defined as "the benefits people obtain from ecosystems" [33]. Ecosystem services can be divided into four types: (1) provisioning, such as food and water; (2) regulating, including diseases and wastes; (3) cultural, providing recreational and spiritual benefits; and (4) supporting, such as soil formation and nutrient cycling. Humans are fundamentally dependent on the flow of ecosystem services. Two examples of ecosystem services that have been degraded are fisheries and freshwater. Important commercial fish stocks are overharvested, and global freshwater usage exceeds long-term accessible supplies [33].

2.7.3 Chemical Pollution and the Release of Novel Entities

Novel entities are considered to be new substances, new forms of existing substances and modified life forms that can cause disturbances to the Earth's systems if they are persistent, have the potential to spread across systems and impact vital functions of those systems. These entities may include chemicals and other new types of man-made materials or organisms not previously known to the Earth's systems as well as naturally occurring elements (for example, heavy metals) mobilised by anthropogenic activities. Emissions of toxic and long-lived substances represent some of the key anthropogenic changes to the natural environment. These compounds can have potentially irreversible effects on living organisms and on the physical environment. In ship operations, chemical pollution can result from the use of antifouling paints or the release of oil into the environment.

When an anthropogenic pollutant ends up in the marine environment, several factors govern its fate, including the chemical characteristics of the pollutant and the features of the ecosystem to which it is discharged. Chemical characteristics, such as volatility and hydrophilicity, determine the compartment (i.e., air, water column, or sediment) in which the pollutant will predominantly exist. *Hydrophilic compounds* are water soluble and will occur in the water column, whereas *hydrophobic compounds* have low water solubility and will, to a greater extent, occur in sediments or bound to particles or organisms in the water column. Another factor that is important for the fate of pollutants is *persistence.* When a compound remains in the environment in an unchanged form for a long time, it is persistent. Inorganic pollutants (e.g., lead and arsenic) are completely persistent, i.e., not degradable, whereas some antifouling biocides have a very low persistence and are rapidly degraded within 24 h [34, 35]. Organic pollutants (e.g., polycyclic aromatic hydrocarbons [PAHs] and polychlorinated biphenyls [PCBs]) are differentially persistent. Half-lives of chemicals are listed to aid in understanding the degradation time of a pollutant. *Chemical degradation* can occur in three ways: hydrolysis, which occurs when a water molecule is added to a substance, initiating the chemical bonds to break; photolysis, which occurs when light or UV radiation breaks the chemical bonds in the molecule; or biodegradation, which occurs when organisms metabolically break down a compound [35]. Furthermore, apart from the chemical characteristics that influence persistence, features of an ecosystem influence the degradation of pollutants. For example, the oxygen content is an important factor that greatly affects the rate of biodegradation. Other factors, such as the water temperature and nutrient levels, also contribute to degradation rates [36].

2.7.3.1 Uptake of Pollutants in Organisms

A central concept for understanding how pollutants are assimilated in organisms is *bioavailability*. When a pollutant is discharged into the environment, a portion of the molecules might not be available for uptake in organisms because a fraction binds to particles or chemical complexes in the water, causing them to become unavailable. Bioavailability is a condition where the extent of the pollutant, that absorbs onto or into and across biological membranes (available for uptake) is expressed as a fraction of the total amount of pollutant to which the biota is exposed [37]. The bioavailability of an anthropogenic pollutant differs depending on the physical, chemical and biological conditions under which the biota are exposed to the compound and contributes greatly to determine the produced toxic effect in the biota [38].

Certain substances have the ability to be assimilated in organisms to a greater extent than in the surrounding water. Hydrophobic compounds do not dissolve in water to a great extent but will instead be readily stored in the lipids of organisms. This results in a higher concentration of the pollutant in the organism even though the concentration in the water might be low. This uptake from the water onto or into organisms is termed *bioconcentration*. In addition to taking up pollutants through the water, animals can also assimilate pollutants through their food. When an anthropogenic substance accumulates over time in an organism through the food or water at a rate greater than it is lost, this is termed *bioaccumulation*. Highly hydrophobic pollutants, such as PCBs, bioconcentrate/bioaccumulate in organisms and become stored in fat deposits. When organisms higher up in the food chain consume prey, each with accumulated toxicants, the predator may, as a consequence, acquire large amounts of the substance. If the substance is persistent and hard to degrade, the toxicant will further concentrate in the predator. This process of enrichment of a pollutant through the food chain is termed *biomagnification*

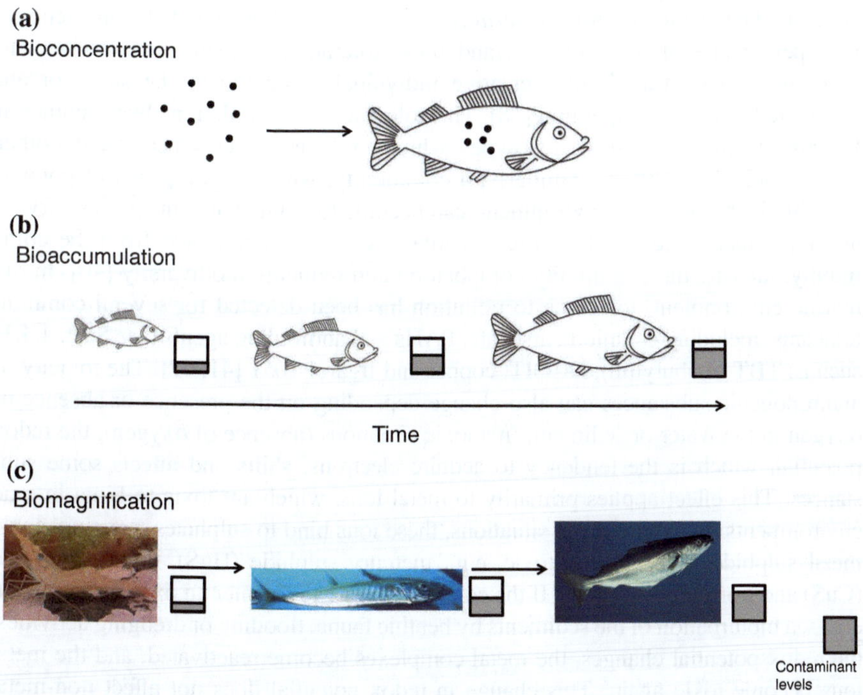

Fig. 2.11 Various ways through which an anthropogenic pollutant can enter a marine food web: **a** bioconcentration, **b** bioaccumulation and **c** biomagnification

(Fig. 2.11) [35]. It is important to note that organisms also have the ability to perform detoxification, where harmful chemicals are degraded and excreted. The ability to degrade and excrete a chemical is an important factor that determines the total effect of a pollutant. These processes occur via several mechanisms; the ability of a species to perform these processes differs widely between groups of organisms [36].

2.7.3.2 Effects of Pollutants

Anthropogenic substances can have different consequences in organisms, ranging from lethal to sub-lethal effects. While lethal effects lead to death in the organism, sub-lethal effects can instead result in reduced growth, changed behaviour, lowered reproductive success, lowered tolerance to other types of stress etc. For example large oil spills commonly lead to lethal effects in organisms due to *narcosis*, which occurs when the chemicals dissolve in the lipids of the cell membranes and cause the cells to leak, whereby the ability to maintain a normal constant internal environment (homeostasis) cannot be upheld, having mortal effects [39]. In contrast, small spills can reduce reproductive success [38]. If an area has been exposed to a pollutant for an extended period of time, populations of organisms can develop a tolerance to the substance through continuous additions or slow degradation. As a result, the negative effects can be masked, providing the appearance of a healthy population. However, the composition of the tolerant population can be changed, with alterations in the genetic diversity. *Tolerance* to toxicants is often encoded in the genome in a percentage of a population, and these tolerant organisms can survive and reproduce; in contrast, more sensitive individuals succumb to the stress or are eliminated through competition [38]. This tolerance in a population always comes at the cost of a loss of sensitive genotypes, which might have been superior under other environmental conditions. Similarly, a community, which is composed of populations of all species in an environment, can become tolerant. Tolerant species survive and reproduce, whereas the more sensitive species are eliminated from the community, altering the community composition and reducing biodiversity [40]. In the marine environment, tolerance to pollution has been detected for several common toxicants, including cadmium, arsenate, PAHs and antifouling agents (see Sect. 4.4.1) such as TBT (Tributyltin), DCOIT, copper and Irgarol 1051 [41, 42]. The toxicity of anthropogenic substances can also change depending on the presence or absence of oxygen in the water or sediment. In anoxic situations (absence of oxygen), the redox potential, which is the tendency to acquire electrons, shifts and affects some substances. This effect applies primarily to metal ions, which are toxic to biota in oxic environments, while in anoxic situations, these ions bind to sulphates, creating stable metal sulphides that are non-toxic, e.g., mercury sulphide (HgS), copper sulphide (CuS) and iron sulphide (FeS). If the complexes were to re-enter an oxic environment, e.g., via bioturbation of the sediments by benthic fauna, flooding or dredging activities, the redox potential changes, the metal complexes become reactivated, and the metal ions become toxic again. This change in redox potential does not affect non-metal substances, such as PAHs and PCBs [38].

The effects of toxicants are generally assessed as if pollutants independently affect an organism in a *dose-response manner*, i.e., higher doses result in larger effects. However, because numerous anthropogenic chemicals will inevitably be discharged into the environment, aquatic organisms are never exposed to one chemical at a time. Instead, these organisms are exposed to a mixture of different chemicals [38]. Streams in the vicinity of agricultural areas were for example found to contain 10–22 different pesticides simultaneously [43]. Instead of separately studying every pollutant to assess toxicity, the summation of toxicities from all individual compounds in the mixture, i.e., the *additive effect*, must be investigated. A clear example of this effect is given by Faust et al. [44], in which the single-cell algae *Scenedesmus vacuolatus* was separately exposed to 18 different herbicides. The concentration that provoked a 1 % toxic effect for each herbicide was established, and the algae were jointly exposed to the 1 % effect concentrations of the different herbicides. Alarmingly, the 1 % effect of 18 toxicants translated into a mixture effect of almost 50 %. This implies that even small discharges, leading to low concentrations in aquatic environments, can have a significant toxic effects on biota, as the summed toxicity arise from chemicals with a shared mode of toxic action (Fig. 2.12) [38]. Another factor that can further enhance the toxicity of a mixture of pollutants is potentiation or a *synergistic effect*. The summed toxicity of a mixture can always be greater than the sum of the individual components, resulting in an increased effect on organisms. This can either occur as a result of the inhibition of detoxification systems in an organism or the activation of the pollutants by generating highly toxic metabolites through the metabolism.

Antagonistic effects in chemical mixtures are also possible, in which the sum of the toxic effects of a chemical mixture is lower than the sum of their individual toxic effects [36]. Currently, knowledge of mixture toxicity is limited because it is experimentally very challenging and time consuming to study, although the importance of the field is growing [38].

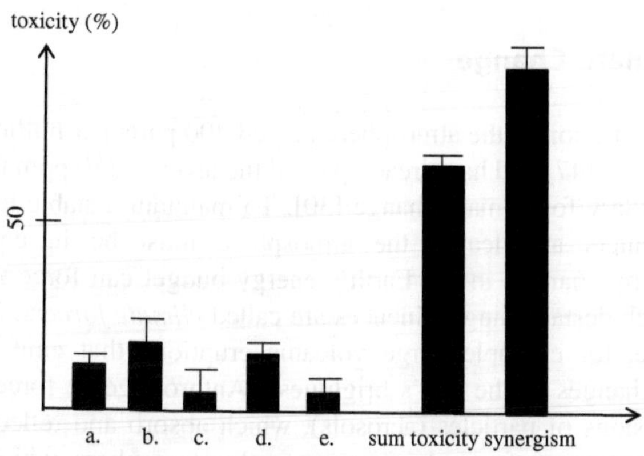

Fig. 2.12 Bars a–e represent the toxicity of various chemical compounds; the summed toxicity and synergism bars represent possible additive or synergistic effects of combined toxicities

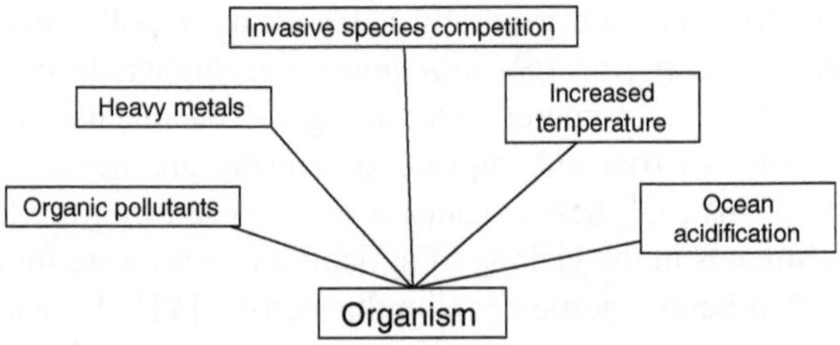

Fig. 2.13 Examples of factors that may contribute to the overall environmental stress to which an organism is exposed

In addition to exposure to a mixture of many different chemicals, communities can also be exposed to multiple types of stressors (Fig. 2.13), ranging from large-scale global stressors, such as climate change, ocean acidification, and water temperature increases, to small-scale stressors, for example, competition from invasive species, increased sedimentation rates from dredging and locally increased nutrient loads [38]. This exposure to multiple stressors can induce additive effects, ranging from a state of no observed or small effects to lethal or sub-lethal effects in individuals or changes in biodiversity and ecosystem functioning [38]. However, the reliable prediction of these effects has been difficult because it has been shown that the stressors may interact in both a non-additive and additive way [45, 46].

Considering the facts above, the discharges from shipping to the marine environment and their resulting effects are not easy to predict. The various conditions in the environment often complicate the estimation of the fate and effects of pollutants on biota. Extensive experimental studies are often needed to elucidate the individual or combined fates and effects of different toxicants in different environments. These uncertainties should be kept in mind when the different types of discharges are studied in more detail. Although the effects of pollutants can be analysed in depth experimentally, the effects on the ecosystem are rarely fully understood.

2.7.4 Climate Change

The CO_2 concentration in the atmosphere passed 400 parts per million (ppm) in the beginning of 2015 [47] and has already passed the level of 350 ppm that is set as the planetary boundary for climate change [30]. To maintain a stable temperature, the energy that enters and leaves the atmosphere must be in equilibrium (see Sect. 2.2.2). Any changes in the Earth's energy budget can force temperatures to rise or fall. Such destabilising influences are called *climate forcers*. Natural climate forcers include, for example, large volcanic eruptions that emit light-reflecting particles and changes in the sun's brightness. Anthropogenic forcers include, for example, emissions of particles (aerosols), which absorb and reflect sunlight, and the rising concentration of greenhouse gases in the atmosphere, which decrease heat radiated to space.

Table 2.5 Global warming potentials of compounds over various time horizons [10]

	Lifetime (Years)	GWP over 100 years ($kg\ CO_2$ eq./kg)[a]	GWP over 20 years ($kg\ CO_2$ eq./kg)[a]
Carbon dioxide (CO_2)	–	1	1
Methane (CH_4)	12.4	28 (34)	84 (86)
Nitrous oxide (N_2O)	121	265 (298)	264 (268)
HFC-134a	13.4	3710 (3790)	1300 (1550)
CFC 11	45	6900 (7020)	4660 (5350)
CF_4	50,000	4880 (4950)	6630 (7350)

[a]GWPs from the IPCC Fifth Assessment report; values with climate-carbon feedback are included in parentheses [10]

Greenhouse gases differ in their warming effects on the global climate system due to their different radiative properties and lifespans in the atmosphere. A common way to quantify the effectiveness of a greenhouse gas is the *global warming potential* (*GWP*). The GWP of a compound is a relative measure comparing the relative amount of heat trapped in the atmosphere to an equivalent mass of CO_2 (GWP = 1). The GWP is calculated over a specific time interval, typically for 20, 100 and 500 years, to account for the residence time of the compound. The time horizon most frequently used is 100 years. In Table 2.5, the GWP is shown for some important greenhouse gases.

The industrialisation of the Earth by humanity in the last two centuries has been fuelled by enormous quantities of fossil fuels, such as oil, coal and natural gas. Through the combustion of these fossil fuels, carbon dioxide is released and added to the natural concentration in the atmosphere. Deforestation, the overgrazing of land by domesticated animals and the altering of ground cover have also contributed to the increasing atmospheric CO_2 levels. Some of the excess CO_2 is taken up by plants or dissolved into the ocean (see Sect. 2.7.5), although 40 % remains in the atmosphere [9]. This increase in the CO_2 levels in the atmosphere has closely followed the anthropogenic CO_2 emissions resulting from the burning of fossil fuels [8]. It is extremely likely that more than half of the observed increase in global average temperature from 1951 to 2010 was caused by the increase in anthropogenic greenhouse gas emissions. The result of this increase in greenhouse gas emissions has been an increased global temperature, with a globally averaged warming of 0.85 °C from 1880 to 2012 [9].

Another gas affecting the climate is methane (CH_4), which is present in much lower amounts in the atmosphere than CO_2, approximately 1.7 ppm (0.00017 %). However, compared with CO_2, methane is more effective (see Table 2.5) at absorbing the radiation emitted by the Earth. Methane is produced by anaerobic bacteria in wet areas, e.g., swamps and wetlands, and in the guts of animals, such as cattle and sheep. The flooding of paddy fields for the cultivation of rice also produces methane. Moreover, the extraction of coal, oil and natural gas is also a source of methane to the atmosphere. Due to the increased human population, the number of rice fields and livestock and the rate of extraction of fossil fuels have increased, increasing methane emissions to the atmosphere.

A third example of a gas that affects the climate is nitrous oxide (N_2O), which is naturally present in the atmosphere as a part of the nitrogen cycle and has several sources. These molecules remain in the atmosphere for up to 120 years before being removed or destroyed. Nitrous oxide is 300 times more potent in warming the atmosphere compared to carbon dioxide. Through human activities, such as fossil fuel combustion, agriculture (the use of fertilisers), wastewater management and industrial processes, the amount of N_2O in the atmosphere is increasing. Approximately 40 % of the global nitrous oxide emissions originate from human activities [48].

Climate change impacts both the natural environment and human health through a variety of environmental mechanisms. The natural environment is affected by the loss of species due to temperature increases, by changes in oceans and seas and by the impacts of extreme weather. Human health, for example, is affected directly by heat waves and indirectly by infectious diseases and malnutrition.

2.7.5 Ocean Acidification

Oceans function as a sink for CO_2, assimilating approximately 28 % of the anthropogenic carbon that has been added to the atmosphere [9]. The emitted CO_2 dissolves in the ocean and forms carbonic acid, which slowly decreases the pH, increasing the acidity of the oceans. Since preindustrial times, the average pH levels in the oceans have decreased by 0.1 units; they are predicted to decrease by another 0.3–0.4 units. Because the pH scale is logarithmic, meaning that a pH decrease of one unit indicates a 10-fold increase in acidity, this decrease of 0.1 units means that the oceans are 26 % more acidic. This acidification decreases the calcium carbonate saturation state, causing it to be more difficult for shell-forming organisms, for example, echinoderms, mussels, corals and plankton, to assimilate calcium carbonate, which completely inhibits or decreases the growth rate of shells and skeletal components and reduces fertilisation success, larval size and developmental rates in the early developmental stages of benthic invertebrates [49]. Determining potential population decreases or species declines is challenging because the science regarding the issue is in its infancy and acidification works in tandem or synergistically with other on-going chemical and physical changes in the ocean [6]. The economic consequences of this process are difficult to determine but will probably be substantial because the effects are both direct, e.g., in the aquaculture industry and tourism, and indirect, e.g., in the fishing industry because fish may find less prey.

2.7.6 Freshwater Consumption and the Global Hydrological Cycle

Freshwater is becoming increasingly scarce; approximately 60 % of the global population risks surface water shortages in 2050 [50]. Freshwater is needed not only for drinking and sanitation but also for irrigated agriculture and industrial use. These activities depend on the withdrawal of a sufficient amount of freshwater from groundwater aquifers, rivers and lakes. Water shortages already exist in some areas, and the expected growth of the human population and economic prosperity will intensify these problems. Human activities are causing stress on the hydrological cycle through, for example, global-scale river flow changes and shifts in vapour flows arising from land-use changes. Climate change poses an additional threat to water security because changes in precipitation and other climatic variables can cause significant changes in the water supply to many regions. The problem of freshwater availability is largely not related to ship operations [50].

2.7.7 Land System Change

The global land system is being changed by humans. Many different land types have been converted to agricultural land, reducing biodiversity and impacting water flows and the geochemical cycling of elements. Forest biomes are especially important because they play a stronger role in land surface-climate linkages than other biomes. Alterations to tropical forests have substantial feedbacks on the climate through changes in evapotranspiration. Moreover, changes in boreal forests affect the albedo of the land surface, leading to changes in the energy balance [30]. Ship operations are not substantially related to land system changes, although this may change in the future due to the use of biofuels in the shipping industry.

2.7.8 Alteration of Biogeochemical Flows

Human changes in biogeochemical flows can affect the biosphere at global, regional and local scales. Changing the carbon cycle is an example of a global-scale impact. The alteration of the nitrogen and phosphorus cycles primarily results in regional- and local-scale problems.

2.7.8.1 Eutrophication

Changes in the nitrogen and phosphorus cycles can cause local- and regional-scale eutrophication. Eutrophication is associated with high levels of nutrients, leading to increased biological productivity of phytoplankton in seas and lakes, e.g., algal blooms. Nitrogen and phosphorus are the most common growth-limiting nutrients; thus, an increase will facilitate increased plankton growth. Different ecosystems are limited by different nutrients, and the sensitivity to an increase in nutrient levels varies geographically. For example, the addition of phosphorus to a limnic environment, where this is limiting, may have large effects and cause a bloom [51].

Algal blooms are followed by algal die-off, which occurs when the excess nutrients are depleted. When the organic material falls to the sea floor, microorganisms begin to decompose the material. This process requires substantial amounts of oxygen. If the amount of decomposing organic material is too large, the oxygen supply in the water will be reduced or even depleted, severely impacting organisms requiring oxygen for respiration. Furthermore, when the oxygen supply is depleted anoxic microbial activities continue the decomposition of organic material. This process is slower, and a by-product, hydrogen sulphide, which produces adverse toxic effects on plants and animals at high concentrations, is released. Furthermore, hydrogen sulphide is also toxic to man and has an unpleasant smell that makes its presence obvious even at low concentrations [51].

Nitrogen oxides are a major pollutant from ship operations as emissions to the air. Nitrogen may also be directly released to the sea via wastewater. The effects on eutrophication from nitrogen compounds are also important to consider when assessing crop-based biofuels because these fuels can be linked to the run off of fertilisers from agriculture.

2.7.8.2 Acidification of Soil and Freshwater

The acidification of soil and freshwater is usually related to changes in the nitrogen and sulphur cycles because sulphuric acid and nitrous acid are strong acids. The characteristic of an acid is its ability to release hydrogen ions in a solution (see Sect. 2.1). Thus, *acidification* means an increased concentration of hydrogen ions, usually expressed in terms of pH. Natural water is generally classified as acidic at a pH of less than 6.2 [3]. Potentially acidifying pollutants released to the air are dissolved in rainwater, therein forming "acid rain". Sulphur dioxide, nitrogen oxide and ammonia are major acidifying pollutants. Carbon dioxide is a weak acid but can also contribute to acidification, for example, in the oceans (see Sect. 2.7.5). The emission of airborne acidifying pollutants thus affects soil, groundwater, surface water, the oceans, biological organisms, ecosystems and materials (buildings). Acidification primarily affects the natural environment by decreasing biodiversity and bio-productivity [3, 59].

"Acid rain" was first discovered after several decades of research on the acidification of lakes in Sweden and Norway. In 1967, it was concluded that *sulphur dioxide* (SO_2) was being transported over long distances in air and that precipitation over Scandinavia had become more acidic, primarily due to SO_2 emissions from industries in the UK and Central Europe. Thus, the problem of acidification was not a local problem; instead, it was discovered that acidification was a large-scale regional problem [52].

Soil and rock as well as sea- and lake-water have the ability to neutralise acidic inputs to a certain degree, namely, they have a buffering capacity. If the capacity is exceeded acid water is produced. The buffering capacity for rock and soil is dependent on the mineralogy, making areas with granitic bedrock and small amounts of limestone more vulnerable [3, 53]. In lakes and oceans, the buffering capacity is dependent on the concentration of other dissolved ions, i.e., the alkalinity; see Chap. 11 [3, 59].

The chemical effects of soil acidification are first observed when acidic deposition has depleted the buffering capacity of the soil. The first effect is significant leaching of mineral nutrients. The second effect is decreased pH levels, followed by increasing concentrations of aluminium ions (Al^{3+}) in lakes and watercourses. The aluminium levels rise sharply in lakes with pH levels below 5.5. Acidification effects have caused substantial decreases in biological diversity in aquatic environments; in particular, a reduction in pH to below 6 is detrimental to many species, e.g., salmonid species. However, whereas many organisms are sensitive to low pH levels, others, for example, some phytoplankton species, are more resilient and benefit from the decline in abundance of other species [3, 36, 53]. The impacts of soil acidification also include the leaching of important nutrients, particularly base cations, such as magnesium, potassium and calcium. The loss of nutrients leads to reduced growth can eradicate sensitive species when combined with low pH level [3, 36, 54].

2.7.9 Air Pollution

Emissions to the air are connected to the weather in two ways. First, weather conditions influence the dispersal and dilution of these emissions. The second relationship is the opposite, i.e., the effects of emissions on weather and climate. Air pollution has existed for a long time, e.g., salt from breaking waves, ash from volcanoes and smoke from forest fires. However, since the introduction of man to the planet, these natural air pollutants have increased in intensity. Anthropogenic air pollutants exist as primary and secondary pollutants. Air pollutants are airborne particles and gases occurring in concentrations that endanger the health of organisms or ecosystem functions. Furthermore, air pollutants can impair visibility and damage materials in cultural heritage and buildings [55]. Primary pollutants are directly emitted from identifiable sources, whereas secondary pollutants are created through chemical reactions of primary pollutants in the atmosphere. Volatile organic compounds, nitrogen oxides, particulates, sulphur oxides and carbon monoxide are the most common primary air pollutants. Secondary pollutants include sulphuric acid and substances that are often formed under exposure to sunlight (UV radiation), e.g., the creation of ground-level ozone or the formation of photochemical smog when nitrogen oxides react with UV light [8, 56].

Air pollution in an area is primarily determined by local emissions to the atmosphere. However, atmospheric conditions, such as wind strength and air stability, can contribute significantly to changes in air quality. Weak winds do not dilute pollution, while stable air conditions with little vertical movement cannot mix the polluted air with clean air at higher altitudes. Strong winds and the substantial mixing of air can instead cause air pollutants to travel great distances, up to hundreds of kilometres [8]. In addition, when constituents from the air closer to the surface, such as particles from volcanic eruptions or persistent chemical compounds, enter the stratosphere, they will be long-lived there due to the lack of vertical movement and precipitation. A long lifetime and efficient horizontal movement via air currents can distribute a compound in the stratosphere globally. This phenomenon is widely acknowledged because organisms in remote areas, such as the Arctic and Antarctic, have been demonstrated to be affected by persistent pollutants [8].

2.7.9.1 Atmospheric Aerosol Loading

Particles, also called particulate matter, can be solid or liquid, and have diameters between 0.002 and 100 μm. The lower limit is not sharply defined, whereas the upper limit is related to the size of very fine sand or fine drizzle. Larger particles are quickly deposited from the atmosphere due to their size. Particles with diameters of approximately 0.002–10 μm are the most important in discussions on atmospheric chemistry and physics. Particles suspended in the atmosphere are often called aerosols. Aerosols consist of solid or liquid particles in a gas [57]. Atmospheric aerosols are a mixture of solid and liquid particles from both natural and anthropogenic sources [58]. The particles in aerosols vary in size and can be directly emitted to the atmosphere, i.e., primary particles, or formed in the atmosphere, i.e., secondary particles [57, 58].

Aerosols have severe human health effects and are affecting the Earth's climate system. Aerosols interact with water vapour, affecting cloud formation and patterns of atmospheric circulation, such as monsoon systems in tropical regions. Aerosols can also reflect and absorb solar radiation, thereby directly affecting the climate. Humans contribute to the formation of aerosols and particles by emitting atmospheric pollution, for example, from different combustion processes and through land-use changes, which increase the release of dust and smoke into the air. In 2007, shipping-related emissions of particulates were estimated to contribute to approximately 60,000 premature deaths worldwide [59].

2.7.9.2 Ground-Level Ozone

Ozone is present not only in the upper regions of the atmosphere, i.e., the stratosphere, but also near the surface in the troposphere. Stratospheric ozone protects us from harmful UV radiation (see Sect. 2.7.1), whereas ground-level ozone is a health hazard to humans. Ground-level ozone can, for example, inflame airways and damage lungs. It also affects sensitive vegetation and ecosystems, such as some crops and forests [60].

Ground-level ozone is not directly emitted to the troposphere. Instead, it is created by chemical reactions. Ozone formation is complex and depends on a number of factors, including the concentrations of NO, NO_2, and volatile organic compounds (VOCs) and ultraviolet radiation. The effect of these various emissions depends on the location and the background NO_x concentration.

Ground-level ozone is formed through the photolysis of NO_2 (2.1), in which atomic oxygen reacts with molecular oxygen to form ozone (2.2) [57]. The NO formed via Eq. 2.1 can rapidly react with ozone and be converted back to NO_2. However, in the presence of volatile organic compounds (VOCs), NO can be converted into NO_2 without consuming ozone, causing a net increase in ozone [57, 60]:

$$NO_2 + h\nu(\lambda < 420\,\text{nm}) \rightarrow NO + O \tag{2.1}$$

$$O + O_2 + M \rightarrow O_3 + M \tag{2.2}$$

2.8 Summary

This chapter discussed the Earth's major systems and provided a basic understanding of the most important environmental issues. The shipping industry plays a role in several of these problems both directly, for example, via the burning of fossil fuel and discharges of chemical compounds, and indirectly by transporting goods, such as fertilisers and pesticides. Governments and international agencies have long since recognised these issues; several conventions to limit the environmental impacts have been introduced (see Chap. 3). However, additional steps must be taken to further limit the impacts. By becoming informed about the various environmental problems that exist and what role the shipping industry plays in these problems, an incentive for future or existing mariners to attempt to be a part of the reduction of the human impact on nature can be created. The following part of this book will focus on ship operations and how they affect the natural environment.

References

1. Nybakken, J. W., *Marine Biology: an ecological approach*. 5th ed. 2001, San Francisco, USA: Addison Wesley Longman. 516.
2. Franklin, M. A., Suckow, S. H. & Weisbroth, C. L., *Chapter 1 - Water, air and ice*, in *Seawater*. 1995, Butterworth-Heinemann: Oxford. p. 4-13.
3. Elvingson, P. & Ågren, C., *Air and the environment*. 2004.
4. Stumm, W. & Morgan, J. J., *Aquatic Chemistry – Chemical Equilibria and Rates in Natural Waters*. 2012, New York: John Wiley & Sons.
5. Franklin, M. A., Suckow, S. H. & Weisbroth, C. L., *Chapter 3 - Salinity in the oceans*, in *Seawater*. 1995, Butterworth-Heinemann: Oxford. p. 29-38.
6. Guinotte, J. M. & Fabry, V. J., *Ocean Acidification and Its Potential Effects on Marine Ecosystems*. Annals of the New York Academy of Sciences, 2008. **1134**(1): p. 320-342.
7. HELCOM. *Atlas of the Baltic Sea*. 2010.
8. Lutgens, F. K. & Tarbuck, E. J., *The Atmosphere, an introduction to meteorology*. 2004: Pearson Education Inc.
9. IPCC. *IPCC Fifth Assessment Synthesis Report*. 2013.
10. IPCC, *Climate Change 2013: The Physical Science Basis. Contribution of Working Group I to the Fourth Assessment. Report of the Intergovernmental Panel on Climate Change.*, T.F. Stocker *et al.*, Editors. 2013: Cambridge, England and USA. p. 1535.

11. Arrhenius, S., *On the Influence of Carbonic Acid in the Air upon the Temperature of the Ground.* Philosophical magazine and journal of science, 1896(41): p. 237-276.
12. Wiscombe, W. J. & Warren, S. G., *A Model for the Spectral Albedo of Snow. I: Pure Snow.* Journal of the Atmospheric Sciences, 1980. **37**(12): p. 2712-2733.
13. IIED, *Breaking New Ground: Mining, Minerals, and Sustainable Development.* 2002, London, UK: Earthscan.
14. IMO. *Revised guidelines for the indentification and designation of particularly sensitive sea areas* 2005, IMO: London, UK.
15. Schulze, E.-D. *et al.*, *Global biogeochemical cycles in the climate system.* 2001, London, UK: Academic Press.
16. Bernhard, A., *The Nitrogen cycle: Processes, players and human impact*, in *Nature Education Knowledge*. 2010. p. 25.
17. Painter, H. A., *A review of the literature on inorganic nitrogen metabolism in microorganisms.* Water Research, 1970. **4**: p. 393-450.
18. Randall, M. J. & Ingraham, J. L., *The denitrifying procaryotes.* The prokaryotes, a handbook on habitats, isolation and identification of bacteria, ed. P. Staar. 1981: Springer-Verlag.
19. Jensen, M. M., Thamdrup, B. & Dalsgaard, T., *Effects of specific inhibitors on anammox and denitrification in marine sediments.* Applied and Environmental Microbiology, 2007. **73**(10): p. 3151-3158.
20. IEA. *World Energy Outlook 2012*. 2012, International Energy Agency: Paris.
21. Rogner H-H *et al.*, *Chapter 7: Energy resources and potentials*, in *Global Energy Assessment: Toward a Sustainable Future*, GEA Writing Team, Editor. 2012, Cambridge University Press and IIASA. p. 423-512.
22. Aitani, A. M., *Oil Refining and Products*, in *Encyclopedia of Energy*, J. Cleveland Cutler, Editor. 2004, Elsevier: New York. p. 715-729.
23. Alfke, G., W., I. W. & Neuwirth, O. S., *Oil Refining*, in *Ullmann's Encyclopedia of Industrial Chemistry*. 2007, Wiley-VCH Verlag.
24. ExxonMobil Australia Pty Ltd. *A Simple Guide to oil Refining*. 2006.
25. Wang, Q., Chen, X., Jha, A. N. & Rogers, H., *Natural gas from shale formation – The evolution, evidences and challenges of shale gas revolution in United States.* Renewable and Sustainable Energy Reviews, 2014. **30**(0): p. 1-28.
26. Vello A. K., Scott H. S. & Keith D. M. *World shale gas and shale oil resource assessment* 2013, Advanced Resources International, Inc. and EIA US.
27. REN21. *Renewables 2013 Global Status Report*. 2013, Renewable Energy Policy Network for the 21st Century: Paris.
28. EIA. *International Energy Statistics, Biofuels Production*. 2014, U.S. Energy Information Administration, 2014-01-23.
29. Rockstrom, J. *et al.*, *A safe operating space for humanity.* Nature, 2009. **461**(7263): p. 472-475.
30. Steffen, W. *et al.*, *Planetary boundaries: Guiding human development on a changing planet.* Science, 2015. **347**(6223).
31. Gonzalez, P. M., Polvani, L. M., Seager, R. & Correa, G. J. P., *Stratospheric ozone depletion: a key driver of recent precipitation trends in South Eastern South America.* Clim Dyn, 2014. **42**(7-8): p. 1775-1792.
32. NASA. *Ozone-depleting compound persists*. 2014: ScienceDaily.
33. Millennium Ecosystem Assessment. *Ecosystems and Human Well-being: Synthesis*. 2005: Washington, DC.
34. Jakobson, A. H. & Willingham, G. L., *Sea-nine antifoulant: an environmentally acceptable alternative to organotin antifoulants.* Science and the Total Environment, 2000. **258**: p. 103-110.
35. Shaw, I. C. & Chadwick, J., *Principles of Environmental Toxicology*. 1998, London, UK: Taylor & Francis. 216.

36. Walker, C. H., Hopkin, S. P., Sibly, R. M. & Peakall, D. B., *Principles of ecotoxicology*. 3rd ed. 2006, Boca Raton, USA: Taylor and Francis.
37. Newman, M. C., *Fundamentals of Ecotoxicology*. 3rd ed. 2009, Boca Raton: CRC Press.
38. Lindgren, J. F. *Evaluating effects of low concentrations of oil in marine benthic communities.* Doctor of Philosophy thesis, Chalmers University of Technology, 2015.
39. Burgess, R., M. *Evaluating ecological risk to invertebrate receptors from PAHs in sediments at hazardous waste sites*. 2009: Cincinnati, OH. . p. 23.
40. Blanck, H. & Wängberg, S.-Å., *Induced community tolerance in marine periphyton established under arsenate stress.* Canadian Journal of Fisheries and Aquatic Sciences, 1988. **45**: p. 1816-1819.
41. Blanck, H., *A Critical Review of Procedures and Approaches Used for Assessing Pollution-Induced Community Tolerance (PICT) in Biotic Communities.* Human and Ecological Risk Assessment, 2002. **8**(6): p. 1003-1034.
42. Lindgren, J. F., Hassellöv, I. M., Landquist, H. & Dahllöf, I., *Low concentrations of PAHs induce tolerance in nitrifying bacteria.* Frontiers in Marine Science, 2015. **2**(35): p. 1-7.
43. Backhaus, T. & Faust, M., *Predictive Environmental Risk Assessment of Chemical Mixtures: A Conceptual Framework.* Environ Sci Technol., 2012. **46**(2564-2573).
44. Faust, M. *et al.*, *Predicting the joint algal toxicity of multi-component s-triazine mixtures at low-effect concentrations of individual toxicants.* Aquatic Toxicology, 2001. **56**(1): p. 13-32.
45. Crain, C. M., Kroeker, K. & Halpern, B. S., *Interactive and cumulative effects of multiple human stressors in marine systems.* Ecology letters, 2008. **11**: p. 1304-1315.
46. Folt, C. L., Chen, C. Y., Moore, M. V. & Burnaford, J., *Synergism and antagonism among multiple stressors.* Limnology and Oceanography, 1999. **44**: p. 864-877.
47. Dlugokencky, E. & Tans, P. *Trends in Atmospheric Carbon Dioxide*. 2014, NOAA/ ESRL, www.esrl.noaa.gov/gmd/ccgg/trends/, last accessed on 20 May 2015.
48. EPA. *Methane and Nitrous Oxide emissions from natural sources*. 2010: Washington DC, USA.
49. Doney, S. C., Fabry, V. J., Feely, R. A. & Kleypas, J. A., *Ocean Acidification: The Other CO_2 Problem.* Annual Review of Marine Science, 2009. **1**(1): p. 169-192.
50. Rockström, J. *et al.*, *Future water availability for global food production: The potential of green water for increasing resilience to global change.* Water Resources Research, 2009. **45** (7): p. n/a-n/a.
51. Ansari, A. A., Sarvajeet, S. G., Lanza, G. R. & Rast, W., *Eutrophication: causes, consequences and control*. 2011: Springer.
52. Lidskog, R. & Sundquist, G. *Regulating european air: The coproduction of science and policy, in Transboundary Air Pollution: Scientific Understanding and Environmental Policy in Europe, Pleijel, H. (ed.)*. 2007: Lund, Sweden.
53. Warfvinge, P. & Bertills, U. *Naturens återhämtning från försurning – aktuell kunskap och framtidsscenarier. Report 5028*. 2000, Swedish Environmental Protection Agency.
54. NEGTAP, *Transboundary air pollution: Acidification, eutrophication and ground-level ozone in the UK* 2001: DEFRA.
55. Masiol, M. & Harrison, R. M., *Aircraft engine exhaust emissions and other airport-related contributions to ambient air pollution: A review.* Atmospheric Environment, 2014. **95**(0): p. 409-455.
56. Elningson, P. & Ågren, C., *Air and the environment*. 2004: Swedish NGO Secretariat on Acid Rain.
57. Finlayson-Pitts, B. J. & Pitts Jr., J. N., *Chemistry of the Upper and Lower Atmosphere—Theory, Experiments and Applications.* 2000, USA: Academic Press
58. Hinds, W. C., *Aerosol technology: Properties, Behavior, and Measurement of Airborne Particles*. 1999, New Jersey, USA: John Wiley & Sons, Inc.

59. Corbett, J. J. *et al.*, *Mortality from ship emissions: A global assessment.* Environmental Science and Technology, 2007. **41**(24): p. 8512-8518.
60. Pleijel, H., *Air Pollution & Climate Change, Two sides of the same coin?* 2009, Solna, Sweden: Swedish Environmental Protection Agency.

Regulating Pollution from Ships

3

Philip Linné and Erik Svensson

Abstract

This chapter begins with a short history of the regulation of ship operations, including the regulation of pollution from ships, and then proceeds to its main focus of explaining the basic international legal framework for regulating pollution from ships, the main actors involved in the international regulatory process, and the process of creating environmental regulations for ships via the International Maritime Organization (IMO). The regulation of ship operations has a long history, although the specific regulation of pollution from ships is a relatively recent phenomenon. Over the course of history, the freedom to use the seas in various ways (the principle of freedom) has been balanced by the interests of sovereign States (the principle of sovereignty). As is demonstrated in this chapter, to explain the regulation of pollution from ships at the international regulatory level, a basic understanding of international law and the international law of the sea is necessary. However, the regulation of pollution from ships should also be viewed as a result of negotiations among States with different interests and with different economic and environmental conditions. In this regard, two divisions can be helpful in understanding how the international instruments that regulate pollution from ships are created via IMO: the first is between *coastal States*, *flag States* and *States with maritime interests*, and the second is between *developed States* and *developing States*.

P. Linné (✉)
Department of Law, University of Gothenburg, Gothenburg, Sweden
e-mail: philip.linne@law.gu.se

E. Svensson
Gothenburg, Sweden
e-mail: e-svensson@outlook.com

K. Andersson et al. (eds.), *Shipping and the Environment*,
DOI 10.1007/978-3-662-49045-7_3

Keywords
Regulation · Pollution · Ships · International law · The law of the sea · LOSC · Jurisdiction · Maritime zones · MARPOL 73/78 · International agreements · International Maritime Organization · IMO · MEPC · Actors

This chapter provides an overview of international regulations on marine pollution from ships. However, because this topic is extensive, it cannot fully be covered in a short chapter. Although later chapters in this book include details of specific international agreements,[1] the current chapter takes a broader perspective. This chapter provides a short history and then examines the basic legal instruments, actors, fora, and processes involved in the regulation of pollution from ships. "Regulation" can have many different definitions. For the purposes of this chapter, regulation should mainly be understood as referring to binding international legal instruments to control pollution from ships, for example, in the form of international conventions drafted with the backing of the International Maritime Organization (IMO). Although softer, formally non-binding instruments, such as guidelines and recommendations, can also be included under a broader understanding of regulation, these are not specifically addressed in this chapter.

The focus of this chapter is on the international regulatory level and on matters that are closer to the *law of the sea* as a part of *international law* than to *maritime law*. Here, lawyers distinguish between domestic and private international *maritime law* and the *international law of the sea* to indicate fields treating different, although overlapping, matters. Maritime law in both its domestic and private international forms typically includes the carriage of goods and passengers, marine insurance and maritime damage claims [2]. In contrast, the international law of the sea, is typically discussed as a part of public international law[2] and concerns the rules and principles governing the rights and duties of States in their marine activities (see Sect. 3.3.2).

In addition, this introduction also provides a short history of the regulation of ship operations, presenting two overarching principles for the law of the sea that are essential for the forthcoming sections of this chapter. However, an additional clarification of the focus of this chapter must be highlighted. The regulation and prevention of marine pollution from ships are closely associated with avoiding various maritime risks. Although maritime safety instruments, such as the 1974 International Convention for the Safety of Life at Sea (SOLAS) [3] and the 1978 International Convention on Standards of Training, Certification and Watchkeeping

---

[1]Henceforth, the common terms "international agreement", "convention" and "treaty" are used interchangeably to indicate the same type of legal instrument, namely "written agreements whereby the states participating bind themselves legally to act in a particular way or to set up particular relations between themselves" (p. 93) [1] (see also Sect. 3.5.1).

[2]A distinction is commonly made between *public* and *private* international law. The former mainly addresses the relations between States, while the latter distinguishes which country's rules should apply, e.g., in a legal conflict between two persons doing business that involves two States' legal systems [1]. Henceforth, when the term "international law" is used in this chapter, it refers to *public international law*. For more information regarding international law, see Sect. 3.3.1.

for Seafarers (STCW Convention) [4] are important for preventing both incidents and pollution, this chapter will not discuss the regulation of maritime safety nor the related area of maritime security (e.g., regulations concerning threats to security at sea, including piracy and terrorism). Furthermore, this chapter is concentrated on the regulation of pollution from ships during *normal ship operations*, i.e., when ships are on international voyages as one of many activities included in shipping.

3.1 A Short History of the Regulation of Ship Operations

Traditionally, the seas have had and still have important functions for sustaining human life and society. For centuries, humans have used the seas in various ways, and a need for regulation has arisen because of these human activities. Remarkably, many of the first marine activities to be regulated have maintained their relevance to this day.

Some of the earliest examples of maritime regulations are believed to originate from shipping in the Mediterranean Sea area and the Old Kingdom of Egypt, i.e., approximately 3000 B.C. Even if no preserved written legal sources from this era have yet been discovered, the Egyptian overseas trade presumably could not have evolved in the way it did in the absence of regulations surrounding river and sea navigation. Later, during King Hammurabi's rule of Sumer, presumably 1711–1669 B.C., regulations regarding various maritime activities were included in the Code of Hammurabi, which is commonly considered to be the oldest preserved legal code in history. In ten of the 282 sections, King Hammurabi's code elaborated on rights and duties regarding matters related to shipwrights, maritime collisions, ship owners, sailors, hiring, payment, and the liability of captains [5].

The law of the sea can be traced back to the infancy of international law [6]. In the beginning, the law of the sea was based on two overarching and conflicting principles: *the principle of freedom* and *the principle of sovereignty*. As will be demonstrated below, balancing these two basic principles is still significant when discussing pollution from ships and what States can legally do to protect their marine spaces from environmental impacts. Historically, traditional maritime States, such as the Netherlands, defended *the principle of freedom* of the seas to safeguard the freedom of navigation. More specifically, the principle of freedom was claimed to ensure that no State could limit another State's access to the oceans. The freedom of the seas was a necessity for expanding trade via ocean routes to other parts of the world. For example, the lawyer Hugo Grotius (1583–1645) argued for the freedom of the seas in his book *Mare Liberum* (The Freedom of the Seas) in 1609 in support of the Dutch East India Company's position against Portugal's claim to exclusive trade in the Far East [6, 7].

If the principle of freedom was used to defend the unhindered navigation of the oceans by maritime States in the early era of the law of the sea, *the principle of sovereignty* (self-governance) represents its balancing counterpart. The principle of sovereignty protects the interests of States along whose coasts other States' ships operate. This principle is the foundation of not only the extension of coastal States' control over different sea areas but also to their claims of control over different sea areas (see Sect. 3.3.2). Traditionally, the principle of freedom had a strong position on the oceans. The territorial sea of States was restricted to only a small strip near their coasts, and the vast majority of ocean areas were considered to be high seas.

After World War II, the principle of sovereignty began to gain strength as a counterbalancing principle, and coastal States gradually began to extend their national jurisdiction over sea areas. The sovereignty over an expanded territorial sea became increasingly important for States to defend because of customs control, security, sanitary regulations and fishing [6]. Moreover, claims over other sea areas for economic purposes also followed with time (see Sect. 3.3.2).

To summarise, this introduction has demonstrated that the regulation of various activities on the seas extends far back into history and that the regulation of these activities is still relevant today. The principle of freedom and the principle of sovereignty are two basic principles that have shaped the law of the sea. On the one hand, the principle of freedom has been used by States to defend the freedom to use the seas for various purposes. On the other hand, the principle of sovereignty has been used by coastal States to secure control over their sea areas. A common point for these principles is that they have been used to defend the economic and political interests of both maritime and coastal States.

3.2 The History of the Regulation of Pollution from Ships

Both early *maritime law* and *the law of the sea* evolved to a considerable extent because of economic interests of States connected to trade, which continues to remain important. However, what are the motivations for specifically regulating the pollution of the marine environment from ships?

Regulation to protect the marine environment from pollution is a relatively recent occurrence, beginning only after World War II.[3] Before this time, international competition among States for living resources, such as fish, fur seals and whales, resulted in several international treaties regarding the seas in the early twentieth century. However, these treaties only focused on the marine environment to the extent that commercial over-exploitation of certain living resources was regulated, not the protection of the marine environment *from pollution.* The slow beginning and marginal attention of States to regulating pollution in the marine environment before World War II may be partially explained by the limited

[3]However, exceptions exist, such as the draft convention on pollution from ships, which was prepared in 1926 but never signed by any State [8, 9].

experience of the more profound effects of marine pollution coupled with the limited scientific understanding of the sea environment at that time [8].[4]

With a growing global economy, greater demand for petroleum products in the 1950's, and the increasing number and size of tankers at sea, oil pollution issues were given increasing attention. In 1954, the International Convention for the Prevention of Pollution of the Sea by Oil (OILPOL 1954) [13], the first broadly accepted multilateral agreement[5] addressing an international pollution problem, was adopted at a conference in London. Among other issues, OILPOL 1954 applied to tanker vessels and prohibited discharges of oil and oily mixtures from ships in specifically defined zones at sea. The convention was primarily created to reduce *operational pollution* from ships, arising from such activities as de-ballasting and tank cleaning [14]. However, OILPOL 1954 never became very successful. It was difficult to negotiate because several maritime States believed it was restricting the freedom of the seas. Furthermore, enforcement problems occurred because it relied to a large extent on the full cooperation of flag States for enforcement (see Sect. 3.3.2). Many flag States were less interested in enforcement beyond their territorial jurisdiction or were not actually parties to OILPOL 1954 [8, 15].

In March 1967, the grounding of the tanker *Torrey Canyon* southwest of the coast of England caused the largest marine oil spill in history at that time. Approximately 80,000 tonnes of its 120,000-tonne crude oil cargo escaped and ultimately polluted the coasts of Britain and France. The incident received extensive media coverage and attracted great public opinion. The fate of the *Torrey Canyon* became a catalyst for a new era of environmental regulation for ships that would lead to the creation of several new conventions. For example, two conventions directly resulting from the wake of the *Torrey Canyon* incident were the 1969 International Convention on Civil Liability for Oil Pollution Damage [16] and the 1969 International Convention Relating to Intervention on the High Seas in Cases of Oil Pollution Casualties [15, 17] (for a perspective on the role of accidents as a catalyst for regulation, see Box 3.1). At this time, incidents similar to the grounding of the *Torrey Canyon* clearly demonstrated the dangers of the daily activities occurring on the seas, with effects that could seriously affect human coastal populations and fish resources [8]. Notably, these incidents also involved polluting substances crossing national borders and even the barrier between the sea and land environments.

With the *Torrey Canyon* still in mind, a few additional important contemporary events raised environmental consciousness that extended beyond the marine environment. Developments in environmental regulation beginning in the late 1960s must also be viewed against the backdrop of a newly awakened environmental movement in the early 1960s that was inspired by various publications, such as Rachel Carson's *Silent Spring* [18]. In the early 1970s, Nordic scientists brought the transboundary environmental problem of "acid rain" (see Sect. 2.7.8) to the

[4]E.g., 1911 Convention for the Preservation and Protection of Fur Seals [10], 1923 Convention for the Preservation of Halibut Fishery of the North Pacific Ocean [11] and 1931 Convention for the Regulation of Whaling [12].

[5]I.e., an agreement involving more than two States or parties.

international agenda, and the first large international United Nations conference on the environment was held in Stockholm in 1972. This conference witnessed a general upsurge in environmental interest that later inspired the negotiation of several new environmental treaties and the establishment of the United Nations Environment Program (UNEP) [19].

By 1973, the growing environmental consciousness spurred by the environmental movement and such events as the Stockholm Conference had provided a foundation for a new international instrument to address pollution from ships. OILPOL 1954 [13] had become out-dated, and with the support of the then Inter-Governmental Maritime Consultative Organization (IMCO), which is currently known as the International Maritime Organization (IMO) (see Sect. 3.4.1), States were able to adopt a new convention with a broader scope that was more modern than OILPOL 1954: the 1973 International Convention for the Prevention of Pollution from Ships (MARPOL 73) [20]. MARPOL 73 did not only regulate oil pollution but also other pollutants, such as chemicals, sewage and garbage. However, the path towards the entry into force of MARPOL 1973 was slow. Several spectacular tanker disasters[6] and another conference hosted by IMCO in 1978 occurred before an addition of a protocol to MARPOL 73 finally made its entry into force possible in 1983 as a combined instrument, which is commonly referred to as MARPOL 73/78 [14, 15, 21].[7]

Box 3.1 The Role of Accidents as a Catalyst for Regulation

It is perhaps somewhat cynical to state that modern regulation of environmental problems and risks has generally required a major disaster or serious impending environmental effects before legislators have reacted and created regulations. However, this is largely true and not only in the context of regulating environmental impacts of ship operations. For example, on land, major industrial incidents, such as those occurring in Seveso in 1976 and Bhopal in 1984, resulted in stricter regulation for the prevention and control of accidents involving chemicals [23]. Again, for the regulation of ship activities, accidents have had an important driving role. Early maritime accidents, e.g., the *Titanic*, stimulated the creation of new safety regulations for ships. Moreover, with regard to marine pollution specifically, the effects of major accidents, such as the *Torrey Canyon* and the *Amoco Cadiz*, cannot be underestimated in their influence on pollution regulation development for ships [24].

[6]The period between December 1976 to January 1977 has been described as "one of the worst periods in shipping history" (p. 127) [15], with accidents involving several large tankers and subsequent environmental impacts.

[7]MARPOL 73/78 is actually "a package" of several documents, including MARPOL 73 [20], MARPOL 78 [21], the 1997 Protocol [22] and six annexes. This chapter focuses on MARPOL 73/78 in its combined form and its six annexes. For a discussion of these protocols in relation to international agreements, see Sect. 3.5.2.

Regarding the approach to regulating ship operations in the wake of a disaster, two points should be discussed. First, despite receiving great attention in the media, especially incidents involving clearly visible pollution in the form of oil spills, a large portion of pollution from ships originates *not from accidents* but as a result of *operational pollution from ships* [14] (see Sect. 4.2.1). Operational pollution occurs as a part of everyday ship operations and is often completely legal. The effects of small, continuously introduced amounts of less visible pollutants seem harder to grasp and are consequently not as likely to increase the progress of regulation in the same way as visible pollutants released in spectacular accidents. Second, although the *prevention* of marine pollution is increasingly the focus of the regulation of pollution from ships, i.e., as in the case of MARPOL 73/78 [25], and less visible pollutants, such as air pollutants, have also recently been regulated, the creation of new regulations for pollution from ships is largely accident- and *reaction-driven*. Moreover, the creation of new regulations remains largely dependent on visible impacts of pollution or impacts made visible in a convincing enough manner, e.g., by scientists.

Much has occurred since MARPOL 73/78 was adopted. Currently, environmental protection of the marine environment, including protection from pollution from ships, is considered to be one of the most important areas of the law the sea [6, 8]. The broadly accepted 1982 United Nations Convention on the Law of the Sea (LOSC) [26][8] is firmly in place and provides the general legal framework for creating new instruments to protect the marine environment from ship pollution (see Sect. 3.3.2) and other sources. Furthermore, the annexes to MARPOL 73/78 have been updated and also include air pollution and greenhouse gas regulations for ships [25]. The relevance of regulating other environmental impacts from ships has also recently been acknowledged. For example, the invasion of alien species from ballast water and the effects of organotin compounds in antifouling systems on ship hulls are two examples of environmental problems that have recently been regulated in dedicated conventions (Sect. 3.3.5).

In summary, although ship activities have been regulated since early human civilisations, pollution from ships has only recently captured the attention of regulators and become included in international agreements. Several motivations for the regulation of pollution from ships exist. Since World War II, the increased transport frequency and volume of potentially polluting substances, such as oil, have led to more profound and visible environmental impacts when such substances reach the marine environment. Notably, several spectacular ship incidents with

[8]Although frequently also used, "UNCLOS" will not be used here as an abbreviation for the United Nations *Convention* on the Law of the Sea because it is easily confused with the three United Nations *Conferences* on the Law of the Sea (UNCLOS I, II and III). Henceforth, the term "LOSC" is used when referring to the convention, as recommended by Edeson [27].

extensive media coverage demonstrated the effects of marine pollution along coastlines on human populations and fish resources, altering opinions in favour of better pollution control. Furthermore, increased scientific knowledge regarding pollution and an upsurge in environmental consciousness since the 1960s boosted awareness of regulating environmental pollution, especially pollution of the marine environment from ships. Currently, protecting the marine environment is considered to be one of the most important components of the law of the sea. The main convention regarding pollution from ships, i.e., MARPOL 73/78, has been extended with new annexes to include regulations on air pollution and greenhouse gases. Moreover, other environmental impacts from ships have been regulated in dedicated international instruments.

3.3 The Legal Framework for Regulating Pollution from Ships

Before discussing the role of IMO in regulating pollution from ships (Sect. 3.4), the context in which regulations are created, in which basic forms regulation appears, and the current status of this regulation must be discussed. Therefore, the following sections are devoted to briefly introducing aspects of international law and the international law of the sea that are necessary for understanding the later sections of this chapter.

3.3.1 An Introduction to the International Law Context

With the long history of ships operating at sea around the globe, the law of the sea is often considered one of the oldest branches of international law [6]. Not only are many marine activities naturally international but also the regulation of marine pollution from ships is generally governed by and based on globally applicable *international agreements* [6] (see Sect. 3.5.1). To describe how pollution from ships is currently regulated, several basic aspects of international law must first be explained. Although a considerable simplification of a complex and extensive legal field, an explanation of *international law* can simply begin with a general description of law as follows:

> a series of rules regulating behaviour, and reflecting, to some extent, the ideas and preoccupations of the society within which it functions. And so it is with what is termed international law, *with the important difference that the principal subjects of international law are nation-states, not individual citizens* (p. 1, emphasis added) [1]

International law can be said to be the legal order of an international society or a global community in which mainly States act and interact with one another. This legal order has grown out of a necessity to address all types of matters and to achieve goals that reach beyond the national borders of a single State, e.g., economy or security [1]. Simultaneously, issues that extend beyond the borders of single

States also affect other existing States. In the case of the environment, the need for international law is clear because many environmental problems transcend national borders and cannot be solved by the actions of a single State. States must cooperate to solve such problems. International law contains the rules and principles governing the rights and duties primarily of States and their relations to one another. Moreover, international law has facilitative and coordinating functions, e.g., when States cooperate or solve conflicts regarding economy, security and the environment.

As was just emphasised, States are the main actors or *subjects* of international law. To be a subject of international law means that a certain entity is regarded as having rights and duties that are enforceable within the system of international law [1]. Apart from States, other important subjects of international law also exist, such as *international organisations*. Section 3.4 provides a more detailed description of an international organisation with essential functions for the regulation of pollution from ships: IMO. However, (public) international organisations mostly consist of States, and similar to States, international organisations are important actors or subjects of international law [1].

Regarding the rules and principles of international law, it is important to note that the rules formulated in, e.g., international agreements, such as the 1997 Kyoto Protocol to the United Nations Framework Convention on Climate Change [28], *do not work* against the same background and according to the same mechanisms as the rules of national law, e.g., a specific State's rules regulating consumer rights.

What are the main differences between international and national law? A national legal system is often built as a hierarchy and normally contains three essential parts: a legislature (a supreme rule maker), a judiciary (a court system) and an executive authority (a government and governmental organs) that can punish individuals (the citizens) in response to breaches of law. This classic three-part division of national legal systems containing law-making, law-interpreting, and law-enforcing functions (performing the functions of the legislature, the judiciary and the executive authority, respectively) is largely absent in international law [29]. Although the United Nations is an international organisation that can formulate resolutions in the General Assembly, *these are not legally binding* in international law in the same sense as a created national law would be within a State. Moreover, no global system of courts exists with compulsory jurisdiction to force States to appear before a certain court. For example, although courts such as the United Nations International Court of Justice exist in international law, they can only rule in cases *where the parties have accepted the court's right to solve a conflict*. Furthermore, the decision to follow a judgement from, e.g., the International Court of Justice is entirely up to the parties when two States have tried to resolve a conflict with the help of the court. Finally, no global government exists to enforce international law within the global community. Although the United Nations was intended to have a governing role, it does not function as a global governing authority and does not have the same powers as a national government or governmental organs [1]. Thus, the decentralised system of international law *lacks executive authorities to enforce the rules*, a weakness of international law that has commonly been noted, especially in international environmental law [19].

International law is not hierarchical, with a division of powers that includes the three basic functions described above; instead, it builds on States' *sovereign* (self-governing) existence and the *formal equality* of States with other States in the global community [1, 29]. States are the supreme entities in international law, and they coexist with other States in a *horizontally* organised system, which differs from the *vertically* organised system associated with national law. In international law, the States serve both as legislators and as those affected by the created rules. Moreover, the States *themselves* decide if they want to comply with these rules or not [1, 29]. In other words, international law has facilitative and *coordinating* functions for States rather than functioning as a national hierarchical *subordinating* system for individuals.

In summary, international law is the context in which the regulation of pollution from ships is created. International law can simply be described as the legal order of an international society or a global community, whereby States act and interact with each other. Thus, this legal order contains the rules and principles governing the rights and duties of States. However, other important actors also exist in international law, such as international organisations. International law differs from national law in that the former is not founded on the law-making, law-interpreting, and law-enforcing functions that are typically found in a national legal system. Instead of being structured around a three-part division of power and having a vertical organisation, international law can be described as a horizontally organised system that comprises coexisting sovereign and formally equal States.

3.3.2 An Introduction to the Law of the Sea Context

If international law is described as the legal order establishing the rules and principles governing the rights and duties of States, *the law of the sea* can be described as the portion of international law that establishes the rules and principles governing the rights and duties of States *in their marine activities*. More specifically, the law of the sea regulates various marine activities, including navigation on the seas, overflight, laying undersea cables and pipelines, the construction of artificial islands, fishing and the conduct of marine scientific research. Additionally, the international law of the sea divides marine spaces into jurisdictional zones and forms the basis for international cooperation among States for protecting the marine environment [6].

Before further discussing how regulations related to marine pollution from ships are created, several general concepts of the law of the sea must be explained because these concepts are not only closely related to the conditions and possibilities of how States can and even must regulate marine pollution from various sources but also related to how regulation can and must be enforced [14]. The international law of the sea contains rules that directly affect how States can and even must create and enforce regulations regarding marine pollution originating from various sources. Under the law of the sea, the 1982 United Nations Convention on the Law of the Sea, i.e., the LOSC, provides the fundamental framework for the most basic matters of the law of the sea [6]. For example, the framework of the LOSC contains rules that balance States' national requirements to control marine pollution from ships in relation to what is possible and allowed within the framework of international law. Moreover, the rules of the LOSC balance free navigation of the seas with States' interests to protect their coastal environments from marine pollution.

3.3.2.1 Legislative and Enforcement Jurisdiction

How a State wants or is obligated to regulate marine pollution from ships is dependent on questions regarding *jurisdiction.* In legal terminology, several types of jurisdiction exist and can be exercised on different grounds [30]. For the purposes of this chapter, jurisdiction can generally be explained as the right or power to exercise various types of (lawful) control. In the case of the law of the sea and the regulation of marine pollution from ships, two types of jurisdiction are relevant: *legislative jurisdiction* and *enforcement jurisdiction.*

Legislative jurisdiction is exercised when a State adopts laws that mandate protection of its coastal environment by its own initiative or as required by international law [14, 31]. In contrast, *enforcement jurisdiction* covers measures in which States can and in some situations are required to ensure compliance with international or national rules and standards, e.g., investigating offences on the seas, including detaining, arresting, prosecuting and sanctioning of offenders [14, 31]. Below, legislative jurisdiction will be the focus, although some discussion of enforcement jurisdiction must be included for explanatory purposes.

3.3.2.2 Flag State, Coastal State and Port State

Apart from the concepts of legislative and enforcement jurisdiction, the LOSC balances maritime and coastal interests of States depending *on the roles in which a State is acting* when using its jurisdiction to create or enforce regulation. Here, a State can act in three types of roles. A State does not necessarily act in only one role at a time. The roles set the conditions for legislative and enforcement jurisdiction. Therefore, these roles determine how regulation to protect the marine environment can and must be created and enforced. With regard to legislative and enforcement jurisdiction, a State can act as a *flag State, a coastal State* and *a port State*, three roles that must be understood *in relation to a ship.*

A *flag State* is a "State which has granted a ship the right to sail under its flag" (p. 152) [6]. Furthermore, a ship has the nationality of the State whose flag it is entitled to fly [26]. The nationality of a ship establishes an important legal relationship between a ship and a particular State, which is connected to the rights and duties of a flag State. On the high seas, the flag State has exclusive jurisdiction over vessels flying its flag [6].

Since the late 1950s, the existence of a "genuine link" between a State and a ship flying its flag has been demanded. This requirement, essentially motivated by ensuring that flag States are effectively following their duties, has resulted in substantial discussion regarding what really constitutes a "genuine link" [2, 32]. Approximately 73 % of the global merchant fleet was foreign flagged in 2014 [33], meaning that the nationality of a ship's owner was different from the flag under which a ship was registered [34]. The practice of States opening up their registries to foreign-owned ships is associated with the discussion of States offering "open registries", which often have little or no connection at all with the ships' actual activities apart from offering an open registry [8, 32]. Although the meaning of "open registries" in relation to "purely national flags" has become increasingly blurred,[9] several benefits attract owners or operators choosing to "flag out" to so-called "flag of convenience" (FOC) States,[10] which are often related to or even equated with "open registry" States (p. 157) [6]. These benefits, which are offered in many cases by developing States that lack any maritime tradition or infrastructure, include lower ship operating costs as a result of reduced tax burdens and cheaper manning. Moreover, laxer regulation or enforcement of regulation of environmental protection, are other appealing benefits in comparison with a ship-owner's home country [6, 8]. Despite international attempts to phase out "open registries", that host ships flying flags of convenience, their frequent use still persists. Because of this trend, efforts to require a "genuine link" between a State and a ship flying its flag have also been complemented by other viable alternatives, including imposing stricter obligations on flag States to legislate and enforce regulation and to increase port State control [32].[11]

This leads to the two other roles in which States can act in relation to a ship, which balances the traditionally strong position of flag States. A *coastal State* may broadly be described as a State in whose maritime zone a ship is situated at a given time [2, 26]. Finally, a *port State* may be described as a State in whose ports and internal waters a ship is situated [32].

Considering the three roles of flag State, coastal State and port State in relation to legislative and enforcement jurisdiction in general, the main power and primary responsibility to create and enforce rules for ships belongs to the flag State [14, 26].

[9]"Purely national flags" are flags used almost exclusively (i.e., greater than 95 %) by ships flying the same flags as the nationality of their owners [34]. However, between the extremes of flags almost exclusively used by national owners or almost exclusively used by foreign owners, other flags host ship owners in other proportions. For example, foreign-to-national ownership in a proportion of 2:1 or 1:1 is also possible. Because of these varied proportions, the use of the term "open registry" has become more problematic because it can mean many different things [34].

[10]For a list of FOC countries, see the list compiled by the International Transport Workers' Federation: http://www.itfglobal.org/en/transport-sectors/seafarers/in-focus/flags-of-convenience-campaign/. Last accessed on 20 May 2015.

[11]Port State control can be described as "a mechanism for verifying whether a foreign vessel itself and its documentation comply with international rules and standards relating to the safety of ships, living and working conditions on board ships and protection of the marine environment set out by relevant treaties" (p. 285) [6].

This main role of the flag State goes back to the history of the law of the sea and still applies because of the historically strong principle of freedom as a foundation for various activities on the seas, except where international law provides for the exercise of coastal State or port State jurisdiction. With regard to legislative jurisdiction, flag States *may* create more stringent regulations than the *generally accepted international rules and standards* (GAIRS).[12] Here, GAIRS represents *the minimum level* for regulatory requirements by flag States [32].

Although the primary jurisdiction over vessels still remains in the hands of flag States [32], the interests of coastal States in preventing pollution from ships and due to their discontent with the inadequacy of flag State jurisdiction has led coastal States to claim increased power to legislate and enforce regulations to protect their coasts [14, 31]. However, in their legislative jurisdiction, coastal States can generally only create regulations *that give effect to* generally accepted international rules and standards. Therefore, they must keep their regulatory requirements *within* the bounds of GAIRS. Here, GAIRS represents *the maximum level* for regulatory requirements introduced by coastal States [32].

Finally, a greater potential exists for port States to legislate and enforce rules. In its legislative jurisdiction, a port State is generally unrestrained in adopting rules and standards for ships voluntarily entering its ports and internal waters [14, 35, 36].

The meaning and content of generally accepted international rules and standards, GAIRS, in relation to States' rights and duties to regulate ship activities is a matter of controversy and is not fixed in time. A common perspective among legal scholars regarding pollution from ships is that GAIRS should be interpreted as including the rules and standards of the legal instruments that have a high level of State acceptance [32]. An often-mentioned example is MARPOL 73/78, where the first two mandatory annexes undoubtedly enjoy a high level of acceptance because States representing greater than 99 % of the gross tonnage of the global merchant fleet have ratified this convention and its first two annexes [37, 38] (see also Sect. 3.3.4). Another perspective regarding GAIRS is that the central factor deciding which rules and standards are generally accepted at the international level is whether they have been widely accepted at the State level and have reached a status of customary international law [37].[13] In any case, the interpretation of the exact meaning and content of GAIRS remains controversial.

3.3.2.3 Marine Spaces and their Division into Maritime Zones

As stated in the introduction to this chapter, the basic conflict between the principle of freedom and the principle of sovereignty has led to the division of marine areas into different zones [6]. Dividing the seas into zones has assisted in balancing the freedom of the seas for flag States with the control and protection of coastal areas by coastal States. The LOSC provides an important framework in which each zone

[12] See immediately below for comments regarding the meaning of GAIRS.

[13] For this chapter, customary international law can be explained as "state practices recognised by the [global] community at large as laying down patterns of conduct that have to be complied with" (p. 6) [1].

has a certain assigned and interest-balanced jurisdiction that is related to the roles of flag States, coastal States and port States [32]. More specifically, to further determine what rights and duties flag States, coastal States and port States have in relation to creating and enforcing rules, the LOSC decides the geographical extent of States' jurisdiction over ships by dividing the seas into several maritime zones.

For the purposes of explaining the regulation of marine pollution from ships, the balance of interests provided by the LOSC regarding five zones is particularly relevant: (1) *the internal waters*, (2) *the territorial seas*, (3) *the contiguous zone*, (4) *the exclusive economic zone* and (5) *the high seas* [32].[14] An over-arching principle for the rights and duties of States in these zones is that coastal States' interests, competence and powers to act in legislative or enforcement roles increase as the distance from the coast decreases. This means the farther away from the coast, the weaker is a coastal State's capacity to act [31, 32].

Beginning at the coastline, the innermost maritime zone is the *internal waters*, consisting of all waters landward of *the baselines*.[15] The baselines are the points where the territorial sea and other maritime zones are measured [26]. Internal waters primarily include bays, ports, estuaries and in some cases waters landward of the straight baselines of a State [2].

The maritime zone directly adjacent to the internal waters of a coastal State is *the territorial sea*. This zone marks the beginning of a State's maritime territory and may comprise an area that is 12 nautical miles wide (measured from the baselines). The territorial sea also includes airspace, seabed and subsoil areas [2, 26].

If claimed by a coastal State, *the contiguous zone* is directly adjacent to the territorial sea [6]. This zone is measured from the same baseline as the territorial sea and may extend up to 24 nautical miles [2, 26]. In this zone, a coastal State is entitled to act to prevent breaches of its "customs, fiscal, immigration or sanitary laws and regulations within its territory or territorial sea" (Art. 33(1)(a)) [26] and to punish infringements committed on its territory or in the territorial sea [14, 26].

The exclusive economic zone (EEZ), which must be declared and legislated by a coastal State, is a zone that is adjacent to the territorial sea and stretches up to 200 nautical miles. This zone is measured from the same baseline as the territorial sea. If declared, the EEZ may overlap with the contiguous zone if a coastal State also claims the latter zone [6, 14, 26]. In the EEZ, a coastal State has limited "sovereign rights" in the waters and seabed for several economic purposes. Among other rights, the coastal State has sovereign rights for the "purpose of exploring and exploiting, conserving and managing the natural resources" and regarding "other

[14]The LOSC further demarcates two additional zones with specific relevance to the exploration and exploitation of the seabed and its subsoil: the continental shelf and the deep seabed or 'the Area'. This chapter does not address these zones (see instead Tanaka 2012 [6] and Churchill and Lowe 1999 [2]).

[15]There is more than one type of baseline. A *normal baseline* is drawn along the coast following the low-water line. A *straight baseline* can be used in situations where it is impractical to draw a normal baseline due to complex coastal shapes [6].

activities for the economic exploitation and exploration of the zone, such as the production of energy from the water, currents and winds" (Art. 56(1)(a)) [26].

The maritime zone situated beyond the EEZ is *the high seas*. This area is open to all States, regardless of whether they have coasts or are entirely landlocked [26]. On the high seas, all States can enjoy the freedom of navigation, freedom of overflight, freedom to lay submarine cables and pipelines, freedom of fishing and freedom of scientific research as specified in the LOSC [26] (Fig. 3.1).

3.3.2.4 Regulating Marine Pollution from Ships in Different Maritime Zones

The possibilities for regulating marine pollution from ships differ between the maritime zones. The easiest way to understand how legislative jurisdiction varies is by examining the differences in the rights and obligations for a coastal State in the various maritime zones. As one author stated, "while the obligations of flag states

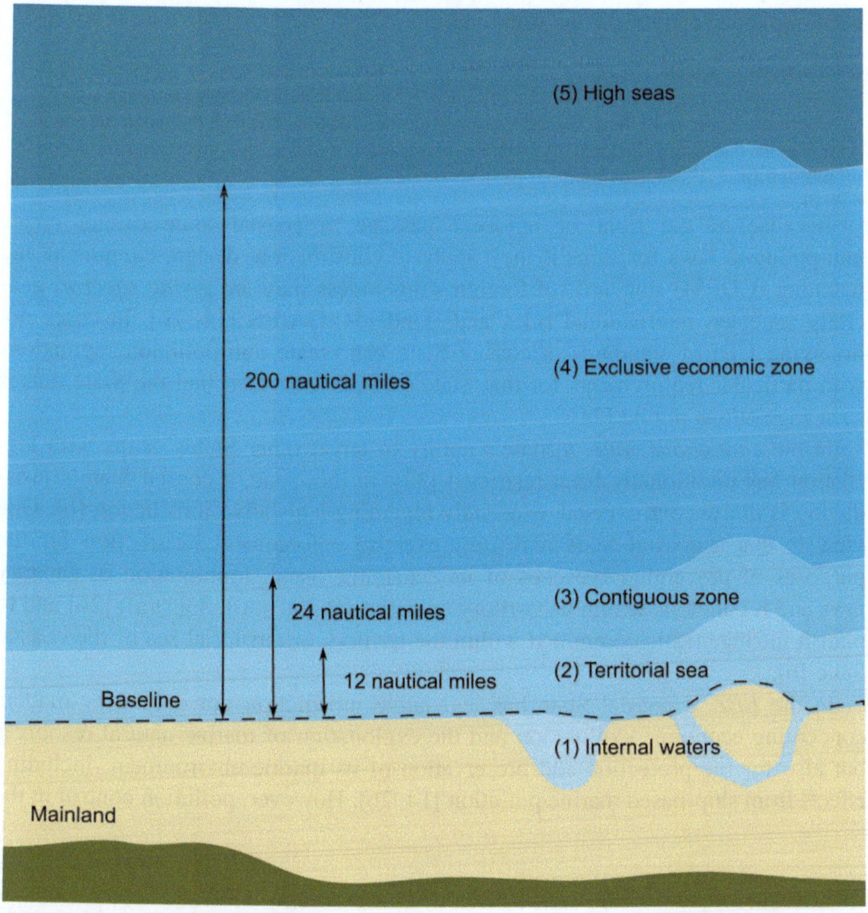

Fig. 3.1 Maritime zones

are the same *irrespective of the sea area concerned*, coastal states' rights depend on whether the (foreign) ship is in the internal waters, territorial sea, exclusive economic zone of the coastal State or in the high seas" (p. 21, emphasis added) [39]. Thus, considering the coastal State perspective in relation to legislative jurisdiction, the jurisdiction in each maritime zone must be examined.

In the *internal waters*, a coastal State enjoys full sovereignty; no general limitations are placed on a coastal State's freedom to legislate and adopt regulations for foreign ships voluntarily entering its internal waters or ports situated in these waters. As a point of departure, a coastal State can therefore legislate as it pleases in this zone, e.g., setting discharge standards for ships [14, 26, 32].

In the *territorial sea*, the general sovereignty and the rights of the coastal State to legislate continue. However, *the right of innocent passage* of ships of other States also applies [32]. According to the LOSC, the passage shall be "continuous and expeditious" (Art. 18(2)) [26] and must not be "prejudicial to the peace, good order or security of the coastal State" (Art. 19(1)) [26]. In practice, innocent passage has been explained to include the following:

> situations in which a vessel is proceeding to or from internal waters, or proceeding to or from a call at a roadstead or port facility outside internal waters. Vessels may even stop for a period of time: only those vessels that interrupt their passage through the territorial sea in a manner that is not 'incidental to ordinary navigation' will lose the protection of innocent passage (p. 17, footnotes omitted) [36].[16]

Because of the right of innocent passage, a coastal State cannot create anti-pollution laws for ships if they apply to construction, design, equipment and manning (CDEM) standards of foreign ships unless they are *giving effect to* generally accepted international rules and standards, GAIRS [14, 26]. In cases not involving CDEM standards, a coastal State can create anti-pollution regulations with particular requirements for that State's coasts, provided that the State makes such regulations public [14].

In the *contiguous zone*, a State's ability to target other States' ships with legislation has traditionally been restricted [32]. In this zone, a coastal State's jurisdiction is not uncontroversial, especially regarding legislative jurisdiction [6]. One view is that a coastal State may *only* exercise enforcement jurisdiction for the purposes of preventing breaches of its "customs, fiscal, immigration or sanitary laws and regulations within its territory or territorial sea" (Art. 33(1)(a)) [26] and to punish infringements committed within the territory or territorial sea of the coastal State [6].

In the *EEZ*, a coastal State has legislative jurisdiction not only over matters concerning economic exploration and the exploitation of marine natural resources but also for the protection and preservation of its marine environment, including effects from ship-based marine pollution [14, 26]. However, pollution control in the

[16] For activities to be considered non-innocent, they must be grave, such as weapons testing or wilful and serious pollution. Therefore, under normal circumstances, foreign ships will enjoy the right of innocent passage through another State's territorial waters [26, 36].

EEZ must conform to and can at a maximum *give effect to* GAIRS, which means that a coastal State must not go beyond what is necessary to create rules mirroring international requirements, e.g., MARPOL 73/78 [14, 26, 32]. An exception to this rule is if a "special area" with stricter rules and standards has been approved by IMO after evaluation. Regardless, anti-pollution standards in such an area may still not have CDEM requirements other than those that conform to GAIRS [14, 37].

On the *high seas*, which extends beyond national jurisdiction [6], the traditional strength of the flag State remains unrestrained. With the exception of unusual situations involving maritime casualties that could affect a coastal State's waters or territory, a coastal State has no right to regulate marine pollution from ships in this zone [14, 32].

In summary, the LOSC recognises three different roles in which States can or are obliged to act according to international law when legislating or enforcing rules and standards related to the law of the sea. These roles exist in relation to a ship and are the roles of the flag State, the coastal State and the port State. The maritime zone divisions in the LOSC provide another dimension in the legal framework, in which varying rights and duties of States using the oceans are established. Each zone has assigned jurisdiction related to a State's desires or obligations (to legislate or enforce rules) and a State's role as a flag State, coastal State or port State. Five maritime zones are primarily relevant for the regulation of marine pollution from ships: the internal waters, the territorial sea, the contiguous zone, the exclusive economic zone and the high seas. In these zones, the powers of coastal States to regulate marine pollution from ships vary. However, in general, coastal States have greater interests, competence and powers to act in zones closer to the coast. Inversely, a State's capacity to act decreases farther from the coast.

3.3.3 Links Between the LOSC and the Role of IMO in the Regulation of Pollution from Ships

This section considers the links between the LOSC and the regulatory activities of IMO, especially the regulation of marine pollution from ships.

Briefly, the LOSC contains fundamental rules regarding the legislative jurisdiction of a coastal State and variations in jurisdiction according to the maritime zone, e.g., when a coastal State can regulate marine pollution from ships. More generally, the LOSC also contains fundamental rules regarding State protection and preservation of the marine environment. These and other obligations are linked to the standard-setting role of IMO via so-called *rules of reference* that are expressed in different forms in the LOSC [32, 40]. To understand the importance of the links between the LOSC and IMO via the rules of reference, some additional comments regarding the LOSC are required.

The rules of reference are formulated in many different ways in the LOSC [32]. Nevertheless, the LOSC includes references that are related to other legal instruments, perhaps most significantly to instruments adopted by IMO [37], including safety and anti-pollution instruments, e.g., SOLAS and MARPOL 73/78.

The LOSC was drafted in an open manner and neither regulates specific pollution control measures nor includes new standards for particular types of marine pollution. Instead, the underlying idea was to create *a general framework* with basic rules by proclaiming "a general regime of powers and duties which builds upon the codification and development of existing and future pollution control conventions" (p. 195) [32]. Thus, by deliberately linking the LOSC via the rules of reference to international standards in other existing conventions, and also by providing a general legal base even for future legal instruments that are *yet to be drafted*, the LOSC became a flexible and dynamic framework that is more likely to stand the test of time. Although environmental standards change over time, by referring to other instruments, the LOSC *itself* can avoid the need to be amended each time environmental standards require updating [37]. Here, IMO is critically important because it is the primary body responsible for specifying more precise anti-pollution standards for ships linked to the LOSC via the rules of reference [40].

To understand the possibilities provided within the LOSC to formulate pollution standards (including future ones), e.g., standards for ships drafted by IMO, one must begin with the LOSC definition of "pollution of the marine environment":

> the introduction by man, directly or indirectly, of substances or energy into the marine environment, including estuaries, which results or is likely to result in such deleterious effects as harm to living resources and marine life, hazards to human health, hindrance to marine activities, including fishing and other legitimate uses of the sea, impairment of quality for use of sea water and reduction of amenities (Art. 1(1)(4)) [26]

Based on this definition, certain circumstances follow. First, "pollution of the marine environment" is defined in an open manner, including *all sources of pollution*, both known existing and future sources, rendering the definition flexible and adaptable over time. Second, even *potentially damaging* effects to the marine environment are included in the definition. Thus, the LOSC allows regulation of not only those sources that result in deleterious effects (e.g., for marine life and human health) but also those that are *likely to result* in deleterious effects. In general, the LOSC established general rules for four main categories marine pollution sources: land-based marine pollution, pollution from ships, pollution by dumping, and pollution from seabed activities in and beyond national jurisdiction [6]. This chapter focuses on pollution from ships, although the other categories of pollution sources can also have potentially damaging effects on the marine environment.

According to PART XII of the LOSC, States have a general obligation to protect and preserve the marine environment [26]. Although the LOSC confirms States' rights to exploit their natural resources, this must be conducted"pursuant to their environmental policies and in accordance with their *duty to protect and preserve the marine environment*" (Art. 193, emphasis added) [26]. Furthermore, the LOSC requires States to take, either individually or jointly, all measures that are necessary to prevent, reduce and control pollution of the marine environment *from any source* [26].

The links to IMO as a drafter of standards for marine pollution from ships via the rules of reference are present in the article of the LOSC that is specifically dedicated to regulating marine pollution from ships. This article generally obligates States to establish international rules and standards to prevent, reduce and control pollution of the marine environment from ships. Moreover, it also obligates States to adopt laws and regulations for the prevention, reduction and control of pollution from ships flying their flag or of their registry. Such laws and regulations must at least have the same effect as that of GAIRS and be "established through *the competent international organization*" (Art. 211(1)-(2), emphasis added) [26]. In this case, the international organisation competent to formulate these standards is IMO [38].

Furthermore, some general comments regarding the LOSC in relation to pollution from ships provide useful context for this section. With the introduction of the LOSC in 1982, some significant changes occurred in the environmental law aspects of the law of the sea.

First, regarding *the principle of freedom* (discussed above) and its relation to the environment, States are not as free to use the seas today as they once were. On the contrary, according to the LOSC, States have a *duty* to protect the marine environment by creating and enforcing regulations [32], including controlling pollution *from all sources* (including from ships). This duty is not only an obligation with respect to the interests of other sovereign States but also a duty for the benefit of the entire marine environment [8].

Second, the LOSC has shifted the balance of power between the historical freedom of flag States to perform various activities on the seas and the *sovereignty* of coastal States to protect their coasts from pollution. More precisely, the LOSC demands at least a minimum level of flag State obligations to regulate ship pollution, and coastal (and port) States have been allowed to expand their ability to protect their coasts [8, 32].

Finally, considering environmental damage and the law of the sea, the LOSC has shifted the focus from traditional State liability and "compensation thinking" to concepts building on international regulation and cooperation among States to protect the marine environment [8]. As discussed above, IMO drafts the standards for protecting the marine environment from ship-based pollution. Simultaneously, the regulatory work performed by IMO reflects the changes induced by the introduction of the LOSC.

To summarise, the LOSC contains fundamental rules that are linked to IMO's standard-setting role via so-called rules of reference. Via these rules, referring to international standards found in other instruments , the LOSC itself can avoid constant amendment as new environmental standards emerge. Here, IMO is critically important because it is responsible for specifying these more precise anti-pollution standards for ships that are related to the LOSC via the rules of reference. With the introduction of the LOSC, some significant changes occurred in the environmental aspects of the law of the sea. According to the LOSC, States have a duty to protect the marine environment by creating and enforcing regulations. Furthermore, the LOSC has shifted the balance of power between the historical freedom of flag States to perform various activities on the seas and the sovereignty of coastal States to protect their coasts from pollution.

3.3.4 An Introduction to MARPOL 73/78 and its Annexes

Although the LOSC is the primary international legal instrument that provides a broadly accepted framework to address fundamental questions related to the law of the sea, such as jurisdiction and marine environmental protection, MARPOL 73/78 is *the primary legal instrument for the prevention of pollution from ships* due to both operational and accidental discharges [6, 20]. Additionally, the more detailed standards contained in MARPOL 73/78 exemplify how the LOSC is linked to other legal instruments created with the support of IMO. In this case, MARPOL 73/78 is related to the more generally formulated environmental requirements in the LOSC via the rules of reference, including those found in Article 211 of the LOSC, the LOSC article dedicated to regulating marine pollution from ships [8].

In MARPOL 73/78, the basic articles primarily address jurisdiction, enforcement and inspection powers in relation to ships [8]. These articles specify that MARPOL 73/78 applies to "ships entitled to fly the flag of a Party to the Convention; and … ships not entitled to fly the flag of a Party but which operate under the authority of a Party" (Art. 3(1)) [20]. However, MARPOL 73/78 does not apply to warships or naval auxiliary [20].

Furthermore, under the basic articles of MARPOL 73/78, any violation of the requirements, wherever it occurs, is prohibited, and sanctions must be established under the law of the administration of the ship concerned [20]. In MARPOL 73/78, *the administration* means the "Government of the State under whose authority a ship is operating" (Art. 2(5)) [20].

As a point of departure, MARPOL 73/78, like other international treaties, only governs the relations between the parties of the treaty (see Sect. 3.5). Thus, the requirements of MARPOL 73/78 apply only to those States that have consented to become parties to this convention. Nevertheless, MARPOL 73/78 can still be applied *to States that are not parties to the convention* in certain situations. A good example of the application of MARPOL 73/78 to non-parties is the application of port State control regimes to *any ship* entering ports included in such a control regime. When a ship is voluntarily in port, a port State can enforce its own (convention-based) laws to ships *regardless of flag* and can control whether the requirements of international regulations, e.g., MARPOL 73/78, are followed [36].

The jurisdiction of MARPOL 73/78, like the LOSC, relies on *flag State* legislative and enforcement jurisdiction, although the convention also partially balances flag State primacy by recognising the jurisdiction of coastal States (and port States) to regulate pollution from ships in their internal waters and territorial seas [8, 20]. In general, MARPOL 73/78 creates an enforcement scheme that builds on the cooperation of flag States, coastal States and port States in a system of certification, inspection and reporting. The intention of introducing such a system was to render the operation of sub-standard ships more difficult or impossible while simultaneously reinforcing the role of flag States to act on violations and to enforce relevant laws [8].

The actual pollution requirements of MARPOL 73/78 are found in the annexes to the convention, where detailed technical rules are contained. Currently, six annexes have been added to MARPOL 73/78 concerning different types of pollution from ships. These annexes have frequently been amended and can be amended via the so-called *tacit acceptance* procedure of IMO (see Sect. 3.5.2). The first two annexes (Annex I and II) of MARPOL 73/78 are mandatory for all parties to the convention, whereas Annexes III-VI are optional [8, 20]. Here, some of the main features included in each annex are briefly discussed.

Annex I, entitled *Regulations for the Prevention of Pollution by Oil,* entered into force on 2 October 1983. This mandatory annex regulates the prevention of pollution by oil or oily mixtures from ship operations and from accidental discharges. Annex I includes obligatory double-hull requirements for new oil tankers and a schedule to phase existing tankers into compliance [41].

Annex II, entitled *Regulations for the Control of Pollution by Noxious Liquid Substances in Bulk,* entered into force on 2 October 1983. This mandatory annex specifies discharge criteria and measures for the control of marine pollution with noxious liquid substances transported in bulk. Noxious liquid substances are classified into four categories: X, Y, Z and other substances. The discharge of the most dangerous noxious substances in category X into the sea is prohibited. However, discharge of substances under categories Y and Z are permitted within certain limits. Other substances may be discharged during, e.g., tank cleaning or de-ballasting [6, 41].

Annex III, entitled *Regulations for the Prevention of Pollution by Harmful Substances Carried by Sea in Packaged Form,* entered into force on 1 July 1992. This optional annex establishes provisions regarding the packing, marking, labelling, documentation, stowage and quantity limitations for harmful substances. "Harmful substances" are those marked as marine pollutants in the International Maritime Dangerous Goods Code (IMDG Code), which is related to Annex III, or those substances that qualify as harmful according to the criteria in the Appendix of Annex III [41].

Annex IV, entitled *Regulations for the Prevention of Pollution by Sewage from Ships,* entered into force on 27 September 2003. This optional annex contains requirements to control sewage pollution from ships. Sewage discharges from ships into the marine environment are prohibited unless the ship has been fitted with an approved sewage treatment plant or the sewage discharge has been comminuted and disinfected with an approved system. In situations in which sewage can be discharged, the annex also specifies minimum distances from the nearest shoreline for discharge [41].

Annex V, entitled *Regulations for the Prevention of Pollution by Garbage from Ships,* entered into force on 31 December 1988. This optional annex regulates the disposal of various types of garbage from ships. Garbage includes domestic and operational waste. In general, the annex prohibits the discharge of garbage into the sea except under specific circumstances. The disposal of plastics into the sea is altogether prohibited [6, 41].

Annex VI, entitled *Regulations for the Prevention of Air Pollution from Ships*, entered into force on 19 May 2005. This optional annex regulates emissions to the air from ships, including sulphur and nitrogen oxides. Under this annex, special so-called Emission Control Areas (ECAs) for sulphur oxides, nitrogen oxides and particulate matter can be designated. Since 2011, Annex VI has also hosted mandatory technical and operational energy efficiency measures for ships to reducing their greenhouse gas emissions. This annex also regulates other emissions to the air, such as deliberate emissions of ozone-depleting substances [6, 41].

In summary, MARPOL 73/78 is the primary legal instrument for the prevention of pollution originating from ships due to both operational and accidental activities. The detailed standards contained in MARPOL 73/78 exemplify how the LOSC is related to other legal instruments created with the support of IMO. In its jurisdiction, MARPOL 73/78, like the LOSC, relies on flag State legislative and enforcement jurisdiction, although the convention also partially balances flag State primacy by recognising the jurisdiction of coastal States (and port States) to regulate pollution from ships in their internal waters and territorial seas. In general, MARPOL 73/78 creates an enforcement scheme that builds on the cooperation of flag States, coastal States and port States in a system of certification, inspection and reporting. The actual pollution requirements of MARPOL 73/78 are currently enumerated in six annexes. The first two annexes (Annex I and II) are mandatory for all parties of MARPOL 73/78, whereas Annexes III-VI are optional. The annexes of MARPOL 73/78 regulate pollution of the marine environment from ships. More specifically, the annexes cover pollution from ships by oil, chemicals, sewage, garbage and air emissions.

3.3.5 Other International Agreements Regulating Pollution from Ships

Although MARPOL 73/78 is the central instrument for regulating pollution from ships, other instruments have been recently drafted to regulate additional aspects of pollution from ships. Table 3.1 shows the important issues regulated by MARPOL 73/78 and its annexes and four other conventions; the adoption and entry into force dates are also shown. The four additional conventions are briefly introduced in this section.

3.3.5.1 International Convention on the Control of Harmful Anti-fouling Systems on Ships

The International Convention on the Control of Harmful Anti-fouling Systems on Ships (AFS Convention) was adopted in 2001 and entered into force in 2008.

Table 3.1 The important issues regulated by MARPOL 73/78 and its annexes and four other conventions; the adoption and entry into force dates are also listed

IMO Convention	Regulates	Adoption (year)	Entered into force (year)
MARPOL 73/78	Pollution from ships	1973/1978	1983
Annex I	Oil & oily mixtures	1973	1983
Annex II	Liquid chemicals in bulk	1973	1983
Annex III	Liquid chemicals in packaged form	1973	1992
Annex IV	Sewage	1973	2003
Annex V	Garbage	1973	1988
Annex VI	Air pollution and GHG emissions	1997	2005
International Convention on the Control of Harmful Anti-fouling Systems on Ships	Antifouling paints	2001	2008
International Convention for the Control and Management of Ships' Ballast Water and Sediments	Invasive species	2004	Not yet in force
Hong Kong International Convention for the Safe and Environmentally Sound Recycling of Ships	Recycling of ships	2009	Not yet in force
Nairobi International Convention on the Removal of Wrecks	Removal of wrecks	2007	2015

The AFS Convention prohibits and restricts the use of organotin compounds in antifouling systems (see Chap. 4) [42]. An antifouling system is defined as "a coating, paint, surface treatment, surface, or device that is used on a ship to control or prevent attachment of unwanted organisms" (Art. 2(2)) [43]. The AFS Convention also establishes a framework to prevent the future use of other harmful substances in antifouling systems [42].

The AFS Convention requires its parties to prohibit and/or restrict "the application, re-application, installation, or use of" (Art. 4(a) and (b)) [43] antifouling systems listed in Annex 1 of the convention. With some exceptions, e.g., warships and naval auxiliary, this convention applies to ships flying the flag of a party, ships operating under a party's authority, and ships that enter a port, shipyard or offshore terminal of a party [43].

More specifically, Annex 1 provides a list of control measures for antifouling systems. This list can be updated when necessary. In its current form, Annex 1 requires that organotin compounds in antifouling systems shall not be applied or re-applied after 1 January 2003. This annex applies to all ships governed by the convention. Since 1 January 2008, all ships, with some exceptions for fixed and floating platforms, must be coated with organotin compounds or bear a protective coating that acts as a barrier to prevent non-compliant compounds from leaching into the marine environment [42, 43].

3.3.5.2 International Convention for the Control and Management of Ships' Ballast Water and Sediments

The International Convention for the Control and Management of Ships' Ballast Water and Sediments (BWM Convention) [44] was adopted in 2004. However, it has not yet entered into force (as of 20 May 2015) [45].[17] The convention will enter into force 12 months after at least 30 States have ratified it, representing 35 % of the world merchant fleet tonnage. In general, the BWM Convention will apply to ships flying the flag of a party and ships operating under a party's authority [44]. Several exceptions to its application are listed, including ships that only operate in the waters of an individual party and do not cause damage to other States' "environment, human health, property or resources" (Art. 2(b)) [44].

The main objective of the BWM Convention is to "prevent, minimize and ultimately eliminate the transfer of Harmful Aquatic Organisms and Pathogens" (Art 2 (1)) [44] from ship ballast water and sediments. More specifically, the convention refers to aquatic organisms and pathogens that, "if introduced into the sea including estuaries, or into fresh water courses, may create hazards to the environment, human health, property or resources, impair biological diversity or interfere with other legitimate uses of such areas" (Art. 1(8)) [44] (see Chap. 4). By "sediments", the convention means "matter settled out of Ballast Water within a ship" (Art. 1(11) [44]). The convention requires management practices for ship ballast water and sediments, i.e., Ballast Water Management (BWM), in accordance with specific standards [45]. In addition, the BWM Convention allows a party to take stricter measures in certain areas in accordance with specified provisions [44, 45].

The specific regulations are established in the Annex to the convention. The Annex includes a requirement to establish and use a ship-specific Ballast Water Management Plan [44]. Such a plan provides details on the on-board implementation of the convention's requirements and supplemental practice [45]. In addition, ships are required to record the uptake, circulation, treatment and discharge of ballast water in a Ballast Water Record Book [44, 45].

In practical terms, the BWM Convention provides a phase-in schedule for ships with two standards depending on the date of construction and the ballast water capacity [44, 45]. The first step is to comply with the Ballast Water Exchange Standard, which requires a volumetric exchange of at least 95 % efficiency [44, 45]. To meet this standard, ballast water must be exchanged from a specified depth and distance from land or as far from land as possible. The second step is the Ballast Water Performance Standard, which contains specific parameters for permitted quantities of organisms and pathogens in discharged ballast water and sediments [44]. To meet the criteria of the second standard, "most ships will need to install an on-board ballast water treatment system" [45]. With some exceptions, the approval of the responsible national authority ("the Administration") is required for the use of such systems [44].

[17]For a list of the status of ratifications, see http://www.imo.org/About/Conventions/StatusOfConventions/Pages/Default.aspx, last accessed on 20 May 2015.

3.3.5.3 Hong Kong International Convention for the Safe and Environmentally Sound Recycling of Ships

Hong Kong International Convention for the Safe and Environmentally Sound Recycling of Ships

The Hong Kong International Convention for the Safe and Environmentally Sound Recycling of Ships (Hong Kong Convention) was adopted in 2009. It has not yet entered into force (as of 20 May 2015) [46].[18] Entry into force will occur 24 months after the date on which 15 States, representing 40 % the world merchant shipping by gross tonnage, have ratified the agreement. Additionally, a third condition is that the combined maximum annual ship recycling volume of the 15 States during the preceding 10 years must constitute no less than 3 % of the gross tonnage of the combined merchant shipping of the same States [47].

The Hong Kong Convention addresses not only the environmental and health risks of ship recycling associated with the hazardous substances contained in the ships but also concerns regarding the working conditions and environmental conditions at many of the world's ship recycling facilities [46]. More details on the environmental risks and impacts of scrapping are provided in Chap. 7. "Ship recycling" is defined by the convention as "the activity of complete or partial dismantling of a ship [...] in order to recover components and materials for reprocessing and re-use, whilst taking care of hazardous and other materials" (Art. 2(10)) [47]. With some exceptions, such as for warships and ships under 500 GT, the Hong Kong Convention will apply to ships flying the flag of party or operating under a party's authority. Moreover, it will apply to ship recycling facilities operating under a party's jurisdiction [47].

The specific regulations are established in the Annex to the convention and cover a broad spectrum of activities of a ship from cradle to grave, including the design, construction and operation of ships, and the operation of ship recycling facilities [46]. Moreover, the convention requires a ship-specific Inventory of Hazardous Materials (IHM) to be carried on ships [47]. For new ships, the IHM is to be established at the shipyard, and an initial survey is required for verification. The IHM must be maintained throughout a ship's operational life, with renewal surveys at intervals not exceeding five years. A final survey is also required before a ship is sent for recycling [46, 47]. For existing ships, the IHM is to be carried on board five years after the convention has entered into force at the latest or earlier if a ship is to be sent for recycling [46, 47].

Appendix 1 to the convention provides a list of hazardous materials. Not only does the convention require that these materials be identified in IHMs but also requires the parties to "prohibit and/or restrict the installation or use" (Reg. 4) [47] of these materials on ships flying the flag of a party, ships operating under a party's authority, and ships in ports, shipyards, ship repair yards or offshore terminals [46, 47]. A further important feature of the Hong Kong Convention is that a ship of a party is only allowed to be recycled in recycling facilities that are authorised in accordance with the convention [47].

---

[18]For a list of the status of ratifications, see http://www.imo.org/About/Conventions/StatusOfConventions/Pages/Default.aspx, last accessed on 20 May 2015.

3.3.5.4 Nairobi International Convention on the Removal of Wrecks

The Nairobi International Convention on the Removal of Wrecks, 2007 (Wreck Removal Convention) was adopted in 2007 and entered into force on 14 April 2015 [48]. The Wreck Removal Convention "fills a gap in the existing international legal framework by providing a set of uniform international rules for the prompt and effective removal of wrecks" [48] that pose a hazard within defined areas [49]. By "hazard", the convention means "any condition or threat that: (a) poses a danger or impediment to navigation; or (b) may reasonably be expected to result in major harmful consequences to the marine environment, or damage to the coastline or related interests of one or more States" (Art. 1(5) [49], see also Chap. 7).

The convention applies to all shipwrecks in the "convention area" except for a few exclusions, such as the wrecks of warships [49]. This area is defined as the EEZ of a party to the convention or, in cases where a party has not established an EEZ, "an area beyond and adjacent to the territorial sea of that State determined by that State in accordance with international law and extending not more than 200 nautical miles from the baselines from which the breadth of its territorial sea is measured" (Art. 1(1)) [49].

In addition, the Wreck Removal Convention provides the possibility for parties to extend the application to include their territorial sea, although this is subject to certain conditions and exclusions with regard to the application of the convention [49]. Another relevant feature of the convention is that it enables measures to be taken in cases when a ship is about to or reasonably expected to sink or strand [48]. Finally, regarding liability, a wreck's owner is liable for the costs of locating, marking and removing a wreck, although this is subject to several exceptions [49].

In summary, alongside MARPOL 73/78, other instruments have been recently drafted to regulate additional aspects of pollution from ships. The International Convention on the Control of Harmful Antifouling Systems on Ships prohibits and restricts the application of organotin compounds in antifouling systems on the hulls of ships and also establishes a framework to prevent the future use of other harmful substances. The International Convention for the Control and Management of Ships' Ballast Water and Sediments (not yet in force) aims to control the spread of alien species from ship ballast water by requiring management practices in accordance with certain standards and allows a party to the convention to take stricter measures in accordance with specified provisions. The Hong Kong International Convention for the Safe and Environmentally Sound Recycling of Ships (not yet in force) addresses the environmental and health risks of ship recycling associated with hazardous substances contained in ships and the working conditions and environmental conditions at ship recycling facilities. This convention requires an inventory of hazardous materials to be carried on ships and provides a list of prohibited and/or restricted hazardous materials. The Nairobi International Convention on the Removal of Wrecks provides international legal possibilities for its parties to remove wrecks that pose a hazard within their EEZ or an equivalent zone of 200 nautical miles. Under certain conditions, this convention also allows its parties to extend the application to include their territorial sea.

3.4 The Role of IMO in the Regulation of Pollution from Ships

As emphasised in Sect. 3.3.1, international organisations are important subjects of international law. The United Nations (UN) represents the majority of the world's States and is one of the most (if not the most) important international organisations in international law [1]. Under the UN's principal organs, programmes and specialised agencies are involved in drafting the regulatory framework for activities at sea and in marine environments. They address issues that include fisheries and aquaculture (the Food and Agriculture Organization, FAO), the working conditions of seafarers (the International Labour Organization, ILO), and the sustainable regional management and use of sea areas (the United Nations Environment Programme, UNEP, including its Regional Seas Programme) [32, 50, 51].

The focus in this chapter is on the most important forum for regulating pollution from ships: the International Maritime Organization (IMO). The role of IMO is central to understanding how regulations regarding pollution from ships are created. This section introduces the functions and structure of IMO, followed by an overview of the actors involved in drafting regulations regarding pollution from ships.

3.4.1 Functions and Structure

IMO is a so-called *specialised agency* of the UN [52]. Such agencies are "autonomous organizations" [51] that have international responsibilities within fields warranting relationship with the UN [53] and are connected to the UN by agreements with the United Nations Economic and Social Council (ECOSOC) that specify the terms of the relationship [53]. The ECOSOC coordinates the work of specialised agencies [51].

IMO is the international body responsible for measures regarding safety, security and pollution prevention for international shipping [50]. Officially, IMO is described as a "global standard-setting authority" [52], although the role of IMO should not be viewed as an "authority" but rather as a forum in which States can discuss, negotiate and make common decisions regarding regulations for ships. The primary purpose of IMO is defined in Article 1(a) of the Convention on the International Maritime Organization, 1948, as amended (IMO Convention) [54]:

> To provide machinery for co-operation among Governments in the field of governmental regulation and practices relating to technical matters of all kinds affecting shipping engaged in international trade; to encourage and facilitate the general adoption of the highest practicable standards in matters concerning the maritime safety, efficiency of navigation and prevention and control of marine pollution from ships; and to deal with administrative and legal matters related to the purposes set out in this Article. (Art. 1(a)) [54]

One of the main functions of IMO is to *develop draft conventions* (or other instruments) that are recommended to States for adoption at international diplomatic conferences [54]. An equally important function of the organisation is to *amend existing conventions* (or other instruments). In addition to conventions, IMO also develops other instruments, such as guidelines and codes that are not formally legally binding, although they still serve important purposes. For example, conventions can refer to guidelines that must be considered, and several codes have become binding after being incorporated into conventions. More than 50 international conventions and hundreds of other instruments have been developed through IMO [32, 50, 55].

IMO was founded in 1948 as the Inter-Governmental Maritime Consultative Organization (IMCO). Its initial role was as a consultative and advisory UN agency for the shipping industry, which had been dominated by major traditional maritime States for centuries, e.g., the UK, France and the Netherlands. IMCO did not have an explicit mandate for environmental issues, and its functions were primarily concerned with maritime safety. Nonetheless, a gradual extension of its mandate and activities to address marine pollution from ships followed with time. IMCO held its first meeting in 1959 because it took ten years before the convention establishing IMCO entered into force in 1958. Thus, IMCO already faced the challenges of a changing world at its beginning. IMCO had to adapt its functions in response to major ship-related oil spills and political pressures related to increased environmental awareness in the 1960s and 1970s, as described in Sect. 3.2. Thus, IMCO began to address environmental issues, such as marine pollution, and liability and compensation for pollution damages. In addition, IMCO had to adapt to significant geopolitical changes. New States emerged and the number of IMCO member States increased, in particular the number of developing States [32, 50, 56].

The organisation eventually became so different that its name was changed to IMO in 1982. The word "consultative" was thought to give the impression that "IMO could only talk, rather than take decisions and act" (p. 6) [56]. Additionally, the purpose of IMO was adjusted in 1984 by adding pollution prevention and control to the text of IMO Convention and by removing the "consultative" and "advisory" limitations [56]. These changes in IMO mandate and activities have been described as historically unique compared with other international organisations [40].

Figure 3.2 shows the current structure of IMO. The *Assembly* is the governing body, which is primarily concerned with the function of the organisation, such as deciding on the rules of procedure and financial arrangements. All member States are represented in the Assembly, which normally meets once every two years. The *Council* is IMO's executive body under the Assembly and is responsible for the task of supervising the organisation's work. The Council has the same functions as the Assembly during the time between Assembly sessions.[19] Another function of

[19]Except for the reserved right of the Assembly to make recommendations on maritime safety and pollution prevention to its member States.

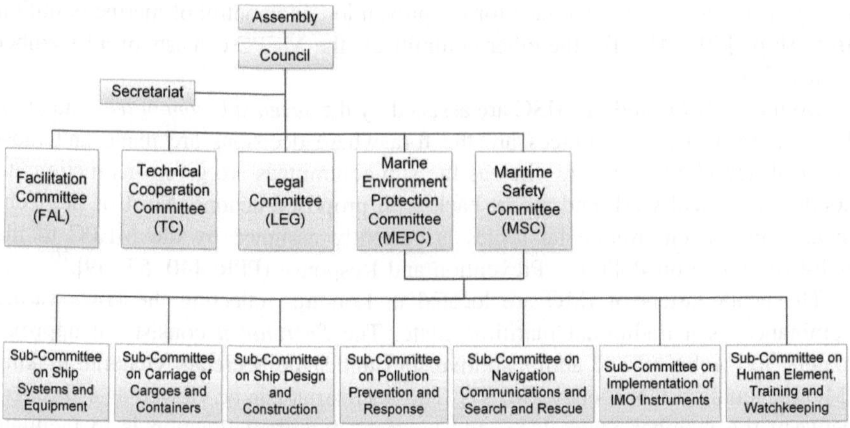

Fig. 3.2 Structure of IMO [40, 57]

the Council is to coordinate the activities of the different IMO bodies shown in Fig. 3.2 [57].

The Council consists of representatives from 40 member States that are elected by the Assembly to two-year terms and are divided into three categories in accordance with criteria of the IMO Convention. Category A consists of 10 member States with "the largest interest in providing international shipping services" (Art. 17(a)) [54]. Category B consists of 10 member States with "the largest interest in international seaborne trade" (Art. 17(b)) [54]. Category C consists of 20 member States with "special interests in maritime transport or navigation, and whose election to the Council will ensure the representation of all major geographic areas of the world" (Art. 17(c)) [54]. Therefore, maritime interests influence IMO through the Council, and coastal States claiming their rights to protect their environments are underrepresented [58]. This underrepresentation is a result of strong pressures to protect the interests of the traditional maritime States in the conference establishing IMCO in 1948. Those states were concerned that the users of shipping services would create safety standards that were "over-constricting and over-expensive" (p. 62) [50]. The chosen solution was the restricted membership and significant powers of the Council. The initial criteria for membership were the supply of shipping services and interests in international seaborne trade. The third category was introduced to increase the democracy of IMO [50].

Figure 3.2 shows the *five committees* that conduct the main work of IMO. The Maritime Safety Committee (MSC) is IMO's highest technical body, which considers all IMO maritime safety matters. The Marine Environment Protection Committee (MEPC) is the main IMO body for addressing environmental issues. This committee considers all issues regarding the prevention and control of pollution from ships. The main task of the MEPC is to develop conventions and other instruments related to marine pollution from ships and to amend existing instruments [57]. Other functions include gathering information and crafting recommendations and guidelines for the prevention and control of marine pollution from ships [40, 54]. Like the other committees, the MEPC consists of all member States [57].

Both the MEPC and the MSC are assisted by the *seven sub-committees* shown in Fig. 3.2 [57]. The committees are the fora where decisions are made and most political negotiations occur, whereas the sub-committees are given instructions to conduct technical work and report back with proposed actions. Most of the technical work on environmental issues is currently assigned by the MEPC to the Sub-Committee on Pollution Prevention and Response (PPR) [40, 57, 59].[20]

The headquarters of IMO are located in London, reflecting the UK's earlier dominance as a traditional maritime State. The *Secretariat* consists of approximately 300 technical and administrative staff under the Secretary General, making IMO the smallest UN agency [50, 57]. The Secretariat can be viewed as a "broker" between the member States [58], and its most important function is to facilitate negotiations and discussions. The Secretariat provides a policy-making arena for the members and has several administrative functions, such as preparing and holding meetings, collecting and distributing documents, informing on legal issues, writing reports and providing interpreters. Another significant function is to finalise and publish the texts of instruments in English, Arabic, Chinese, French, Russian and Spanish. The Secretariat is also responsible for publishing manuals containing guidance and interpretations, e.g., *MARPOL: How to Do It* [60]. The sales of these publications assist in financing IMO [50, 58, 61].

In summary, IMO is the main forum for regulating pollution from ships. At its headquarters in London, a small Secretariat provides a policy-making arena for the IMO members. One of the main functions of IMO is to develop draft conventions for adoption by States at international diplomatic conferences, although an even larger role is related to amending existing conventions. The organisation's initial role in 1948 was consultative and advisory, primarily regarding maritime safety for a shipping industry that had been dominated by traditional maritime States. Today, developing States are the majority, although traditional maritime States are still influential. Maritime interests have a particular influence in the IMO Council, which is underrepresented by coastal States. The Council supervises the organisation's work under the Assembly, which is the governing body of IMO. Major ship-related oil spills and political pressures related to increased environmental awareness in the 1960s and 1970s resulted in the addition of pollution prevention and control to IMO's functions. Thus, the Marine Environment Protection Committee (MEPC) was established, and it considers all matters regarding the prevention and control of pollution from ships. The Maritime Safety Committee (MSC) is the highest technical body. Both the MEPC and the MSC are assisted by seven sub-committees, and most of the technical work on environmental issues is assigned to the Sub-Committee on Pollution Prevention and Response (PPR).

---

[20]The structure of the sub-committees has changed over time. From 1995 to 2013, this sub-committee was called the Sub-Committee on Bulk Liquids and Gases (BLG).

3.4.2 An Overview of Actors at IMO

Before further discussing the drafting of conventions for regulating pollution from ships (Sect. 3.5), this section discusses the view that these conventions are the result of negotiations to find common ground among the different priorities of the IMO member States. Moreover, the positions of the member States are influenced by other actors, such as inter-governmental and non-governmental organisations. This section provides an overview of these actors and a simplified perspective of the diversity of interests involved in the drafting of conventions regulating pollution from ships through IMO.

3.4.2.1 Member States

Currently, IMO has 171 *member States* [62]. The IMO Convention allows all States to become members, although it sets a few provisions for membership that favour member States of the UN. A common requirement for both UN members and non-UN members is to become a party of the IMO Convention [40, 54]. In addition, territories or groups within member States can become *associate members*, which have no formal decision-making powers and are excluded from membership in the Council [40, 54]. Currently, Hong Kong, Macao and the Faroe Islands are associate members of IMO [62].

The crafting of conventions regulating pollution from ships can be described as a political contest between *maritime interests* and *coastal interests*. The latter group of interests includes, but is not limited to, environmental interests (e.g., see Tan [32]). However, looking solely at the aforementioned division between flag States and coastal States is not sufficient to convey an understanding of how regulations for pollution from ships are drafted through IMO. For example, the term "flag State" represents the State that has granted a ship the right to sail under its flag and not necessarily the ship owners or other maritime industries. Other categories that can help explaining the dynamics in IMO should thus be considered. Two divisions of States regarding interests and conditions for regulating pollution from ships are highlighted: the first is between *coastal States*, *flag States* and *States with maritime interests*, and the second is between *developed States* and *developing States*.

As a simplification of what was described in Sect. 3.3.2, *coastal States* assert the right to environmental protection of their coasts and waters and seek strict regulation of ships through IMO. Historically, coastal States have been the primary initiators of regulating pollution from ships, e.g., Canada and Australia. However, many developed States that were previously traditional maritime States, e.g., western European States and the US, have joined the traditional coastal States in prioritising environmental protection [32].

In contrast with environmental protection, the *flag States* have traditionally emphasised their freedom to use their ships in various ways [32]. As described in Sect. 3.3.2, the possibility of ships flying flags of nationalities that are different from the owners' home-countries, and so-called "open registries", have led to the problems associated with Flags of Convenience (FOC). Among the top five flag

States in 2014, Panama had the most ships with owners registered in different States, representing approximately 21 % of the world fleet based on deadweight tonnage (dwt). Panama was followed by Liberia (approximately 12 %), the Marshall Islands (approximately 9 %), Hong Kong in China (approximately 8 %), and Singapore (approximately 6 %). Together, these five States represented approximately 57 % of the world fleet by dwt [33]. Regarding the role of flag States as actors in IMO, all funding for IMO is provided by its member States and is calculated in proportion to the size of their merchant fleets. Thus, IMO's budget is largely financed by the major flag States, where a majority of the registered ships are foreign-owned [32, 57]. The top three contributors to the IMO budget for 2012 were Panama, Liberia and the Marshall Islands [57].

The category *States with maritime interests* include the interests of several industries, such as ship owners, shipbuilders, cargo owners, shippers, brokers and insurers. Another characteristic of this category of States is an economic dependence on seaborne trade and transport. For example, Japan is heavily dependent on imports by sea transport due to its lack of natural resources. These industries have traditionally been located in developed States, such as the US, the UK, Greece, Germany, Norway, the Netherlands and Japan. However, the proportion of major ship owners located in Asia has increased [32]. Among the top five ship-owning States in 2014, Greece was the largest, with ownership of 15-17 % of the world merchant fleet by tonnage, followed by Japan, China, Germany and the Republic of Korea. The US, the UK and Norway were still among the top ten (numbers seven, eight and ten, respectively), as were Singapore (number six) and Taiwan (number nine) [33]. The influence of Asian States in IMO has traditionally been weak compared with European States [32], although a trend shift due to the recent increase in Asian shipping industries is possible.

In regard to the division of State actors involved in the crafting of regulations on pollution from ships, it should however be noted that many States can have simultaneous and differing interests, e.g., coastal, flag and maritime interests [32, 40]. For example, Greece is a large flag State with a fleet representing 4.6 % of the world fleet tonnage in 2014 [33]. However, as the above figures indicate, Greek ship owners have registered most of the fleet in other States. Thus, in IMO negotiations on pollution from ships, Greece would be considered a State with maritime interests. Simultaneously, Greece can be viewed as having coastal State interests in protecting its environment from ship operations in the Mediterranean. Thus, when dividing States within defined categories, it should be considered that "a State may consider one of these roles more important than the other due to its economic, geographical and environmental interests" [40]. A State with a large merchant fleet may prioritise and represent shipping industry interests due to the shipping sector's importance in the national economy. In contrast, other States have previously prioritised coastal and environmental interests [40].

Finally, the division of the member States into *developed* and *developing* States is important. Although the dynamics between developed and developing States have existed since the early years of IMCO, the tensions between developed States in "the north" and developing States in "the south" have recently become more apparent with the introduction of new issues to the agenda, e.g., climate change, resulting in the division of developed and developing States into opposing groups in negotiations within the MEPC [40]. In the past, most developing States had few maritime interests. However, they have begun to dominate as flag States due to the opening of registries and the flag of convenience phenomenon [32]. In 2014, more than 75 % of the world fleet was registered in developing States [33]. A parallel can be drawn to the aforementioned financing of IMO. When viewed as a division between developed and developing States, IMO is financially dependent on the contributions of these developing flag States, which has previously led to temporary shortages in IMO's budget [58].

Developing States form the majority of the IMO membership, although the influence of developed States is significantly greater. A primary reason for the influence of the developed States is their capacities to supply large delegations and high levels of expertise to participate in meetings. In contrast, developing States are often unable to be fully represented during IMO meetings. Due to lack of resources and high travel costs, developing States often have small delegations, affecting their abilities to cover multiple issues on the agenda or to participate in the different groups that constitute a substantial portion of the meetings [32, 40] (see Sect. 3.5.2 regarding the arrangements of meetings). In addition, their delegations often lack the technical expertise and the accumulated experience and influence that the developed States have gained through many years of attendance [32].

Finally, factors other than those related to the capacities and resources of the member States should be discussed. Section 3.4.1 briefly highlighted historical explanations for the particular influence of traditional maritime States. Their influence as western developed States is primarily gained through participation in the Council and from their leadership and participation in drafting essential policies, e.g., in working groups at meetings [32]. Council membership is an important component of the influence of developed States. Although Council members in Category C (see above) are currently different than in 2006 and 2007, the election of Council members in 2005 is worth highlighting regarding the tensions between developed and developing States. Despite an amendment to the IMO Convention that widened the geographic representation to all regions of the world, the election did not include West Africa, Central Africa, Latin America, or Eastern Europe on the Council. Notably, Liberia was not included in any of the three Council member categories [63]. Currently, Liberia is included in Category C [57]. Yet another factor that could disfavour developing States is the frequent use of the English language without interpretation in different groups at IMO meetings [32].

As mentioned earlier, the States' interests are influenced by other inter-governmental organisations and non-governmental organisations. The roles of these organisations are briefly outlined below.

3.4.2.2 Inter-Governmental Organisations

In addition to cooperation between IMO and any specialised agency of the UN, the IMO Convention allows for cooperation with Inter-Governmental Organisations (IGOs) if their activities and interests are related to the purpose of IMO [54].

Currently, 63 IGOs have entered agreements of cooperation with IMO [62]. Depending on the areas of interests, IGOs are represented at IMO meetings as observers, which means that they can participate in discussions and submit documents, although they do not have any formal decision-making powers in IMO [64].

However, IGOs can have powerful influences on the positions of the IMO member States. The observer IGOs are primarily regional organisations representing the common interests of States in specific regions. Examples of regional organisations that have important roles in developing regulations on pollution from ships are the European Union (EU) and the Helsinki Commission (HELCOM) regarding the protection of the Baltic Sea. Other examples of regional IGOs are the African Union, the League of Arab States, and several Memoranda of Understanding (MoUs) between States on the matter of port State control. These regional IGOs commonly act as coordinators for their respective member States at IMO. For example, the EU is represented not only by its member States but also by the European Commission, which is an observer IGO [40, 62]. For several years, the European Commission has strived for more effective EU participation in IMO by increased cooperation and by seeking to enhance its observer status to full membership [65, 66]. Such membership would involve the same rights as an IMO member State, although it is currently not allowed under the IMO Convention [40].

3.4.2.3 Non-Governmental Organisations

The term Non-Governmental Organisation (NGO) encompasses diverse actors, including civil society movements, multinational corporations, trade associations, and research-oriented associations [67]. NGOs involve themselves in many fora in their efforts to shape environmental policy, e.g., international negotiations and domestic policy-making. As an actor in forming international policies, an NGO can be defined as "any international organization which is not established by inter-governmental agreement" (p. 8) [68]. The phrase "non-State actor" is often used. No formal decision-making powers are given to NGOs within the UN. The States of a UN organisation or conference decide which NGOs can participate and the terms of their participation. The level of NGO participation and their opportunities for influence vary among international organisations [67, 69]. IMO has been described as "more generous with respect to the opportunities for NGOs to be active" (p. 4) [69]. NGOs play an important role in the work of IMO regarding the drafting of conventions [37, 40].

The IMO Convention is not specific regarding the terms for membership and participation of NGOs at IMO meetings. The convention does permit IMO to "make suitable arrangements for consultation and co-operation with non-governmental international organizations" (Art. 67) [54] on matters that are within the scope of IMO. Nevertheless, rules adopted by the Assembly and guidelines developed by the Council provide the possibility for an international NGO to be granted so-called "consultative status" "if it can be reasonably expected to make a substantial contribution to the work of IMO" (p. 1) [70]. The conditions for granting consultative status include a requirement to demonstrate "a considerable expertise as well as the capacity to contribute, within its field of competence, to the work of IMO" (p. 1) [70]. Statuses for applicants are granted by the Council, subject to approval by the Assembly and are regularly reviewed by the Council. A possibility also exists for provisional consultative status up to a maximum of four years [70]. Currently, 77 NGOs have consultative status in IMO [62]. Most represent the interests of various industries related to maritime transport and trade, although environmental interests are also represented. The following are several examples of NGOs participating at IMO meetings regarding matters of pollution from ships [40, 50, 62]:

- Ship owners and ship operators
  - For example, the International Chamber of Shipping (ICS), the Baltic and International Maritime Council (BIMCO), the International Association of Independent Tanker Owners (INTERTANKO) and the International Association of Dry Cargo Shipowners (INTERCARGO)
- Shippers and cargo owners
  - For example, the European Chemical Industry Council (CEFIC) and oil industry organisations, such as the Oil Companies International Marine Forum (OCIMF) and the International Petroleum Industry Environmental Conservation Association (IPIECA)
- Shipbuilders and equipment manufacturers
  - For example, the Community of European Shipyards' Associations (CESA), the Institute of Marine Engineering, Science and Technology (IMarEST) and the European Association of Internal Combustion Engine Manufacturers (EUROMOT)
- Ports, terminals and port services
  - For example, the International Association of Ports and Harbors (IAPH) and the Society of International Gas Tanker and Terminal Operators Limited (SIGTTO)
- Classification societies and industry standards
  - For example, the International Association of Classification Societies (IACS) and the International Organization for Standardization (ISO)
- Environmental NGOs
  - For example, the Friends of the Earth International (FoEI), Greenpeace, the World-Wide Fund for Nature (WWF) and the Clean Shipping Coalition (CSC)

Consultative status does not give NGOs formal decision-making powers in IMO. However, they have the right to submit and receive documents, to have a representative present at sessions of IMO bodies and to speak on any agenda item of interest [70]. They participate in formal discussions and have informal talks with delegates [69]. Organisations representing ship owners and ship operators have a particular influence in IMO due to their ability to influence States with maritime interests [40]. Another very active and influential group of NGOs are organisations representing oil industries [40]. Environmental NGOs are also active participants [40], and an increased importance of their roles in IMO has been observed [37], especially in the MEPC. However, their influence has been partially restricted due to the lack of financial resources to cover the plethora of issues on the agenda at meetings of the different bodies of IMO [32].

In summary, the member States of IMO are the actors with formal decision-making powers. Regulating pollution from ships is the result of negotiations among different States with different economic and environmental conditions and different interests and priorities. Two divisions of State actors based on their different interests and conditions to regulate pollution from ships were discussed: (1) between coastal States, flag States and States with maritime interests and (2) between developed States and developing States. These divisions provide a simplified perspective of the dynamics involved in drafting international regulations on pollution from ships. In addition, the member States are greatly influenced by inter-governmental organisations (IGOs) that are represented at IMO and by non-governmental organisations (NGOs) that are given the opportunity to influence IMO negotiations. Maritime interests are represented by flag States, States with other maritime interests and a large group of influential industry NGOs, whereas environmental interests are represented by coastal States and a few environmental NGOs. Developed States also have a dominating influence, although developing States represent the majority of both the IMO membership and the flag State tonnage.

3.5 An Introduction to the Crafting of International Agreements

Although many international agreements include and are based on unwritten but widely recognised State practices in the form of *customary law*, in the field of marine environmental protection, specifically marine pollution from ships, customary law can only provide general rules that are not practically useful for addressing the problem of marine pollution [6]. Therefore, most marine pollution regulations are currently found in written sources in the form of international agreements, such as the LOSC and MARPOL 73/78, which were negotiated with the backing of IMO [32]. Here, the process associated with the creation of an international agreement between States is presented by examining important steps that are commonly included in this process. A closer examination of the adoption and amendment of IMO conventions follows. In Box 3.2, some perspectives are provided regarding the creation of agreements at the international versus the regional level for the regulation of pollution from ships.

Box 3.2 International versus Regional Regulation of Pollution from Ships

Ships traversing the seas with goods form the backbone of a particularly international business. Thus, it would quickly become untenable for ships to navigate and comply with innumerable different rules throughout the world if States persisted in maintaining their own special national regulations. For example, trade would be severely hindered. Even regional regulatory differences could potentially distort competition in the truly international business of shipping. In addition to the effects on trade, environmental protection would also be rendered inefficient from pollution and conservation perspectives [39] because both pollution and species travel across the artificial borders of States.

To avoid a patchwork of different national rules, shipping regulators have supported a harmonisation of regulation for ships *at the highest international level* in the form of international agreements for a long time. Simultaneously, globally applicable environmental regulations for ships are not always possible or even particularly desirable. In the case of pollution from ships, *regional regulation* in the form of regional agreements to protect the marine environment may also be supported for several reasons [6, 8].

- Regional rules are ecologically defensible given the diverse ecology and differing susceptibilities to pollution of different parts of the world. All States will not experience environmental problems with the same intensity and are not equally interested in regulating the same phenomena.
- Regulation at the regional level can better coordinate with integrated ecosystem and coastal zone management, which is increasingly relevant for protecting sensitive regions, such as the Baltic Sea.
- Regional cooperation in regulation may be more practical, especially in emergency situations where rapid response is needed.
- Regional cooperation in regulation can also result in fewer compromises, more stringent requirements and higher environmental standards than when the regulation is formulated at the highest international level.
- Regional regulation can also provide a "testing room" for regulations. Successful attempts to regionally regulate pollution may become precursors to global regulation.

3.5.1 Basic International Agreement Terminology

In the crafting of international treaties/agreements/conventions, there are no steps and formalities that are applicable to all situations. In principle, an international agreement can be concluded in any way that the parties prefer [1]. Regardless, certain underlying and basic conditions normally apply to the conclusion of international agreements. One such condition is the importance of the *consent* to be bound by an international agreement. When States are negotiating and considering becoming parties to an international agreement, the process must begin with consent.

How do States express their consent to be bound by an international agreement according to international law? The 1969 Vienna Convention on the Law of Treaties [71] is a widely accepted framework that provides guidance on several matters regarding international agreements, including how States normally give their consent to become parties to international agreements [1]. Based on this framework, some of the important terms and steps that are typically present in the conclusion of an international agreement are presented below. However, these steps are also preceded by several crucial steps, which are specifically described in relation to the work of IMO in the next section (see Sect. 3.5.2). Nevertheless, when States have already agreed to begin working on the text of an international agreement and are willing to convene a meeting or meetings in a certain forum, the following steps are of particular interest in the process of concluding a binding international agreement.

(1) *Adoption* is the term used for the formal establishment of the form and content of the proposed text in an international agreement. In general, States participating in the crafting of an international agreement adopt the text of the agreement by expressing their consent. When an international agreement is negotiated with the backing of an international organisation, the adoption (and expression of consent) typically occurs via a *resolution* in a representative body of that particular international organisation, e.g., by a United Nations General Assembly resolution [1, 72]. However, an international agreement can also be adopted at an international conference convened for the purpose of concluding the international agreement. At international conferences, adoption generally occurs when a two-thirds majority of those States present and voting votes in favour of adoption, unless, by the same majority, the States have decided to apply a different rule for adoption [72]. However, a visible trend at international conferences, inspired by the work of the third United Nations Conference on the Law of the Sea, i.e., UNCLOS III, which led to the adoption of the LOSC, is that it has become common to attempt to reach adoption of international treaties through consensus before the negotiating parties proceed to a vote [1]. Thus, no voting will occur "until all efforts to reach agreement by consensus have been exhausted" (p. 910) [1]. Here, *consensus* refers to a situation in which the opinions of the negotiating parties have begun to intersect; opinions may still differ, although not strongly. An adoption by consensus *without any voting* differs from *unanimity*, which "requires the affirmative vote of all negotiating states" (p. 43) [37].

(2) *Signature* refers to the step that often follows adoption. Situations exist when the signature of an international agreement is immediately binding, although a State's signature commonly indicates *only a preliminary acceptance* of being bound to an international agreement. This distinction is important because a State can still withdraw from being bound by an agreement *after it has signed the international agreement* [19]. For a State to be fully bound to an international agreement, signature is often combined with the next step.

(3) *Ratification* is the step in which a State formally agrees to be bound to a treaty. When an international agreement has many parties, the common procedure is for a so-called *depositary* to collect the ratifications from the States. The United Nations is an example of a depositary of ratifications [1]. The function of ratification is that a State gains additional time for consideration before it is fully bound by an international agreement. Moreover, this step allows for public opinion regarding the agreement to be voiced by representatives in a particular State's national parliament. Thus, a State can withdraw from being bound to an agreement if strong public disagreement is voiced [1]. In some situations, a State may wish to join an international agreement after a particular signature deadline has passed. If a particular international agreement enables this, consent to be bound can instead be given by *accession*, which has the same function as *ratification* [1].

After ratification, i.e., when States have been formally bound, the agreement must (4) *enter into force* to become operative. Entry into force conditions can vary between agreements. Some agreements set a specific date when they enter into force. In other cases, in which many parties are involved in an international agreement, a predefined number of States are often required to have expressed their consent to be fully bound in the form of a predefined number of ratifications. Some agreements use combinations of additional conditions for entry into force [72]. For example, an agreement can enter into force after a defined period and when a predefined number of States have ratified it; these States must simultaneously represent a group that is vital for the agreement to have a sufficient effect (see Sect. 3.5.2). For clarity, in the context of IMO, agreements can specify "effective dates", which is not the same as the entry into force of an agreement. An effective date (or dates) means a specified date within an agreement from which certain regulations apply in practice. This date occurs after entry into force and allows additional time before certain regulations take effect [73].

In summary, in the field of marine environmental protection, specifically marine pollution from ships, most regulations can be found in written sources in the form of international agreements, such as the LOSC and MARPOL 73/78, which were negotiated with the backing of IMO. In the creation of international agreements, certain basic conditions apply. When States are negotiating and considering becoming parties to an international agreement, the process must begin with consent. States can express their consent to be bound to an international agreement in several ways. Commonly, when the subject of an international agreement has been decided and negotiations have begun at an international conference, the following steps are included in the crafting of an international agreement: adoption, signature, ratification or accession and entry into force.

3.5.2 The Crafting of IMO Conventions on Pollution from Ships

As briefly noted above, crucial preparatory steps precede the adoption of an international agreement. For example, before negotiations are initiated, issues must be brought to the attention of the States, e.g., an international transboundary pollution problem. The establishment of such issues on the international agenda as a sufficiently pressing problem to necessitate States crafting an agreement can require many years [19]. The preparatory steps are particularly important for understanding IMO's role in the crafting of conventions that regulate pollution from ships. Because of the involvement of IMO, some differences in the steps compared with the aforementioned general description can occur. In this section, the preparatory work and adoption procedures are outlined. Next, the entry into force criteria that are common in IMO conventions are addressed. Finally, the procedure of amending existing IMO conventions is described.

3.5.2.1 Preparatory Work and Adoption Procedures

In general, a convention developed by IMO originates from one or several member States raising an issue to IMO [50] (e.g., see Svensson [74]). However, restrictions exist for placing new issues on the agenda of MEPC sessions, which is the forum that is used for drafting environmental regulations in IMO. First, the MEPC's workload "is growing enormously"(p. 25) [40]. Second, the overall strategic plan of IMO provides the main strategic objectives for a six-year period, and a "High-level Action Plan" addresses how these objectives should be effectively achieved by the different bodies of IMO over a two-year period [59]. Third, the current guidelines regarding the work of the MEPC [59] restrict proposals considering new issues and the development of measures, such as conventions. A proposal for developing a new convention must include a demonstrated "compelling need" for the proposed measure or measures, an analysis of the costs and benefits to the shipping industry, and the legislative and administrative burdens of the member States. Finally, when the MEPC considers such proposals, it should conduct "a comprehensive and thorough assessment" (Para. 4.14) [59] that considers the aforementioned criteria (with some exceptions in cases of urgent matters) [59].

Aside from these restrictions, what happens if the MEPC agrees on a proposal for a new convention? After endorsement by the Council, the MEPC begins a detailed consideration and often instructs a sub-committee (such as the Sub-Committee on Pollution Prevention and Response, PPR) to perform the main technical work for drafting a convention [50, 59].

Committee or sub-committee sessions are held during a normal workweek. All decisions are made in *plenary* sessions in the main hall of the IMO headquarters, and all participating delegations and observers are represented. Each agenda item is considered in the plenary; submissions that require plenary consideration are also discussed. In general, the technical consideration of specific issues on the agenda is assigned to *working groups* before decisions are made. Additionally, *drafting groups* are given instructions to conduct editorial work for already-decided requirements in a draft text [74]. Because it is difficult for developing States to fully participate in these groups, as discussed in Sect. 3.4.2, the number of working groups is limited to three, and the total number of groups (working groups and drafting groups) is five. If additional work is needed to meet the deadlines, working groups can be assigned to meet between the sessions or members can comment on an issue via email in *correspondence groups* [59]. In addition, *informal talks* represent a significant portion of policy-making within IMO. Informal groups are established, and much discussion and negotiation occur between the formal working hours of a session [74].

An important difference from the general description of international organisations is that *IMO does not have the mandate to adopt international conventions on its own*. The adoption of an IMO convention must occur through an international diplomatic conference that is specifically convened for that purpose. When agreement on a final draft convention has been reached by the MEPC, it is sent to the Council and the Assembly with a recommendation to convene a diplomatic conference for adoption. The work of the MEPC is then complete [50].

If the draft convention is agreed upon, a resolution is then adopted by the Council or the Assembly. The resolution calls for a diplomatic conference and invites all IMO member States and all member States of the UN. The draft convention is circulated for comments before the conference. At this stage, the mandate of IMO ends. The time required to reach this stage varies, although it can require several years [50]. The IMO process has been described as a naturally "slow negotiating process" (p. 183) [75], and the extent of IMO's work should not be underestimated. Any excessive haste could result in an inadequately prepared draft convention, which could result in failure at the conference [50]. In summary, the time-consuming preparatory work performed by IMO plays an important role in the crafting of conventions by defining and resolving the critical issues with a resultant draft convention that provides the conditions for later consensus among governments.

Once the conference is held, it becomes an international body with its own arrangements and procedural rules, even if it is held at the IMO headquarters. In addition to IMO member States and other UN member States, NGOs can also participate without formal decision-making powers [50, 74]. Although participating States are generally equal at a conference [50], voting rights can differ depending on the situation, e.g., when adopting a new protocol under an existing convention. For example, the addition of Annex VI to MARPOL 73/78 was adopted by the parties present and voting (a two-thirds majority was required) at the 1997 Conference of Parties to MARPOL 73/78 [76].

3.5.2.2 Entry into Force and Amendments

As described in Sect. 3.5.1, the adoption of a convention is merely the first stage of an often long process that requires signature, ratification or accession by States before its entry into force and even more time before the regulations become effective. Generally, a specified number of States representing a specific percentage of the world tonnage (merchant fleet) must express their consent to be bound by an IMO convention [32]. A current example is the 2009 Hong Kong International

Convention for the Safe and Environmentally Sound Recycling of Ships (see Sect. 3.3.5), which will enter into force 24 months after the date on which 15 States, representing 40 % of the world merchant shipping by gross tonnage, have ratified the agreement. Additionally, a third condition is that the combined maximum annual ship recycling volume of the 15 States during the preceding 10 years constitutes no less than 3 % of the gross tonnage of the combined merchant shipping of the same States [47].

Once a convention has entered into force, amendments are possible; however, the restrictions on introducing new measures described above must still be taken into account [59]. The work of *amending conventions* follows the preparatory work towards a new convention [59], with the important difference that *no conference must be held.* The ability of the MEPC to adopt amendments to technical standards through *resolutions* is part of an important additional mandate given to IMO by amendment procedures incorporated into IMO conventions, such as MARPOL 73/78 [56].

The amendment procedures of the early IMO conventions required ratification or acceptance by a two-thirds majority of the parties. However, these procedures were so slow that some amendments never entered into force. Therefore, the *tacit acceptance procedure* was introduced in several IMO conventions to hasten the amendment process [40, 50]. By "tacit acceptance" of an amendment, a specific date is set for its entry into force; this date applies if no objections are received from "a specified number of Parties" [55] before that date [40, 55]. For example, the revised 2008 MARPOL Annex VI was set to enter into force on 1 July 2010 unless no "less than one-third of the Parties or Parties the combined merchant fleets of which constitute not less than 50 % of the gross tonnage of the world's merchant fleet" (Preamble, para. 2) [77] *had objected* prior to 1 January 2010.

In summary, this section highlighted the particular importance of the preparatory work of IMO in the crafting of conventions for regulating pollution from ships. However, many restrictions must be met for the MEPC of IMO to address new measures, e.g., a compelling need for regulation must be demonstrated. If agreement is reached that a convention should progress, the development of a draft text of a convention is the scope of its work because IMO does not have the mandate to adopt international conventions. Adoption of an IMO convention must occur through an international diplomatic conference that is specifically convened for that convention. The often time-consuming preparatory work by IMO provides important conditions for consensus among governments at a diplomatic conference. After adoption, a predefined number of States representing a specific percentage of the world tonnage generally have to express their consent to be bound by an IMO convention for it to enter into force. Once a convention has entered into force, amendment is possible. In this process, the MEPC has the ability to adopt amendments on its own. By the tacit acceptance procedure, a specified date applies for the entry into force of the amendments if no objections are received from a set number of parties.

Recommended further reading for the different areas described in this chapter are presented in Box 3.3.

Box 3.3 Recommended Further Reading

International Law:

Cassese, A., *International Law*. Second Edition ed. 2005, Oxford University Press.

Shaw, M. N., International Law. Sixth Edition ed. 2008: Cambridge University Press.

International Environmental Law:

Birnie, P., Boyle, A. & Redgwell, C., *International Law & the Environment*. 3rd ed. 2009, Oxford: Oxford University Press.

Bodansky, D., *The Art and Craft of International Environmental Law*. 2010: Harvard University Press.

The International Law of the Sea:

Tanaka, Y., The International Law of the Sea. 2012: Cambridge University Press.

Churchill, R. R. & Lowe, A. V., *The law of the sea*. Third edition ed. 1999: Juris Publishing and Manchester University Press.Harrison, *Making the Law of the Sea: A Study in the Development of International Law*, Cambridge University Press 2011.

The International Framework Regulating Pollution from Ships:

De la Rue, C. & Anderson, C. B., *Shipping and the Environment*. 2nd ed. 2009: Informa Law.

IMO webpage, *List of IMO Conventions*

http://www.imo.org/About/Conventions/ListOfConventions/Pages/Default.aspx, last accessed on 20 May 2015.

The Role of IMO and the Dynamics of Actors and Interests:

Boisson, P., *Safety at Sea: Policies, Regulations & International Law*. 1999, Paris: Edition Bureau Veritas.

Karim, M. S., *Prevention of Pollution of the Marine Environment from Vessels: The Potential and Limits of the International Maritime Organization*. 2015, Switzerland: Springer International Publishing.

Tan, A. K. J., *Vessel-Source Marine Pollution: The Law and Politics of International Regulation*. 2006, Cambridge: Cambridge University Press.

IMO. IMO 1948-1998: a process of change. Focus on IMO, September 1998. 1998, International Maritime Organization: London, http://www.imo.org/KnowledgeCentre/ReferencesAndArchives/FocusOnIMO(Archives)/Documents/Focus%20on%20IMO%20-%20IMO%201948-1998,%20a%20process%20of%20change%20(September%201998).pdf, last accessed on 20 May 2015

References

1. Shaw, M. N., *International Law*. Sixth Edition ed. 2008: Cambridge University Press.
2. Churchill, R. R. & Lowe, A. V., *The law of the sea*. Third edition ed. 1999: Juris Publishing and Manchester University Press.
3. *International Convention for the Safety of Life at Sea* (SOLAS), 1 November 1974, entered into force 25 May 1980 1184 UNTS 2, 1974.
4. *International Convention on Standards of Training, Certification and Watchkeeping for Seafarers (STCW)* 7 July 1978, entered into force 28 April 1984, 1978.
5. Khalilieh, H. S., *Admiralty and Maritime Laws in the Mediterranean Sea (ca. 800–1050)*. 2006, The Kitāb Akriyat al-Sufun vis-à-vis the Nomos Rhodion Nautikos: Brill.
6. Tanaka, Y., *The International Law of the Sea*. 2012: Cambridge University Press.
7. Onuma, Y., *Hugo Grotius., last accessed on 20 May 2015, accessed via University of Gothenburg, Gothenburg University Library*, in *Encyclopædia Britannica Online*. 2015.
8. Birnie, P., Boyle, A. & Redgwell, C., *International Law & the Environment*. 3rd ed. 2009, Oxford: Oxford University Press.
9. *Draft Convention of the Preliminary Conference on Oil Pollution of Navigable Waters* 1926.
10. *Convention Between the United States of America, the United Kingdom of Great Britain and Ireland, and Russia, for the Preservation and Protection of Fur Seals* 7 July 1911, entered into force 15 December 1911, 8 IPE 3682, 1911.
11. *Convention for the Preservation of Halibut Fishery of the North Pacific Ocean* 2 March 1923, entered into force 21 October 1924, 32 LNTS 93, 1923.
12. *Convention for the Regulation of Whaling* 24 September 1931, entered into force 16 January 1935, 155 LNTS 349, 1931.
13. *International Convention for the Prevention of Pollution of the Sea by Oil* (OILPOL 1954), 12 May 1954, entered into force 26 July 1958, 327 UNTS 3, 1954.
14. De la Rue, C. & Anderson, C. B., *Shipping and the Environment*. 2nd ed. 2009: Informa Law.
15. Gold, E., *Gard Handbook on Protection of the Marine Environment*. 2006: AS Gard.
16. *International Convention on Civil Liability for Oil Pollution Damage* 29 November 1969, entered into force 27 January 1980, 973 UNTS 3, 1969.
17. *International Convention Relating to Intervention on the High Seas in Cases of Oil Pollution Casualties* 29 November 1969, entered into force 6 May 1975, 9 ILM 25, 1969.
18. Carson, R. L., *Silent Spring*. 1962, New York: Houghton Mifflin Company.
19. Bodansky, D., *The Art and Craft of International Environmental Law*. 2010: Harvard University Press.
20. *International Convention for the Prevention of Pollution from Ships* (MARPOL 73), 2 November 1973, not in force, 12 ILM 1319, 1973.
21. *Protocol Relating to the International Convention for the Prevention of Pollution from Ships, 1973* (MARPOL 78), 17 February 1978, entry into force 2 October 1983, 17 ILM 546, 1978.
22. *Protocol of 1997 to amend the International Convention for the Prevention of Pollution from Ships, 1973, as modified by the Protocol of 1978 relating thereto* (1997 Protocol), 1 January 1998, entry into force 19 May 2005, Doc. MP/CONF. 3/34 1997.
23. European Commission. *Industrial accidents: The Seveso Directive - Prevention, preparedness and response*. 2015, European Commission: Brussels, http://ec.europa.eu/environment/seveso/ , last accessed on 20 May 2015.
24. Gold, E., *Learning from Disaster: Lessons in Regulatory Enforcement in the Maritime sector*. Review of European Community & International Environmental Law, 1999. **8**(1): p. 16–20.
25. *International Convention for the Prevention of Pollution from Ships as amended by the Protocol of 1978 (MARPOL 73/78)*. 1973/78: International Maritime Oranization.
26. *United Nations Convention on the Law of the Sea* (LOSC), 10 December 1982, entered into force 16 November 1994, 21 ILM 1261, 1982.
27. Edeson, W. R., *Confusion over the use of "UNCLOS", and references to other recent Agreements*. International Journal of Marine and Coastal Law 2000. **15**(3): p. 415–421.

28. *Protocol to the United Nations Framework Convention on Climate Change* (Kyoto Protocol), 11 December 1997, entered into force 16 february 2005, 37 ILM 22, 1997.
29. Cassese, A., *International Law*. Second Edition ed. 2005, Oxford University Press.
30. Law, J. & Martin, E. A., *Jurisdiction. A Dictionary of Law*. 7th ed. 2009: Oxford University Press Published online: Current Online Version: 2014.
31. Bodansky, D., *Protecting the Marine Environment from Vessel-Source Pollution: UNCLOS III and Beyond.* Ecology Law Quarterly 1991. **18**(4): p. 720–777.
32. Tan, A. K. J., *Vessel-Source Marine Pollution: The Law and Politics of International Regulation*. 2006, Cambridge: Cambridge University Press.
33. UNCTAD. *Review of Maritime Transport 2014*. 2014, United Nations Conference on Trade and Development (UNCTAD): New York and Geneva.
34. UNCTAD. *Review of Maritime Transport 2013*. 2013, United Nations Conference on Trade and Development (UNCTAD): New York and Geneva.
35. Marten, B., *Port State Jurisdiction in New Zealand: The Problem with Sellers.* Victoria University of Wellington Law Review, 2013. **44**(3/4 - 60th Anniversary Issue): p. 559-572.
36. Marten, B., *Port State Jurisdiction and the Regulation of International Merchant Shipping*. 2014, Switzerland: Springer International Publishing.
37. Harrison, J., *Making the Law of the Sea: A Study in the Development of International Law.*. 2011, Cambridge: Cambridge University Press.
38. IMO Secretariat. *Implications of the United Nations Convention on Law of the Sea for the International Maritime Oranization. LEG/MISC.8*. 2014, International Maritime Organization: London.
39. Ringbom, H., *Preventing Pollution from Ships – Reflections on the 'Adequacy' of Existing Rules.* Review of European Community & International Environmental Law, 1999. **8**(1): p. 21–28.
40. Karim, M. S., *Prevention of Pollution of the Marine Environment from Vessels: The Potential and Limits of the International Maritime Organization.* 2015, Switzerland: Springer International Publishing.
41. IMO. *International Convention for the Prevention of Pollution from Ships (MARPOL).* 2015, International Maritime Organization.: London, http://www.imo.org/About/Conventions/ListOfConventions/Pages/International-Convention-for-the-Prevention-of-Pollution-from-Ships-%28MARPOL%29.aspx, last accessed on 20 May 2015.
42. IMO. *Anti-fouling systems.* 2015, International Maritime Organization.: London, http://www.imo.org/OurWork/Environment/Anti-foulingSystems/Pages/Default.aspx, last accessed on 20 May 2015.
43. *The International Convention on the Control of Harmful Anti-fouling Systems on Ships 2001* (AFS Convention), 5 October 2001, entered into force 17 September 2008, Doc. AFS/CONF/26 2001.
44. *The International Convention for the Control and Management of Ships' Ballast Water and Sediments* (BWM Convention), 1 June 2004, not entered into force, Doc. BWM/CONF/36 2004.
45. IMO. *International Convention for the Control and Management of Ships' Ballast Water and Sediments (BWM).* 2015, International Maritime Organization: London, http://www.imo.org/About/Conventions/ListOfConventions/Pages/International-Convention-for-the-Control-and-Management-of-Ships%27-Ballast-Water-and-Sediments-(BWM).aspx last accessed on 20 May 2015.
46. IMO. *Recycling of ships.* 2015, International Maritime Organization: London, http://www.imo.org/OurWork/Environment/ShipRecycling/Pages/Default.aspx, last accessed on 20 May 2015.
47. *Hong Kong Convention for the safe and environmentally sound recycling of ships, 2009* (Hong Kong Convention), 1 September 2009, not entered into force, Doc. SR/CONF/45 2009.

48. IMO. *Nairobi International Convention on the Removal of Wrecks*. 2015, International Maritime Organization: London, http://www.imo.org/About/Conventions/ListOfConventions/Pages/Nairobi-International-Convention-on-the-Removal-of-Wrecks.aspx, last accessed on 20 May 2015.
49. *Nairobi International Convention on the Removal of Wrecks* (Wreck Removal Convention), 19 November 2007, in force 23 May 2007, 46 ILM 697, 2007.
50. Boisson, P., *Safety at Sea: Policies, Regulations & International Law*. 1999, Paris: Edition Bureau Veritas.
51. United Nations. *The United Nations System*. 2013, United Nations Department of Public Information: New York, http://www.un.org/en/aboutun/structure/pdfs/UN%20system%20chart_lettercolor_2013.pdf, last accessed on 20 May 2015.
52. IMO. *Introduction to IMO*. 2015, International Maritime Organization: London, http://www.imo.org/About/Pages/Default.aspx, last accessed on 20 May 2015.
53. *Charter of the United Nations* 26 June 1945, entered into force on 24 October 1945, 1 UNTS XVI.
54. *1948 Convention on the International Maritime Organization* (IMO Convention), 6 March 1948, entered into force 17 March 1958, Amendments by Assembly Resolutions in: Basic Documents, Volume I, 2010 Edition. 2010, International Maritime Organization. 1948.
55. IMO. *Introduction: Adopting a convention, Entry into force, Accession, Amendment, Enforcement, Tacit acceptance procedure*. 2015, International Maritime Organization: London, http://www.imo.org/About/Conventions/Pages/Home.aspx, last accessed on 20 May 2015.
56. IMO. *IMO 1948–1998: a process of change. Focus on IMO, September 1998*. 1998, International Maritime Organization: London, http://www.imo.org/KnowledgeCentre/ReferencesAndArchives/FocusOnIMO(Archives)/Documents/Focus%20on%20IMO%20-%20IMO%201948-1998,%20a%20process%20of%20change%20(September%201998).pdf, last accessed on 20 May 2015.
57. IMO. *Structure of IMO*. 2015, International Maritime Organization: London, http://www.imo.org/About/Pages/Structure.aspx, last accessed on 20 May 2015.
58. Campe, S., *The Secretariat of the International Maritime Organization: A Tanker for Tankers*, in *Managers of Global Change: The Influence of International Environmental Bureaucracies*, Frank Biermann & Bernd Siebenhüner, eds. 2009, MIT Press: Cambridge.
59. IMO. *Guidelines on the Organization and Method of Work of the Maritime Safety Committee and the Marine Environment Protection Committee and Their Subsidiary Bodies. MSC-MEPC.1/Circ.4/Rev.2*. 2012, International Maritime Organization: London.
60. IMO. *MARPOL - How to do it*. 2013, International Maritime Organization: London.
61. IMO. *Publications catalogue: July 2014*. 2014, International Maritime Organization: London.
62. IMO. *Member States, IGOs and NGOs*. 2015, International Maritime Organization: London, http://www.imo.org/About/Membership/Pages/Default.aspx, last accessed on 20 May 2015.
63. Ademun-Odeke, *From the "Constitution of the Maritime Safety Committee" to the "Constitution of the Council": Will the IMCO Experience Repeat Itself at the IMO Nearly Fifty Years On?-The Juridical Politics of an International Organization*. Texas International Law Journal, 2007. **43**(1): p. 55–113.
64. IMO. *Rules of Procedure of The Marine Environment Protection Committee, in Basic Documents, Volume I, 2010 Edition.*. 2010, International Maritime Organization.
65. European Commission. *Strategic goals and recommendations for the EU's maritime transport policy until 2018. COM(2009) 8 final*. 2009: Brussels.
66. European Commission. *Keep Europe moving: Sustainable mobility for our continent: Midterm review of the European Commission's 2001 Transport White Paper. COM(2006) 314 final*. 2006: Brussels.
67. Betsill, M. M. & Corell, E., *NGO diplomacy: the influence of nongovernmental organizations in international environmental negotiations*. 2008, Cambridge: MIT Press.

68. Ahmed, S. & Potter, D. M., *NGOs in International Politics*. 2006, Sterling, VA: Kumarian Press.
69. Peet, G., *The role of (environmental) non-governmental organizations at the Marine Environment Protection Committee (MEPC) of the International Maritime Organization (IMO), and at the London Dumping Convention (LDC)*. Ocean & Coastal Management, 1994. **22**(1): p. 3–18.
70. IMO. *Rules and Guidelines for Consultative Status of Non-Governmental International Organizations with the International Maritime Organization*. 2015, International Maritime Organization (IMO): London, http://www.imo.org/About/Membership/Documents/RULES%20AND%20GUIDELINES%20FOR%20CONSULTATIVE%20STATUS.pdf, last accessed on 20 May 2015.
71. *Vienna Convention on the Law of Treaties* 23 May 1969, entered into force 27 January 1980, Treaty Series 1155 331, 1969.
72. UN Treaty Collection. *Glossary of terms relating to treaty actions*. 2015, United Nations Office of Legal Affairs: New York, https://treaties.un.org/pages/Overview.aspx?path=overview/glossary/page1_en.xml, last accessed on 20 May 2015.
73. IMO. *List of amendments expected to enter into force this year and in the coming years* 2015, International Maritime Organization: London, http://www.imo.org/About/Conventions/Pages/Action-Dates.aspx, last accessed on 20 May 2015.
74. Svensson, E. *Sulphur Regulations for Shipping - Why a Regional approach?: Scientific and Economic Arguments in IMO Documents 1988–1997* Doctor of Philosophy thesis, Chalmers University of Technology, 2014.
75. Okamura B., *Proposed IMO Regulations for the Prevention of Air Pollution from Ships*. Journal of Maritime Law and Commerce, 1995. **26**(2): p. 183–195.
76. *1997 Conference of Parties to the International Convention for the Prevention of Pollution from Ships, 1973, as modified by the Protocol of 1978 relating thereto*. Doc. MP/CONF.3/2. 1997.
77. *Resolution MEPC.176(58). Amendments to the Annex of the Protocol of 1997 to Amend the International Convention for the Prevention of Pollution from Ships, 1973, As Modified By the Protocol of 1978 Relating Thereto (Revised MARPOL Annex VI). Adopted on 10 October 2008, MEPC 58*. International Maritime Organization. 2008.

Part II
Environmental Impacts

Discharges to the Sea

4

J. Fredrik Lindgren, Magda Wilewska-Bien, Lena Granhag, Karin Andersson and K. Martin Eriksson

Abstract

In this chapter, various environmental issues from the shipping industry which ends up in the oceans are described. Oil pollution, wastewater, antifouling paint, ballast water and litter are all described in detail. Various sources of oil pollution exist, ranging from large accidents to small continuous leakages from, e.g., propeller shaft bearings. The behaviour of oil when it enters the sea can differ, ultimately affecting the environment. Wastewater from ships is divided into sewage and grey water, and different regulations can affect their characteristics. Fouling on ship hulls affects the drag on the ship, which increases fuel consumption when maintaining a constant speed. The various antifouling paints used today to combat fouling are described herein, and a review of the environmental implications of using these paints is provided. Ballast water contains organisms that can become invasive if released into a new geographical area. Invasive species can entail costs on the order of millions of euros. Finally, litter is discussed in this chapter. Litter is deposited in the ocean via several sources and can affect organisms over long periods of time. Plastic causes the

J.F. Lindgren (✉) · M. Wilewska-Bien · L. Granhag · K. Andersson · K.M. Eriksson
Shipping and Marine Technology, Chalmers University of Technology, Gothenburg, Sweden
e-mail: f.lindgren@chalmers.se

M. Wilewska-Bien
e-mail: magda.w.bien@chalmers.se

L. Granhag
e-mail: lena.granhag@chalmers.se

K. Andersson
e-mail: karin.andersson@chalmers.se

K.M. Eriksson
e-mail: martin.eriksson@chalmers.se

K. Andersson et al. (eds.), *Shipping and the Environment*,
DOI 10.1007/978-3-662-49045-7_4

largest litter-related problem because it does not biodegrade; such material only becomes smaller, ultimately reaching a microplastic state. Hence, litter can affect organisms in different ways.

Keywords
oil pollution · accidental spills · operational spills · wastewater · antifouling · ballast water · litter · invasive species · fouling organisms

The shipping industry contributes to the addition of anthropogenic pollutants to the ocean via several pathways. The restricted usage, discharging and handling of different products from ships are regulated by IMO convention MARPOL 73/78 (The International Convention for the Prevention of Pollution from Ships), which entered into force on 2 October 1983 [1] (Table 4.1). Several of the environmental issues that originate from the discharges of ships into the sea are regulated by the MARPOL 73/78 convention. For example, in Annex I (Regulations for the Prevention of Pollution by Oil), both accidental and operational spills are addressed. The regulations for handling sewage are described in Annex IV (Prevention of Pollution by Sewage from Ships), whereas the rules regarding garbage, including specifications for the distance from land and the manner in which garbage can be

Table 4.1 Various IMO conventions controlling the discharge, handling and usage of different hazardous products on ships. Conventions discussed in this chapter are marked in bold

IMO convention	Regulates	Adoption (year)	Entered into force (year)
MARPOL 73/78	Pollution from ships	1973/1978	1983
Annex I	**Oil & oily mixtures**	**1973**	**1983**
Annex II	Liquid chemicals in bulk	1973	1983
Annex III	Liquid chemicals in packaged form	1973	1992
Annex IV	**Sewage**	**1973**	**2003**
Annex V	**Garbage**	**1973**	**1988**
Annex VI	Air pollution and GHG emissions	1997	2005
International convention on the control of harmful anti-fouling systems on ships	**Antifouling paints**	**2001**	**2008**
International convention for the control and management of ships' ballast water and sediments	**Invasive species**	**2004**	**Not yet in force**
Hong Kong International Convention for the Safe and Environmentally Sound Recycling of Ships	Recycling of ships	2009	Not yet in force
Nairobi International Convention on the removal of wrecks	Removal of wrecks	2007	2015

Annex I-VI refers to annexes in MARPOL 73/78

disposed, are described in Annex V (Prevention of Pollution by Garbage from Ships). Furthermore, other types of pollution from the shipping industry exist that are not regulated by MARPOL 73/78, e.g., heavy metals, biocides, biological discharges in the form of invasive species and anthropogenic noise [1]. Heavy metals originate from different sources, such as fuels, antifouling paints, and the reactivation of metal-sulphide complexes in dredging operations. Biocides are chemical substances used to prevent the establishment and/or growth of organisms or to eradicate organisms that disturb human activities. These chemicals can be included in antifouling paints used to coat ship hulls to prevent the attachment of biological fouling and subsequent drag on the hull [2]. The discharge of biological material can result in the spread of non-indigenous biota to new geographical areas, which can initiate species invasions. Such alien species originate from ships ballast water and can also be attached to ship hulls that lack antifouling paint (Table 4.1) [3]. The IMO conventions regulating the environmental issues that are addressed in this chapter can be found in Table 4.1 (marked in bold).

4.1 Oil

The global need for energy, e.g., for heating, electricity production and propulsion, has entailed that enormous quantities of various forms of petroleum are needed annually. Because petroleum sources are only found at specific locations around the globe, a large need exists for its transport to consumers. During the transport and consumption of petroleum, discharge can occur that immediately or subsequently ends up in marine systems. Because petroleum is an anthropogenic stressor to biota, these releases, depending on several factors, may result in acute lethal to long-term sub-lethal effects. Thorough investigations have been completed on the effects of large petroleum spills, whereas less is known regarding the effects of continuous small-scale discharges on marine biota. Indeed, such studies of continuous small-scale discharges are needed because the number of large spills has decreased, whereas small-scale discharges have remained frequent and may even be increasing. This increase comes from the growing demand for energy from oil and the subsequent increase in transport.

Crude oil is a naturally formed substance that originates from decaying plant and animal material. Over millions of years, organic material becomes incorporated into the sediment and is exposed to heat and pressure with increasing depth. Under optimal conditions, it is eventually transformed into petroleum and either makes its way to the surface or gets trapped beneath solid rock, forming large oil reservoirs [4]. The content of crude oil varies slightly depending on its source. Crude oil can be divided into four main categories: saturates (alkanes and cycloparaffins), aromatics (mono-, di- and polycyclic), resins (aggregates of pyridines, quinolones, carbazoles, thiophenes, sulfoxides, and amides) and asphaltenes (aggregates of polyaromatics, naphthenic acids, sulphides, polyhydric phenols, fatty acids, and metalloporphyrins) [5]. These main groups constitute approximately 90 % of the total content, whereas the remaining 10 % is composed of nitrogen, oxygen, sulphur and different metals;

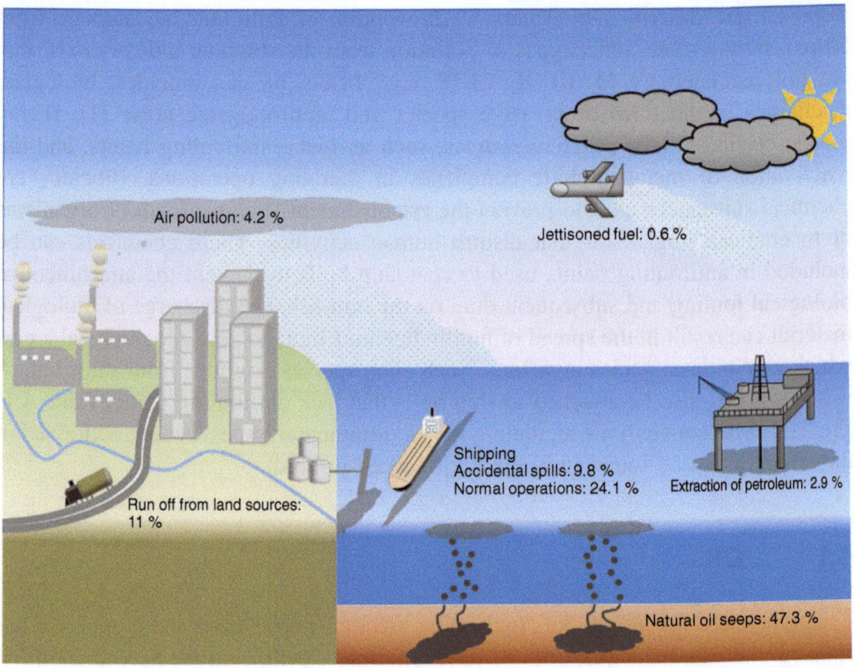

Fig. 4.1 Sources of marine oil pollution (modified from Farrington and McDowell 2004, with the kind permission of Oceanus magazine) [11]

vanadium and nickel are the only elements to exist at concentrations exceeding 1 ppm [6]. Large quantities of crude oil are pumped to the earth's surface by the oil industry and transported to various places around the globe every year. The main system for transportation is the shipping sector, which transported approximately 1.9 billion tonnes of petroleum in 2013, corresponding to 62 % of the total amount of oil that was transported [7]. In deep water travels usually Very large crude carriers (VLCC) or ultra large crude carriers (ULCC) are used for transport, with *Jahre Viking* being the largest tanker ever built, with a deadweight of 565,000 tonnes [8]. These ships are capable of transporting several hundred thousand cubic metres of oil in their storage tanks. At the destination, the crude oil is offloaded and distilled in an oil refinery into different end products, e.g., gasoline, diesel, lubricating oils, and heavy fuel oil (HFO) (see Box 2.4) [8].

Crude oil and its different distillates commonly end up in the natural environment. On average, 1,250,000 tonnes of oil are released annually (calculated from 1988 to 1997) into the marine environment from sea-based sources [9]. The sources of these discharges can include natural seepage, industrial and urban run-off from land, offshore production and shipping (Fig. 4.1). Natural seepage occurs constantly and simultaneously in thousands of places across the sea floor, rendering it difficult to assess the amount of leakage. With new remote sensing technologies, the seepage has been measured in the American portion of the Gulf of Mexico (i.e.,

70,000 tonnes per year). These measurements have been used to extrapolate annual global seepage, resulting in a value of approximately 600,000 tonnes, which is nearly half the total discharge [10]. Petroleum discharge from industrial run-off represents 11 % of the total addition to the oceans [11]. Storm water drainage and industrial outlets are also sources of oil pollutants that eventually end up in the sea [12]. The offshore production industry constitutes a small portion of the total input (2.9 %) of oil into the oceans. Nonetheless, when these rare discharge events occur, the consequences can be widespread, and the amounts of oil released are often enormous. This fact is emphasised by the blowouts of the *Ixtoc I* oil platform in 1979 near the coast of Mexico, where 475,000 tonnes of oil leaked into the ocean, and the discharge from the *Deepwater Horizon* oil platform in 2010 off the coast of the USA, where 750,000 tonnes of oil spilled [13].

4.1.1 Discharges of Oil from Shipping

Accidents, groundings, operational discharges and shipwrecks are all sources of petroleum in the oceans that originates from shipping; these sources are responsible for approximately 34 % of the total input.

4.1.1.1 Accidents/Groundings

The shipping accidents that have occurred over the years that were caused by explosions/fires on-board, collisions between vessels or the grounding of vessels are the most spectacular [14]. These accidents started occurring in the 1960s with the accidents of the *Sinclair Petrolare* near Brazil, the *Heimvard* near Japan and the *Torrey Canyon* near Cornwall, where hundreds of thousands of tonnes of crude oil leaked into the surrounding waters. The *Exxon Valdez* accident in Prince William Sound, Alaska, in 1989 received the most media attention; 35,500 tonnes of crude oil were released during this accident [15]. Remarkably, the amount of oil released at that time was fairly low compared with that of other disasters. In addition to these shipping accidents, several others have occurred, resulting in the release of enormous amounts of oil into the ocean, i.e., more than 5,500,000 tonnes, often with devastating consequences (Table 4.2) [16]. These accidents represent a large portion of the amount of oil spilled by shipping, although few accidents have occurred. Approximately 410 oil spills greater than 700 tonnes have occurred since 1960, whereas the annual number of accidents has decreased since the 1970s [8].

4.1.1.2 Marine Governance and Regulations

Marine governance to reduce the number of accidental spills has been conducted on several levels: global (e.g., IMO), regional (e.g., HELCOM, EU), and regional regulations from, e.g., port authorities and municipalities. These actions can be grouped into four different oil spill governance categories. The first category includes *precautions and new vessel designs*, such as a more modern tanker fleet with sectioned storage tanks, regulations introducing tankers with dual skin hulls

Table 4.2 Major shipping disasters during the period 1960–2014

Year	Name of vessel	Location	Oil lost (t)
1960	Sinclair Petrolare	Brazil	60,000
1964	Heimvard	Japan	50,000
1967	Torrey Canyon	United Kingdom	130,000
1969	Julius Schindler	Portugal	96,000
1971	ABT Summer	Angola	51,000
1971	Texaco Denmark	Belgium	107,000
1971	Wafra	South Africa	69,000
1972	Sea Star	Oman	129,000
1974	Yuyo Maru No. 10	Japan	54,000
1975	Jakob Maersk	Portugal	83,000
1976	Urquiola	Spain	96,000
1978	Amoco Cadiz	France	234,000
1979	Atlantic Express 2	Trinidad and Tobago	145,000
1979	Atlantic Express 2	Barbados	141,000
1979	Independentza	Turkey	98,000
1979	Exxon Valdez	USA	36,000
1980	Irenes Serenade	Greece	125,000
1983	Castillo de Bellver	South Africa	267,000
1983	Assimi	Oman	54,000
1985	Nova	Iran	73,000
1988	Oddessey	North Atlantic	147,000
1989	Khark 5	Marocco	70,000
1991	Haven	Italy	144,000
1992	Katrina P.	South Africa	51,000
1992	Aegean Sea	Spain	74,000
1993	Braer	United Kingdom	85,000
1996	Sea Empress	United Kingdom	72,000
2002	Prestige	Spain	63,000
2003	Tasman Spirit	Pakistan	28,000
2007	Hebei Spirit	Republic of Korea	11,000

Data from GESAMP, 2007 and ITOPF, 2015

(OPA90), the establishment of sea lanes with traffic separation, the institution of protected areas without vessel traffic and the introduction of digital positioning systems, such as global positioning systems (GPS) and automatic identification systems (AIS). The second is *monitoring*, including air surveillance of certain coastal areas (for example, in the Baltic Sea) and inspections of ships by port state control officers. The third category includes the *enforcement* of existing regulations, performed by, for example, flag state control offices and coast guard patrols. The fourth is *remedial action plans* that exist between countries, which can include the pooling of oil remediation equipment or joint exercises of oil combat operations [17].

Some international marine governance exists, i.e., laws and conventions regarding accidental pollution control; these regulations have had a profound impact, decreasing the number of large oil spills. This governance began with the MARPOL 73/78 convention in 1978, which defined measures to eliminate oil pollution from ships and eliminated problems in the system by providing compensation for accidents at sea [18]. This was followed by the Oil Pollution Act of 1990 (OPA 90), which was established in the USA in response to the *Exxon Valdez* spill in Alaska. This act placed liabilities on the responsible parties for future accidental oil spills and enforced new regulations that initiated the phase out of single hull vessels, new technology, and safety procedures [19]. In 1998, the International Management Code for the Safety of Ships and for Pollution Prevention (ISM code) was implemented. This code includes a set of guidelines that describes actions ship owners must undertake to implement safety management systems on vessels and in their on-shore organisations [20]. Finally, a revised chapter of the SOLAS (The International Convention for the Safety of Life at Sea), i.e., Chap. 5 (Safety of Navigation), began to be enforced in July 2002. This chapter not only introduced mandatory equipment for accident investigations, i.e., voyage data recorders (VDRs), but also required the use of the AIS system, which automatically provides ship information to other ships and coastal authorities [14, 21].

4.1.1.3 Operational Discharges

Regardless of the often spectacular accidents that occur, these tanker accidents are not the main source of anthropogenic oil inputs into the oceans due to shipping [22, 23]. Operational discharge, originating from routine operations, bilge water, illegal cleaning of tanks, and bunkering, is the cause of 50 % of the spills with less than 700 tonnes of released oil and 91 % of the spills with less than seven tonnes of released oil [24]; these spills have been estimated to total approximately 200,000 tonnes annually (1988–1997) [25]. Similar to large shipping disasters, actions have been taken by IMO to limit these smaller spills, e.g., the implementation of separated ballast tanks, bilge water treatment equipment, and mandatory crude oil washing (COW) of tanks instead of using hot water. Another example of these actions is the load-on-top procedure, in which the oily mixture that is created when cleaning the tanks with hot water is allowed to separate in an isolated slop tank for the return voyage; then, the relatively clean water can be released into the ocean [17, 26].

Despite the enforcement of international regulations, improvements in vessel design and on-board safety installations, problems still exist that ultimately lead to petroleum spills of various sizes. Statistics demonstrate that approximately 47 % of all vessel accidents in the Baltic Sea are caused by human error, whereas only 13 % are caused by technical factors [27]. Inadequate training of lower-level crews and compliance with labour organisations continue to be problematic and contribute to this high statistics in human error. Another problematic area is the so-called flags of convenience. Flag states have the responsibility to ensure that ships flying their flags abide by international regulations regarding the handling of petroleum. These flags of convenience states in several occasions have a few or non-existent means of enforcing that the regulations is followed. In addition, indications of problems regarding port state control have been suggested. The lack of inspections (only approximately 25 % of all visiting vessels are inspected) and non-identical inspection routines are issues of concern [17].

4.1.1.4 Bilge Water

The lowest inner part of a ship hull to where liquids drain from the interior spaces and upper decks is referred to as the bilge. The main sources of the liquids draining into the bilge are the main engine room and the auxiliary machine room. Other areas that collect fluid drainage into the bilge are the shaft alley and the air conditioning, refrigeration, steering gear and pump rooms. The liquid collected in these various bilge areas is called bilge water [28]. Bilge water is a mixture of various substances used in the machine room and often consists of hydrocarbons, grease, hydraulic fluids, oil additives, cleaning and degreasing solvents, detergents and various metals [29]. For ships over 400 GT, the MARPOL 73/78 regulations state that the hydrocarbon content cannot exceed 15 ppm in the treated bilge water for overboard discharge [30]. There are a number of various bilge water treatment techniques on the market, e.g., membrane technologies (ultrafiltration), chemical oxidation, flocculation or biological treatment combined with oily water separators (see Sect. 11.7.3).

There have been known examples of illicit changes to ship bilge water treatment equipment to circumvent the equipment, so-called "magic pipes" and records "oil book", to be able to discharge bilge oil at higher concentrations than allowed. However, with more strict controls and several high profile cases in which the responsible crew and management were fined or even incarcerated, this phenomenon has likely been partially resolved [31].

A 20000 HP engine has been estimated to generate approximately 19 l of bilge oil per day. Assuming 350 days of operation per year and the number of tankers from Lloyds registry (1999), the total volume of bilge water production from tankers is 19,200 tonnes per year. Assuming bilge water oil content of 15 ppm, less than 0.2 % of bilge oil is discharged into the ocean, resulting in an annual bilge oil discharge of 38 tonnes from the tanker fleet. However, if it is assumed that not all ships are following MARPOL 73/78 regulations, i.e., 1 % of tankers >125,000 DWT, 5 % of tankers 20,000–125,000 DWT and 10 % of tankers <20,000 DWT, and if they discharge their bilge oil without treatment, the total annual discharge would be 1130 tonnes. If similar calculations are applied to the non-tanker fleet, 19 l of bilge oil produced per 20000 HP engine on the 79,547 non-tankers exceeding 100 GT in operation in 1999 results in a total discharge of 15,600 tonnes per year. Moreover, vessels between 100 and 400 GT, which constitute 54 % of the non-tanker fleet, are not obligated to have on-board bilge water treatment equipment. These vessels are not allowed to discharge the bilge water overboard, although it is assumed that 15 % of commercial vessels and 30 % of non-commercial vessels do not comply with the MARPOL regulations. The annual total bilge oil discharge to the ocean from the shipping transport sector thus comes to 16,730 tonnes. However, the input from vessels less than 100 GT cannot be included in this estimate due to the lack of relevant information [19].

4.1.1.5 Shipwrecks

Oil pollution from shipwrecks has recently gained more attention, where reports from different sources indicate concern regarding leaking oil from wrecks. The sea floor is littered with wrecks of different sizes and ages; the primary source of shipwrecks is World War II. The total number of wrecks greater than 400 GT has been estimated to be nearly 8600, and approximately 1600 of these wrecks are tankers [32]. The Swedish maritime administration reported 31 potentially polluting shipwrecks along the Swedish coast in 2011 [33]. Most of the ships contained petroleum products when they sank to the bottom of the sea floor, often close to shore and in shallow waters. Eventually, these wrecks will begin to leak their contents due to corrosion and affect the environment. Shipwrecks have been estimated to contain between 2.5 and 20.4 million tonnes of different oil products [32].

Remediation of shipwrecks is often a time-consuming and costly process and depends on the type of oil contained inside the wreck, water depth, temperature, and weather conditions. A complex remediation operation with water depths >50 m, highly viscous oil, open waters and a wreck in poor condition can involve costs of $5 to > $100 million. However, operational clean-up and the ecological and socioeconomic costs of large- or small-scale continuous discharges are expected to be several times more costly than remediation operations [34]. Therefore, effective risk assessment methods, the determination of which wrecks pose the largest risk to the environment, and appropriate remediation actions are all essential [35].

4.1.1.6 Leisure Boating

Another source of operational spills that is often forgotten is leisure boating. This is not the source of most small-scale spills; however, these spills occur within a limited geographic area, primarily in shallow coastal waters. Most leisure boating vessels use two-stroke engines [36]. For example, in Sweden, the fleet comprises 570,000 leisure boats propelled by engines, of which approximately 30 % have traditional two-stroke engines [37]. Because approximately 20–30 % of the fuel is not combusted and is washed out in the exhaust of these engines, 6500–10,000 m^3 of oil is released annually in Swedish waters based on previously published gasoline consumption values [38, 39]. Therefore, in some areas, these operational spills can be substantial.

4.1.1.7 Propeller shaft bearings

Propeller shaft bearings can be a substantial contributor to the operational oil spills from shipping, as they can leak between two and six litres of oil daily. When a ship is propelled with a propeller there need for an axis from the engine to the propeller. This involves a passage through the ship's hull below the waterline. To keep water from entering the ship, to keep the axis straight and to prevent the passagng material from not wearing down the axis, there is a need for a bearing with a sealing. This part of the axis is called a propeller shaft bearing, and it contains bearings and seals.

These are regularly lubricated with a special oil for that purpose, but common engine- or transmission oils are occasionally used. The oil is pumped into the bearing and passes the seals through small canals. It is then returned to a cooler and pump. This is made through a small increase in pressure (0.1- 0.3 bars) compared to the outside of the ship, to keep water from entering the bearing. This is performed through gravity; the oil tank to the system is located higher in the ship compared with the bearing. Under optimal conditions the seals in the bearing work perfectly, no water enter the ship and no oil is lost to the marine environment. However, because seals harden over time, the bearings wear down, the propeller axel becomes skewed or lines from fishing gear become tangled and damages the bearing; consequently, the propeller shaft bearing begins to leak oil. When the oil begins to run out in the bearing oil is refilled in the tank. It is no simple operation to repair a seal or replace a propeller shaft bearing; this most often has to be performed when the ship is dry docked. Then it may take up to years before the leakage can be stopped [23] Some temporary minor adjustments to reduce the leakage can be made while at sea. A lubricant oil with higher viscosity can be used, but because the lubricating quality is lowered and the friction is increased with a subsequent lower speed of the ship, this is not a long-term solution. Another solution is to lower the oil pressure in the system, thereby reducing the amount of oil escaping from the system. However, this increases the risk of water entering the system, which could induce corrosive damage on the bearings and dramatically increase the wear. The annual global additions of oil from propeller shaft bearings are estimated to range between 30,000 – 100,000 m^3. The estimations were based on the assumption that 80–90 % of all ships emits oil from their propeller shaft bearing, that a ship sails 300-330 days per year, that a ship emits between two and six litres of oil per day and that the world fleet consists of 45,000 – 70,000 ships [23] There are environmental friendly alternatives to standard propeller shaft bearings. Non-mineral oils, which are non-toxic and easily degradable in the marine environment, can be used to lubricate the shaft. However, recent data has shown that these oils are not as readily biodegradable and non-toxic as claimed. Another alternative is to use a special propeller shaft using water lubrication, from the company Thordon bearings. At the end of 2013, the US EPA enforced a regulation that bans the use of non-environmentally friendly oils for propeller shaft bearings within US waters. Because regulations from the USA have previously been adapted in other parts of the world (OPA 90), it can be assumed that this commendable enforcement will be adapted by other countries in the future [23].

4.1.1.8 Illegal discharges

The most common sources of illegal discharges of oil into the marine environment include the washing of tanks and water hosing of engine equipment and rooms, with the subsequent release of the contaminated water into the ocean [22, 23]. Another addition, which is currently uncommon, is the combined use of non-segregated ballast water tanks as fuel storage. Because these actions are connected to a cost, if the oily water is to be delivered ashore or purified in a bilge water separator there is an incentive to attempt to avoid these costs i.e. by illegally discharging the water into the ocean [17]. As described in Sect. 4.1.1.3, new regulations were introduced in the 1960s and 1970s that lessened those incentives. Demands on Load on Top and Crude oil washing procedures have resulted in a dramatic decrease of these types of discharges, both because money could be saved (crude oil washing) and because there was a risk of being fined if the regulations were not met. In addition to these rules, tankers must have filters installed in the engine rooms, to prevent oily water from being washed out if hydrocarbon levels become too high [17]. Calculations on the global annual addition of these type of discharges to the marine environment have been estimated at 36 000 tons. This is a large decrease from earlier estimations in 1975, 1985 and 1990 in which discharge sizes were estimated at 1 080 000, 710 000 and 158 600 tons annually, respectively [19]. Within the participants of The Baltic Marine Environment Protection Commission (HELCOM), which focuses on minimizing the anthropogenic stress on the Baltic Sea, a "no special fee" system has been introduced. This requires large ports within the area to provide facilities that can handle, without charging a fee, oily water and other residual oil products [17]. In addition, the countries are cooperating with satellite and flight surveillance to deter ships from illegally discharging oily water and to connect ships to oil spills. Sweden, Poland and Germany are responsible for the majority of these flight hours. In response to this initiative the detected illegal oil spills have decreased since 1988, from approximately 300 to approximately 100 annually. In 2013, 130 spills were detected, and the estimated volume of these was 11 m^3. Using the flight surveillance, AIS-data and an oil spill drift forecast tool called Sea Track Web the registered oil spill can be deducted to originate from the ship responsible. However, even with all those techniques for surveillance the ship responsible for the discharge can be identified in only in 4.5 % of cases [17]. Furthermore, it is legally complicated in Sweden to sentence a person who was on-board the ship as being responsible for the illegal discharge, even if the ship is within Swedish territorial waters. To complicate things further, ship operators planning to perform this illegal action knows that at night, in bad weather and in particular geographical areas, the surveillance is less comprehensive; hence they choose those opportunities to release the oil. Therefore, the hidden statistics of this illegal operational discharge are likely substantial [17, 23].

4.1.2 Behaviour of Oil Spills

When oil enters the marine environment there are several processes that govern its fate (Fig. 4.2). Immediately after entering the water, the *spreading* of the oil leads to a thin film layer on the water surface that is <10 μm to several centimetres in thickness [40]. The rate at which this occurs depends on the volume and the viscosity of the oil, the water temperature, the wind speed and the currents. Furthermore, *evaporation* of the most volatile products in the oil, such as light hydrocarbons, begins immediately. Approximately 40 % of the oil released by the

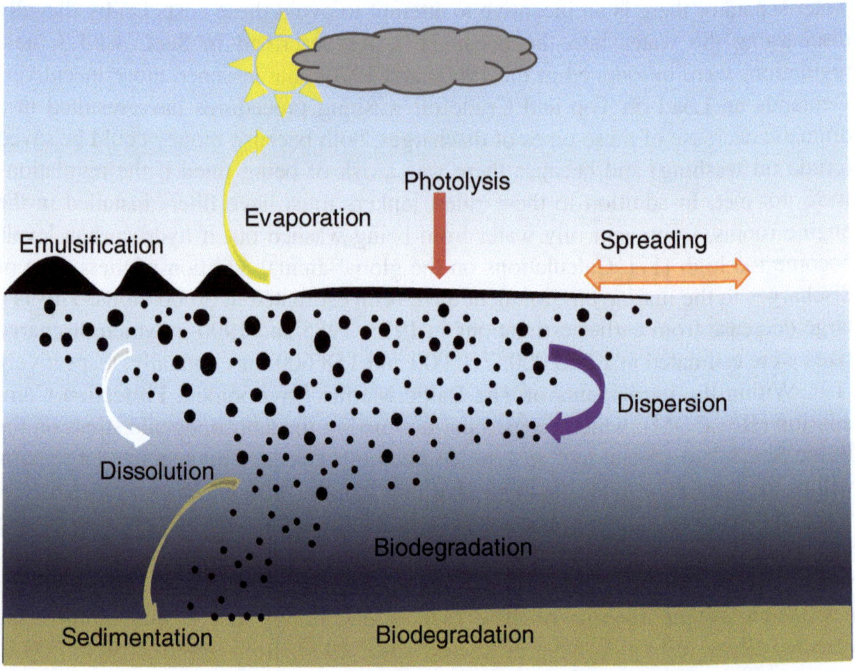

Fig. 4.2 Pathways of how oil enters the marine ecosystem (modified from ITOPF 2002)

Amoco Cadiz spill and 30 % of that released during the *Exxon Valdez* spill are estimated to have evaporated in this manner, whereas heavy fuel oil (HFO), as in the case of the *Prestige*, decreased by only approximately 5–10 % of its initial volume. The harmful products that evaporate are translocated by winds to other locations [8]. Some of the oil will be degraded by UV light through a process called *photolysis*. This process creates photo-oxidised products of oil components such as anthracene, benzo(a)pyrene, fluoranthene, phenanthrene, and pyrene that may produce products that are more toxic than the original hydrocarbons.

Emulsification is a process where oil droplets are mixed into the surrounding water by means of wind and wave power, which expands the volume two to five times. The density and viscosity also increase, rendering it more difficult to mechanically remove the oil from the water via standard oil treatment procedures [41, 42]. A small fraction becomes dissolved in seawater, i.e., less than 1 % depending on the composition, salinity, water turbulence and water temperature of the oil. This fraction quickly becomes diluted and degraded in the water column. A larger portion is *dispersed* into the water column by wave forces. The oil breaks into small droplets (0.01–1 mm in diameter) and is transported and retained until it is *degraded* [4]. This process increases the bioavailability and toxicity of the oil to fish. Finally, the oil may sink to the sea floor. This can occur as either oil or specific compounds in the oil adsorb to organic particles in the water, increasing the weight of the complexes and thereby inducing *sedimentation* or because some oil fractions are more dense than water [43].

4.1.3 Impacts of Oil

The pathways in which oil enters the marine ecosystem can lead to different types of impacts. Large discharges, i.e., so-called black tides from shipping catastrophes and marine oil well blowouts, affect biota in a different way than small but constant discharges from urban/industrial run-off and operational discharges from shipping. The effects can range from *acute toxic* to *sub-lethal responses*, and the magnitude of the effect depends on several factors, such as the oil composition, released volume, geographic location, seasonality, affected habitats and the sensitivity and possibilities for the recolonisation of habitats by the affected species. The exposure of biota can occur via three different pathways: direct contact or ingestion, uptake of bioavailable components through the water or by ingestion of petroleum-contaminated pray [44].

Effects from single large discharges often have fatal consequences for biota and have been examined in numerous studies [4]. These large discharges cause narcotic effects in animals because the chemicals dissolve in the lipids of the cell membranes and cause the cells to leak, whereby normal homeostasis cannot be upheld, resulting in mortal effects [45]. Furthermore, oxygen deficiency from the organic enrichment and subsequent degradation of the oil is commonly triggered. Seabirds are especially vulnerable because only small amounts of oil can result in the loss of insulation capabilities, leading to hypothermia. Larger amounts cause them to lose their buoyancy and flight abilities [8]. For example, during the *Exxon Valdez* spill, approximately 250,000 seabirds succumbed to the harsh conditions in Alaska because their insulation capabilities were lost [46]. This phenomenon can also affect other animals that use fur or blubber for thermoregulation, e.g., seals, sea otters, sea lions, whales and dolphins [28]. Other effects from large spills include the physical hindering of oxygen and the transfer of sunlight into the water caused by the oil slick. This effect leads to anoxia in shallow waters and lowered capabilities for photosynthesis [47], which can reduce the densities of marine plants and benthic fauna [48].

Small, continuous, non-lethal releases of petroleum products are also a great cause for concern in the marine environment; such effects are somewhat less examined. These releases can originate not only from shipwrecks and operational discharges but also from the remnants of large oil spills trapped in anoxic sediment layers, i.e., the *Exxon Valdez* [49]. The effects of these discharges are commonly caused by the most toxic oil components, the *polycyclic aromatic hydrocarbons* (PAHs) [8], which exist in petroleum products in variable concentrations. PAHs represent a group of hydrophobic substances containing two or more fused benzene rings in linear, angular or clustered forms [50, 51] (Fig. 4.3). The consequences of exposure to PAHs can be cancer, decreased species diversity in communities, reduced fecundity and

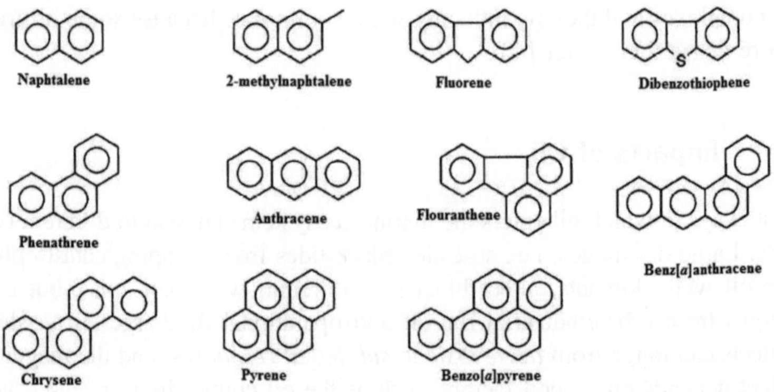

Fig. 4.3 Commonly measured polycyclic aromatic hydrocarbons found in oil (modified from Harayama 1997)

growth and lowered tolerance to other stresses [52, 53]. It is also noticeable that these effects might not be evident until several generations later [54]. Several factors, mainly divided into biotic and bioavailability factors, influence the effects of PAHs. Biotic processes, such as organism size, ingestion rate, growth rate, membrane permeability, ventilation rate, gut residence time and osmotic regulation, influence the uptake of PAHs. Bioavailability, which is expressed as a fraction of the total amount of pollutants that the biota is exposed to, is affected by temperature, oxygen levels, sediment characteristics, pH and salinity [55, 56].

The recovery time of an ecosystem, which is also called *resilience*, varies substantially as a function of several factors. The type of habitat affected has a large effect on the recovery time. For example, exposed shores recover more quickly than salt marshes, estuaries and subtidal ecosystems. The latter may take up to fifteen years to recover [57]; on rare occasions, more than forty years may be required for full recovery [58, 59]. Other factors that influence the recovery time include the climate at the affected site, the time of year, the availability of recolonising species, biological interactions, and the availability of oil-degrading organisms [4].

Species of bacteria, fungi and algae have all been found to degrade oil and PAHs [55], and microorganisms that readily degrade hydrocarbons are widespread in nature [60, 61]. However, the degradation of hydrophobic PAHs is slow; the speed of this process depends on the bioavailability of PAHs, oxygen availability, temperature and whether PAHs exist in concentrations that are toxic to the degraders. In some cases, especially for PAHs with high molecular weights, e.g., benzo(a)pyrene, degradation is very slow or non-existent [55].

In many instances, anthropogenic stressors, such as oil, heavy metals and organochlorides, are eventually deposited on the sea floor and affect benthic systems instead of being transported to shores via processes that include sedimentation and the deposition of faecal pellets. Hence, many of the effects arising from oil pollution can be found in these systems.

4.1.4 Costs Related to Petroleum Contamination

Several factors affect the economic costs of an oil spill. The type and volume of the spilled oil, the geographical location of a spill, the weather and water currents in the area, the time of year, and the efficiency and costs of clean up are several of the factors that affect the final cost [62]. A light refined petroleum product or light crude oil will not remain on the sea surface for a long time but will instead quickly evaporate, minimising the amount of oil that must be cleaned up. An example of this process is the *Braer* incident, which occurred near the Shetland Isles, UK, in 1993. The ship spilled its entire cargo of 85,000 tonnes of light Gullfaks crude oil. Due to the severe weather conditions and the evaporation and dispersion of the light crude oil, minimal shoreline contamination occurred, and the total clean-up cost was only $0.5 million. A highly viscous crude oil or heavy fuel oil can remain at the surface for a long time, with little evaporation, and can enter near-shore areas, coating beaches, anthropogenic structures, fishing gear, aquaculture equipment and other structures. Oil that sinks to the sea floor can become remobilised following storms, washing up on shorelines and causing problems for a long period of time. Moreover, this type of oil is more difficult to remove using common clean-up techniques. These factors lead to a higher total clean-up cost, i.e., usually more than 10 times that of spills with lighter oils [63, 64]. An example of this type of spill is the *Erika* spill, which occurred near the coast of France; 20,000 tonnes of oil were spilled. Weather conditions spread the oil over a large area, leading to extensive coastal contamination, and resulting in one of the most costly clean-up operations in history.

The amount of oil spilled indubitably affects the clean-up costs, although the effect is not linear [62]. The size of the clean-up operation increases with the volume of the oil spill, resulting in naturally higher costs. However, smaller spills are always more costly per tonne because substantial costs are associated with the clean-up response, mobilisation of equipment and personal, and evaluation expertise [63, 64].

The geographical location of a spill, i.e., near-shore or far from the coast, also affects the costs because response actions are usually only undertaken when spills approach land. Near-shore or in-port spills are four to five times more costly to clean up than offshore spills [64]. An example of this type of spill is the *Atlantic express* accident, which occurred in Tobago, West Indies, in 1979; 287,000 tonnes of oil were spilled. This incident occurred far from land and resulted in very low clean-up costs because very little oil reached the nearest coastlines [62].

The weather and water currents in the area of a spill can have large effects on the amount of oil that may reach nearby shorelines. The type of area affected by a spill and the time of year can also affect the clean-up costs. For example, spills occurring near popular recreation beaches in the middle of tourist season or in an estuary in spring during bird migrations are of high priority for society, demanding a rapid, thorough and thereby costly clean-up process [63].

The type of costs associated with a spill can be divided into three main categories: direct clean-up costs, socioeconomic costs and environmental costs.

4.1.4.1 Clean-up Costs

The direct clean-up costs are mainly related to the cleaning, recovery, and storage of spilled oil and the disposal of collected oil waste, although the collection of oil remaining in the vessel can also be considered a direct clean-up cost [65, 66]. Expensive oil removal equipment, vessels, aircraft, trained operators and personnel are examples of the costs included in this category [62]. Personnel costs and technical equipment represent a large portion of this cost, although labour costs vary depending on the amount of voluntary personnel assisting in the clean-up. In the case of the *Prestige* oil spill in 2002, when approximately 60,000 tonnes of oil were spilled, the total clean-up costs came to €509 million [66].

4.1.4.2 Socioeconomic Costs

The socioeconomic costs are primarily related to the fisheries, aquaculture industries and the tourism sector that suffer income losses and property destruction due to an oil spill [65]. Fish stocks can decline due to the lethal effects of the petroleum; the affected fish may also contain higher levels of contaminants than permitted by regulatory authorities, rendering them inedible. Tourists may reject contaminated areas, such as damaged beaches or un-swimmable waters, due to oil contamination; this is a common economic losses for stakeholders associated with the tourism industry in affected areas. Losses in the Spanish fishing sector after the *Prestige* spill in Galicia, Spain (2001–2004), resulted in socioeconomic costs of approximately €113 million. The losses in the mussel cultivation sector were €13 million, whereas the losses in the tourism sector were €110 million [66].

4.1.4.3 Environmental Costs

An oil spill could lead to environmental costs in the form of temporary degradation of natural resources and services [65]. Environmental costs are the most difficult to assess among the three cost factors related to oil spills because most of the goods and services produced by the environment are non-market items that do not have an established market value. For example, it is very challenging to assess the total economic value of a coral reef because it produces many different ecosystem services, which have values that are also difficult to assess. Coral reefs produce value through tourism, as a habitat for various fish species that have a value both as food and in ecosystem functions and as part of the biogeochemical cycling in the sea. Although environmental costs are difficult to assess, several models exists to evaluate these goods and services [67]. The realism of these models can be debated because only three types of damaged resources are generally considered when calculating environmental damages: beaches, birds, and seals. No attention is given to the effects on other habitats, such as coastal shelf areas, deep sea waters or marine benthic systems, or to other effects on animals, for example, decreases in reproduction, offspring survival, productivity, and organic matter degradation, as well as food web disturbances [65]. In the case of the *Prestige* spill, the number of killed birds was 115,000–230,000, which had an estimated economic value of €6.4 billion [66].

4.2 Wastewater

On land, sewage that originates from households, shops and less polluting industrial activities is usually handled by a municipal wastewater system. Here, water from toilets, showers, kitchens and other sources is combined. Occasionally outdoor sources are also added in the form of storm water, i.e., rainwater collected from streets and roofs, which increases the capacity requirements of sewage treatment facilities. On ships, the availability of fresh water and the possibility of storing water that is not permitted to be disposed into the sea are limited. Therefore, the sewage water from different sources is not often mixed. Wastewater streams on ships include *sewage* (black water) and *grey water* (non-sewage wastewater). These wastewaters have different sources and characteristics and are also treated differently by the discharge regulations.

4.2.1 Origin and Characteristics of the Wastewater Streams

Sewage generally includes wastewater from toilets, medical facilities and premises for live animals, according to MARPOL 73/78. The amount of sewage produced on a ship is related to the number of people on board and to the technical parameters of the sewage system. The generation rate of sewage is estimated to be approximately 70 l per person per day for a conventional flushing system and 25 l per person per day with a vacuum system installed on-board [68]. Sewage is more contaminated than grey water and may only be discharged into the sea under specific conditions. It can contain several components of concern, e.g., pathogens (bacteria, viruses and eggs of intestinal parasites), organic matter and nutrients [69, 70]. In addition, sewage may contain heavy metals and residues from pharmaceuticals [71]. Compared with sewage produced on land, black water is more concentrated because less water is used for flushing [69].

Grey water is non-sewage water that originates from dishwashers, showers, laundry machines, baths and washbasins [72]. A single person on a cruise ship is estimated to generate between 120 and 300 l of grey water per day, which constitutes largest type of wastewater generated on cruise ships [73]. This waste stream is not restricted by international discharge regulations if it does not contain components that are regulated in Annex V of MARPOL 73/78 protocol. However, the use of port reception facilities or a sewage treatment plant is also recommended for grey water. Grey water often contains a wide range of pollutants, e.g., bacteria, suspended solids, metals, detergents, oil, grease, and food particles [74]. Grey water may also contain microplastics that originate from the usage of certain types of cosmetics, e.g., facial scrubs and exfoliating hand cleansers, or from washing polyester or acrylic textiles [75–77].

4.2.1.1 Passenger Ships and Pleasure Crafts

Because the generation of both sewage and grey water is naturally associated with persons travelling on ships, passenger ships (especially cruise ships) are regarded as significant sources of these pollutants. This category of ships produces considerable amounts of sewage that may pose a threat to the marine environment if discharged to the sea. In this regard, the regulations are stricter to protect areas visited by passenger ships. In accordance with Annex IV of MARPOL 73/78, the Baltic Sea has been classified as a special area. The stricter regulations will possibly reduce the nutrient load of this inland sea, which is particularly vulnerable and suffers from anthropogenic eutrophication. The amended Annex IV regulation for the discharge of sewage from passenger ships operating in the Baltic Sea set new criteria for on-board sewage treatment plants by monitoring the nutrient concentrations in the discharged effluent. In addition to the Baltic Sea, the state of Alaska has also introduced stricter rules regarding the discharge of black and grey water from passenger ships.

Large cruise ships are not the only vessels that may pose a threat to the marine environment. Sewage discharge from leisure boats, especially in coastal areas, can also have a significant impact on the marine environment. In those areas where heavy leisure traffic exists, local areas with high nutrient levels and low dissolved oxygen levels may form. Restrictions on sewage discharge from leisure boats have already been introduced or will be introduced soon in countries around the Baltic Sea. However, the restrictions may differ slightly between countries [78]. In Sweden, the regulations that came in force in April 2015 prohibit the discharge of sewage from pleasure crafts. The regulation applies to all pleasure crafts, regardless of flag, throughout Swedish territorial waters and to Swedish pleasure crafts in other territorial waters [79].

4.2.2 Environmental Effects

A discharge of untreated or insufficiently treated wastewater from ships to the sea can be aesthetically unpleasing, especially in attractive tourist destinations, i.e., coastal areas. The release of pathogens to water increases the risk of (human) diseases, which can be contracted by swimming in contaminated water or eating contaminated shellfish. The discharge of nutrients and organic matter leads to marine eutrophication and can increase the risk of blooms of algae (and of cyanobacteria that can produce toxins), followed by the decomposition of organic matter. The amount of nutrients released with sewage has been estimated at 12–15 g of nitrogen per person per day and 3–5 g phosphorous per person per day [80]. The decomposition of organic matter can lead to hypoxia (<2 mg O_2/litre) or anoxia (absence of dissolved oxygen) at the sea floor due to the oxygen depletion, creating so-called "dead zones" [81]. This process can be very important in sensitive areas, e.g., the Wadden Sea in the south-eastern North Sea and the Baltic Sea. Under anoxic conditions, phosphorous accumulated in the sediment can be released back into the water column, further increasing the eutrophication effect [81]. Furthermore, under anoxic conditions, other types of bacteria begin to use sulphate as an energy source rather than oxygen when decomposing available organic matter. This results in the formation of hydrogen sulphide (H_2S), which is a highly toxic compound [82]. This process poses a threat to marine life, resulting in decreased biodiversity because only the most resistant species can survive. Moreover, eutrophication reduces the water quality for bathing; consequentially, the recreational value of the coastal environment may decrease [82].

4.2.3 Regulations

Annex IV of MARPOL 73/78 states the rules regarding the discharge of ship-generated sewage into the sea. These rules apply to vessels exceeding 400 GT or carrying more than 15 passengers and also contain principles regarding ship equipment and systems for controlling sewage discharge and the requirements for surveying and certifying the equipment. Requirements regarding the facilities at ports and terminals for the reception of sewage are also included. In 2011 the amendments were adopted by resolution MEPC. 200(62) to Annex IV. The Baltic Sea was introduced as a special area under Annex IV, and new discharge requirements for passenger ships operating in special areas were added.

Table 4.3 Regulations in Annex IV of MARPOL 73/78 on the discharge of untreated sewage from ships

		Discharge of untreated sewage	
		Special areas	Outside special areas
Passenger ships	New ships[a]	Prohibited after 1 January 2016[b]	Allowed at a minimum distance of 12 NM from the nearest land
			En route
	Existing ships	Prohibited after 1 January 2018[b]	Allowed at a minimum distance of 12 NM from the nearest land
			En route
Other ships	New and existing	Allowed at a minimum distance of 12 NM from the nearest land	Allowed at a minimum distance of 12 NM from the nearest land
		En route	En route

[a]Building contract in place on 1 January 2016 or other criteria stated by Annex IV of MARPOL 73/78

[b]The dates that have been postponed lately

Table 4.4 Regulations in Annex IV of MARPOL 73/78 on the discharge of treated sewage from ships

<table>
<tr><th rowspan="2"></th><th rowspan="2"></th><th colspan="2">Discharge of treated sewage</th></tr>
<tr><th>Special areas</th><th>Outside special areas</th></tr>
<tr><td rowspan="3">Passenger ships</td><td rowspan="2">New ships[a]</td><td rowspan="2">After 1 January 2016, ships must use an approved sewage treatment plant that meets the operational requirements in regulation 9.2.1 of Annex IV; the effluent can neither produce visible floating solids nor cause discoloration of the surrounding water</td><td>Comminuted and disinfected sewage can be discharged at a minimum distance of 3 NM from the nearest land using a system approved by the governing administration in accordance with regulation 9.1.2 of Annex IV. En route</td></tr>
<tr><td>No minimum distance is required if a ship uses an approved sewage treatment plant that meets the operational requirements in regulation 9.1.1 of Annex IV; the effluent can neither produce visible floating solids nor cause discoloration of the surrounding water</td></tr>
<tr><td>Existing ships</td><td>After 1 January 2018, ships must use an approved sewage treatment plant that meets the operational requirements in regulation 9.2.1 of Annex IV; the effluent can neither produce visible floating solids nor cause discoloration of the surrounding water</td><td></td></tr>
<tr><td rowspan="2">Other ships</td><td rowspan="2">New and existing</td><td colspan="2">Comminuted and disinfected sewage can be discharged at a minimum distance of 3 NM from the nearest land using a system approved by the governing administration in accordance with regulation 9.1.2 of Annex IV. En route</td></tr>
<tr><td colspan="2">No minimum distance is required if a ship uses an approved sewage treatment plant that meets the operational requirements in regulation 9.1.1 of Annex IV; the effluent can neither produce visible floating solids nor cause discoloration of the surrounding water</td></tr>
</table>

[a]Building contract in place in 2016 or other criteria stated by Annex IV of MARPOL 73/78

Farther than 12 nautical miles from the nearest land, sewage may be discharged at a low rate in most regions (Table 4.3). Stricter discharge requirements for passenger ships operating in special areas were to enter into force in 2016; however, the implementation has been postponed until port reception facilities are prepared to handle the discharge. In general, the discharge of sewage from passenger ships will be prohibited except when the ship has an approved sewage treatment plant, and the effluent cannot produce visible floating solids nor cause discoloration of the surrounding water. Ships may discharge treated sewage closer than 12 nautical miles from the nearest land if certain conditions are met (Table 4.4) [83].

Information on the effluent standards for the approved sewage treatment plants can be found in Chap. 11.

Annex IV of the MARPOL 73/78 regulations regulates only sewage, whereas grey water is sometimes regulated at the national and regional levels. For example, Alaska introduced regulations for passenger ships regarding the monitoring of discharges of both grey water and black water in 2000 [84]. Moreover, grey water from ships is regulated to different extents in the Great Lakes (USA) and European inland waterways [85, 86].

4.3 Fouling, Ship Hull Penalties and Antifouling Paint

The problem of fouling on ship hulls and marine installations has been known for a long time, and the use of antifouling systems is not a modern phenomenon. Throughout history, seafarers have attempted to protect their ship hulls from organisms that increase drag. Indications exist that people used antifouling systems to protect ships even before 1000 B.C. (The Bible, Genesis 6:14). Reports as early as 700 B.C. indicate that the Phoenicians used pitch and possibly copper sheathings to protect their ships. By approximately 300 B.C., the Greeks used tar, wax and even lead sheathings for the same purpose. The Vikings occasionally used seal tar to prevent fouling; from the 13th century to the 15th century, pitch was extensively used. In 1625, William Beale patented an antifouling paint that included a mixture of copper, cement, powdered iron and arsenic [87]. Copper was used until the introduction of iron ships during the late 18th century, when the antifouling effect could not be assured because copper initiates corrosive effects on iron. In the mid-1800s, many different paints were developed. Copper, arsenic, and mercury oxide were some of the additives used in these paints, which were applied over a layer of anticorrosive shellac. Unfortunately, these paints were ineffective, relatively expensive and had a short lifespan. After World War II, changes took place in the antifouling paint industry. Synthetic petroleum-based paints entered the market and improved the mechanical characteristics of paints, and new health regulations were also implemented. In the early 1960s, the excellent antifouling properties of organotin were discovered [88].

Fouling can be defined as "*the undesirable accumulation of microorganisms, algae and animals on artificial surfaces immersed in seawater*" [88]. When a surface is first immersed in the sea, fouling processes immediately commence. Within minutes, organic molecules, such as proteins and polysaccharides, become attached [89]. This process is governed by relatively weak chemical forces, such as van der Waals forces and electrostatic interactions. The fouling process proceeds in an ecological succession, meaning that different types of organisms attach in different phases. The first organisms to attach to the surface are the primary colonisers, such as bacteria and diatoms, typically within 24 h. When they adhere to the surface, they

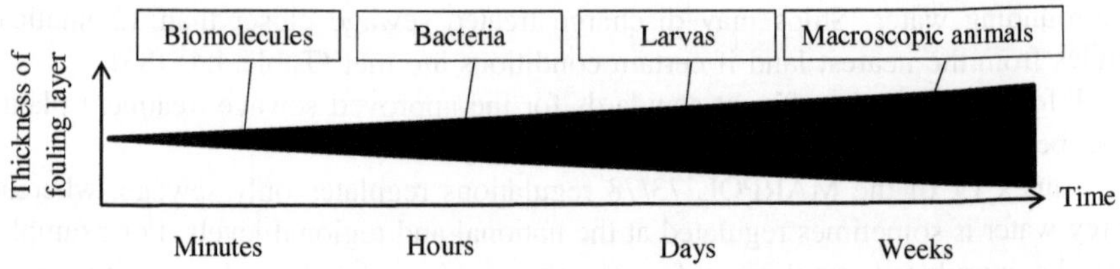

Fig. 4.4 Schematic illustration of the fouling process on a surface immersed into sea-water

form a microbial biofilm. The organisms in the biofilm begin to exude extracellular polymeric substances (EPSs), such as lipids, polysaccharides and proteins. Within approximately a week, the EPSs attract additional colonising organisms, such as spores of macroalgae and protozoans. In the final stage of the fouling process, i.e., after two to three weeks, larva of macrofoulers, such as barnacles, bryozoans, molluscs, and polychaetes, become adhered (settle) and begin to grow on the surface (Fig. 4.4). The amount of biofouling on the immersed surface depends on several factors, e.g., water temperature, salinity, solar radiation, water depth and interactions within the fouling community [88].

The colonisation of surfaces is a universal and fundamental characteristic of the biota in marine environments. No such thing as an (non-living) free surface exists in an aquatic environment. A fierce competition exists for the space between sessile organisms in rocky shore communities [90], and man-made structures immersed into the sea will therefore be targets for fouling. Structures such as ship hulls, offshore oil and gas structures, piers and heat exchange tubes are examples of artificial structures that are exposed to the sea and will get attention by fouling organisms [91]. Approximately 4000 species have been identified on fouled structures worldwide [92]. Numerous marine invertebrates have freely swimming larvae that, to various extents, explore potential surfaces before settling and metamorphosing into their adult life stage [87]. Fouling organisms in the marine environment can be divided into two categories: hard and soft foulers. Hard foulers are organisms with solid skeletons, tubes or shells, such as barnacles, mussels, tube building polychaetes, bryozoans and corals (Fig. 4.5). Soft foulers are organisms that lack hard structures, such as ascidians, macroalgae, hydroids and sponges [93].

Both hard and soft fouling organisms are major problems for the shipping industry because they cause an increase in friction on ship hulls, which results in increased weight, reduced speed and decreased manoeuvrability, effects which must be compensated for by increased fuel consumption. Fuel consumption has been estimated to increase by 6 % for every 100 μm increase in hull roughness due to fouling when the same speed is maintained [94]. Furthermore, a higher frequency of dry-docking is also needed because fouling increases the corrosion of boat hulls. Another consequence of fouling is the introduction of marine species to areas in which they are not native [88].

It has been calculated that fouling costs the US navy approximately $1 billion per year [95]. When assuming that no effective self-polishing copolymer (SPC) antifouling paint is applied to ships, global calculations for 1989 have indicated that an additional 7.4 million tonnes (an increase of approximately 4 %) of fuel will be

Fig. 4.5 Examples of adult barnacles attached to a ship hull. Traces of old barnacle base plates are also visible on the hull. *Photo* Fredrik Lindgren

added to the global usage of 184 million tonnes. The 85–86 % carbon content of the fuel results in approximately 3.1 tonnes of the greenhouse gas CO_2 per every tonne of fuel that is consumed. By using an effective antifouling paint, approximately 22.8 million tonnes of CO_2 can be prevented from being released into the atmosphere each year [2]. Thus, the use of toxic antifouling compounds is an environmental dilemma. It results in adverse effects to the marine environment, in which the biota are affected by the toxic compounds and the marine ecosystem can be severely disturbed, although reduced CO_2 emissions into the atmosphere and a relative reduction in climate change effects are consequences of these antifouling agents. This trade-off between environmental concerns can be difficult to address and further emphasises the need for efficient and non-toxic antifouling systems.

Box 4.1 Common foulers in Swedish waters

In Swedish waters, barnacles, mussels, ascidians and macroalgae are common foulers. The fouler responsible for the most severe problems on human structures are bay barnacles, specifically *Amphibalanus improvisus* (Crustacea, Cirripedia, Balanidae) [96], which were introduced to Swedish waters during the 19th century and are now found in the Baltic Sea and along the western Swedish coast [97]. This barnacle is a euryhaline balanoid that colonises substrates from the water surface to depths of approximately three metres and can be found on a variety of substrata, e.g., rocks, macroalgae, ship hulls and aquaculture equipment [98], although it has a clear preference for smooth substrates [99]. *Amphibalanus improvisus* spawns continuously from June through September. It primarily sexually reproduces, which is likely the cause of the high numbers of *A. improvisus* that can be found living closely together on populated substrates. After fertilisation and rearing, nauplius larvae are released into the water column. These are freely swimming planktonic larvae that feed in the water and pass through six larval stages before moulding into the final larval stage, i.e., the cyprid larvae (Fig. 4.6). At this stage, the larvae are non-feeding and instead obtain energy from stored lipids and proteins. Their dispersal is maintained by near-shore currents. When the larvae are transported into shallow water, they begin a surface-bound exploratory behaviour to search for suitable substrata to settle and metamorphose on [100]. In this exploratory substrate phase, the larvae use their antennae for surface examination. They are believed to respond to several physical and chemical cues, such as surface texture, hydrodynamic forces, biofilm composition, chemical composition and conspecific adults

Fig. 4.6 Microscopically enhanced image of a cyprid larva of the barnacle species *Amphibalanus improvisus*. *Photo* Fredrik Lindgren

[101, 102]. Cyprid larvae from several barnacle species, such as *Semibalanus balanoides*, *Balanus amphitrite* and *Elminus modestus*, are reported to be strongly influenced by conspecifics, which may be the strongest cue for settlement among these species. When a surface is selected, cyprid larvae permanently attach to the surface using a secreted adhesive and metamorphose into juvenile barnacles. The attachment process for *A. improvisus* lasts approximately 8–16 h. Glenner and Hoeg [103] estimate that the juvenile barnacles are able to stretch their cirri (long featherlike limbs used to filter food from the water) for the first time 24 h after the cyprid carapace has been shed.

Another consequence of fouling on ship hulls is the introduction of new species into areas where they are not naturally present because organisms can travel attached to the ship hull. Alien species are considered to be introduced in similar proportions with ballast water and ship hull fouling [104]. When the hull is cleaned at port or when sloughing of the fouling biofilm occurs, the organisms enter a new habitat. The attached organisms release reproductive stages, e.g., spores and larvae. A great number of the introduced species will not survive the environmental conditions in the new habitat, although a history of problematic species invasions confirms that there can be severe effects from those that do survive [105]. Together with pollutants, the overexploitation of resources and the destruction of habitats, invasive species have been identified as one of the four largest threats to the marine environment [106].

Other consequences of fouling for submerged structures include increased weight of oil drilling rigs and pipelines. The insides of heat exchange tubes in electrical power plants can be fouled, resulting in decreased water exchange [107]. Fish farm nets can also become fouled, requiring more frequent replacement [108]. Both the environmental and economic consequences of fouling generate a need for antifouling methods that counteract fouling on man-made substrates immersed in the sea.

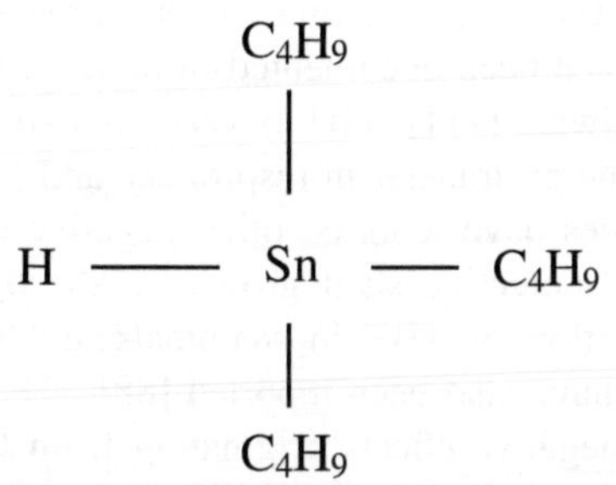

Fig. 4.7 Schematic of a Tributyltin molecule with three butyl groups attached to a central tin atom

4.3.1 Antifouling Paints

4.3.1.1 Tributyltin (TBT)

Tributyltin was originally used to combat the tropical disease Bilharzia [109]. However, in the early 1960s, the substance was discovered to also have excellent antifouling properties, and it was commercialised as an antifouling biocide by incorporating it into paint. Tributyltin works as a toxicant with primary activity on shell (e.g., barnacles) and vegetative fouling.

This new biocide and the SPC paint formulation technology called self-polishing copolymer (SPC) have revolutionised the antifouling industry. TBT-SPC paint is a methyl methacrylate polymer bonded by TBT groups (Fig. 4.7). When introduced into the water, soluble pigment particles in the paint begin to dissolve, exposing TBT groups at the paint surface; this is followed by a slow leakage of TBT that is caused by the breaking of molecular bonds in the matrix by the alkaline seawater environment. Simultaneously, hydrophobic copolymers prevent seawater from penetrating the paint matrix. These two abilities form a stable paint technology with a constant slow leakage of the active substance. After a sufficient amount of TBT is leached, new pigment particles are exposed and dissolved, exposing new TBT molecules at the paint surface [88]. The leaking is more extensive in rough spots, causing a self-polishing effect that decreases the hull roughness, resulting in fuel savings and lowered emissions. The polishing effect can be manipulated to maximise the effective lifetime of the paint, which means that most vessels painted with TBT can delay the dry-docking period by five years. The toxicant release rate is approximately constant for the entire life of the paint [88]. When dry-docking and repainting are needed, the old paint does not require removal; repainting can be conducted directly on the old paint, saving both time and money [95]. In 1996, 70 % of the world fleet was estimated to use TBT-SPC paint [110].

However, TBT has another potential side effect that was first discovered in the early 1970s. A scientist in the U.K. reported numerous dog whelks (*Nucella lapillus*) exhibiting unnatural development. The dog whelks developed *imposex*, and a great number of females had developed male genitalia, which, among other things, resulted in reproductive failure. In the late 1970s, new technology traced this phenomenon to the usage of TBT on boat hulls. This sterilisation of female dog whelks occurs at TBT concentrations of 3–5 ng/l. During summers in the 1980s, TBT values exceeded 100 ng/l, sometimes even surpassing 1000 ng/l in U.K. marinas. Imposex has since been documented in more than 150 species of marine prosobranch snails worldwide [111]. TBT is very lipid soluble and accumulates in cells, where it inhibits energy transfer in respiratory and photosynthetic processes. In addition, several studies have reported other negative effects of using TBT on marine animals, such as defective shell growth in the oyster *Crassostrea gigas* (20 ng/l). The accumulation of TBT in mammals and negative effects on fish immunological systems have also been reported [88].

Due to the multiple negative effects originating from the use of Tributyltin in antifouling paints, IMO decided in October 2001 to place a global ban on the use of this product. By 2003, no ship was permitted to be repainted with TBT, and beginning 1 January 2008, all vessels were required to be free from TBT [112]. Several countries, including Sweden, introduced legislation on TBT in the mid- to late 1980s, i.e., before the IMO ban was enacted [113].

4.3.1.2 Copper

After the TBT paint ban, copper became the primary active ingredient in antifouling paints [114]. Copper is an essential metal that commonly occurs in the environment. Moreover, living organisms use it for growth [94]. Due to weathering, 250,000 tonnes of copper enter the oceans naturally each year [88], versus 15,000 tonnes from the use of antifouling paints [114]. However, copper can be toxic to organisms at high concentrations [94] and can occur in several different states. The most toxic form of copper to living organisms is Cu^{2+}. In this ionic form, copper easily passes through cell membranes. The bioavailability of copper depends on several biotic and abiotic factors, rendering it challenging to estimate [114]. The usage of copper in antifouling paints is often in the form of cuprous oxide, copper thiocyanate or metallic copper, leaking copper in ionic form, in contrast to the copper leaking due to natural weathering. This have an antifouling effect on barnacles, tube worms and several species of algae [94]. The leakage of copper from the paint is governed by several factors, e.g., the paint formula, the percentage of copper in the paint, the age of the paint, the water temperature and the pH. Some copper-containing antifouling paints can have a copper content of up to 75 % (Pettit Trinidad©) [114]. In several cases, algae species are significantly tolerant to high levels of copper, e.g., the common fouling alga *Ulva spp*. Due to this tolerance, paint companies have begun adding other biocides, i.e., so-called *booster biocides*, which are often used in combination with copper to guarantee sufficient effects; in some cases, two substances can have synergistic effects. Booster biocides generally exhibit effects against bacteria, algae and/or fungi, and they are divided into two categories: metal-based and non-metal-based biocides. Globally, approximately 18 different types of booster biocides exist on the market [115]. In the following section, the most commonly used booster biocides are described.

4.3.2 Non-metal-Based Booster Biocides

4.3.2.1 Irgarol 1051

Because several algal species have exhibited tolerance to high copper levels, the substance Irgarol 1051 (2-methylthio-4-tert-butylamino-6-cyclopropylamino-s-triazine) has been added to copper-based antifouling paints to boost the effects against algal foulers [115]. Irgarol 1051 is an herbicide that is primarily toxic to algae and cyanobacteria. Its toxicity is derived from its ability to inhibit photosynthetic electron transport in chloroplasts and cyanobacteria. For these groups of organisms, Irgarol 1051 has a very high toxicity. Irgarol 1051 has been found to inhibit growth at concentrations of 50 ng/l in non-target algal species [88]. For periphyton communities, i.e., microbial communities attached to underwater surfaces, which are very similar to ship fouling, levels as low as 182 ng/l have been found to result in substantial effects [116]. Furthermore, the seaweed *Ulva intestinalis* exhibits acute effects from Irgarol 1051 at 10 μg/l, and the sea grass *Zostera marina* has an EC50 value of 2.5 mg/l. Irgarol 1051 biodegrades slowly compared with other booster biocides, with a lifetime in seawater of approximately 350 days. It is predominantly found in water, although it can accumulate in sediment if introduced as paint particles. Under anoxic sediment conditions, the degradation time is considerably longer than 350 days. The primary metabolite of Irgarol 1051 is called M1 (2-methylthio-4-tert-butylamino-s-triazine), which has toxic effects and is commonly found near marinas, although always at lower concentrations than Irgarol 1051 [117].

In 1993, researchers found high levels (i.e., concentrations as high as 1700 ng/l) of Irgarol 1051 in the port of Côte d´Azur, France. Following this discovery, the substance was also found in harbour areas in the USA, Australia, Japan, Bermuda and Sweden [117]. Because the growth inhibition effects of Irgarol have been documented to affect more than the algae on boat hulls legislation now exists regarding this product in several countries, for example in Denmark, the UK and Sweden. In Sweden, antifouling paints containing copper combined with Irgarol 1051 have not been approved by the chemical inspectorate for usage along the Baltic coast on ships less than twelve metres in length since 2002 [118, 119]. This ban has led to a decrease in Irgarol 1051 levels in the Baltic Sea from the pre-legislation levels of 10–100 ng/l [118, 119]. Little is known regarding the long term effects of this molecule on marine life, although several scientific studies have been performed [88].

4.3.2.2 Diuron

Diuron (3-(3,4-diclorophenyl)-1,1-dimethylurea) is an agricultural product that has been used for many years. Although its primary usage is as a vegetation control substance in non-crop areas, Diuron has also been found to work well as a booster biocide in antifouling paints [88, 94]. Diuron works in the same way as Irgarol 1051, i.e., as a photosynthetic inhibitor, and is also primarily toxic to algae and cyanobacteria. It is commonly found in areas of high leisure boating activity, such as marinas, and its occurrence has been reported in the UK, Sweden, Spain, the Netherlands, and Portugal, with the highest reported levels in Japan (3.05 μg/l) [120]. Diuron exhibits limited bioaccumulation [88], is relatively persistent in seawater and is stable to UV radiation exposure. The degradation products of Diuron have been reported to be up to 215 times more toxic than the original substance [118]. Studies of the detrimental effects of Diuron when present at concentrations lower than those found in the environment on non-target organisms, such as corals, sea urchin embryos, sea grass, the bacteria *Vibrio fisheri*, the water flea *Daphnia magna*, marine invertebrate embryos and larvae, have been performed and have resulted in the ban of this product from use on all vessels of any size in the UK, France, Denmark and Sweden [120].

4.3.2.3 DCOIT

Compared with other non-metallic booster biocides, which primarily function by inhibiting photosynthetic activity, DCOIT (4,5-dichloro-2-*n*-octyl-4-isothiazolin-3-one) has broad-spectrum antifouling activity and is highly toxic to many organisms. DCOIT has a very strong effect on barnacles [121], has high anti-microbial activity against bacteria, diatoms, fungi and algae [88] and has been detected in marinas in Spain, Greece and Denmark [120]. Compared with other antifouling compounds, it has high degradation rates in seawater and sediment, requiring less than 24 h to reach non-toxic levels [88, 121]. Moreover, the degradation products of DCOIT are four to five times less toxic than those of the parent compound, although they are still highly toxic to non-target species, e.g., sea urchin eggs and embryos [88].

4.3.3 Metal-Based Booster Biocides

Several metal-based booster biocides exist. The most commonly used boosters are the polyvalent salts of the substance pyrithione. Zinc and copper are the metals most commonly used with pyrithione. These biocides are effective against soft fouling, such as algae, fungi and bacteria. They have low solubility in water, low sediment accumulation, high biodegradation rates, and the degradation products have low toxicity. Consequently, they might have less of an environmental impact. Copper pyrithione generally has an advantage over zinc pyrithione because it has lower water solubility, a shorter half-life, and a longer biocidal effect [88].

4.3.4 Regulations

The International Convention on the Control of Harmful Anti-fouling Systems on Ships (AFS convention), which was adopted in 2001 and entered into force in 2008, regulates antifouling paints. This convention primarily regulates and prohibits the use of organotin compounds (primarily TBT) in antifouling systems. Usage on ship hulls, floating storage units and floating production storage and offtake units has been included in this ban since January 2008. Furthermore, the convention provides for the establishment of a technical group that reviews proposals for substances used in antifouling systems that may be candidates for restricted or prohibited use in the future [112].

4.4 Ballast Water

4.4.1 Background and History

Ships use ballast for stability, especially during non-laden voyages. In ancient times, sand, stones and other solid materials were placed in the keel of a ship as

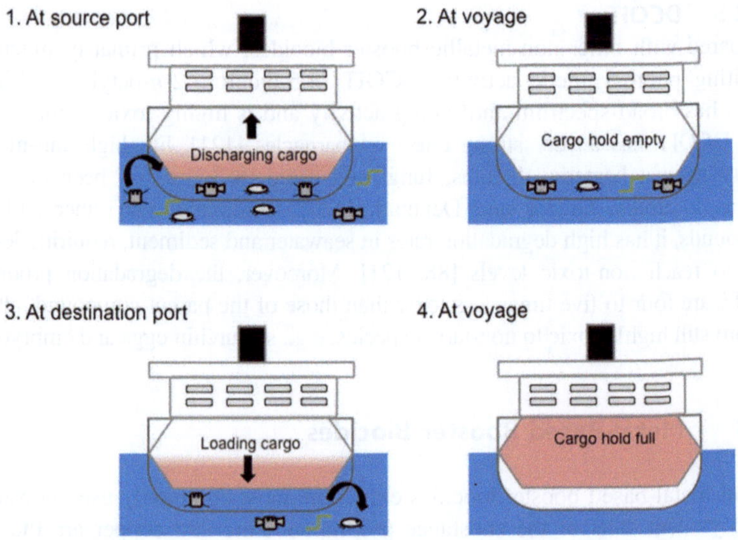

Fig. 4.8 The different stages in the cargo loading-ballast water cycle. In the source port, ballast water containing organisms is loaded into the ballast water hold. In the destination port, the ballast water containing organisms is discharged into a new marine environment

ballast. Since the mid-19th century, when ships changed from wooden hulls to metal hulls with ballast tanks, seawater has been used as ballast. The increased amounts of goods carried at sea during the last century have resulted in large volumes of ballast water being transferred between coastal areas and ports. When ballast water is collected, aquatic organisms are also transferred into the ballast water tanks. The ballast tank constitutes a "home away from home" for plankton and various aquatic organisms in their young stages, e.g., fish or crab larvae (Fig. 4.8). Frequent sea traffic and rapid transportation have enhanced the survival rate of organisms in the tanks because the reduced shipping times cause organisms to spend less time in darkness and under low-oxygen conditions. Shipping is considered the major vector for the transfer of organisms between different ecosystems around the globe and the largest vector for the introduction of alien species to different sea areas [122]. Marine organisms can be transferred either by being attached to the hull, as described earlier in this chapter, or by being contained within the ballast water.

The largest ballast water volumes are used by tankers, followed by container ships; ferries use much smaller volumes of ballast water. Approximately 4000 species have been estimated to be transferred by ships each day and can cross natural barriers, e.g., water currents, due to this process. The organisms can also be transported over long distances that they would have been unable to swim or move across on their own. When ballast water organisms are discharged, their tolerance to the physical conditions in the new area, e.g., temperature and salinity, determine whether they become established. The following species characteristics are important for successful establishment: tolerance to broad salinity and temperature ranges, generalist concerning food preferences, and quick and vast reproduction. Other prevailing factors, e.g., nutrient supply, light and oxygen, also affect the survival and establishment of organisms in new areas. Light conditions are important for organisms using sunlight for photosynthesis, whereas oxygen conditions can be important for organisms living within the sediment. When species are introduced to new areas and are tolerant of the new physical conditions, they can affect existing ecosystems. For example, when organisms that feed on a new species are absent, the new population will not be controlled by grazing/predation, allowing it to expand rapidly. When organisms spread and increase in number without control, they are considered to be *invasive*. The effects of aquatic invasions or biological pollution are difficult to estimate. However, numerous examples exist of disturbed ecosystems with high economic costs and severe human health problems.

The first aquatic invasion was reported in the early 1900s by the Danish professor Ostenfeld after the vast occurrence of the Asian phytoplankton *Odontella biddulphia* was reported in the North Sea [123]. Following this event, several examples of transfer of organisms from native to invaded areas have occurred, with profound consequences. Examples of these events include the Chinese mitten crab, which is now found in the North Sea; the comb jellyfish, *Mnemiopsis leidyi*, which is native to the American east coast and is now found in the Black Sea and the Swedish west coast; the zebra mussel from the Ponto-Caspian area, which invaded the Great Lakes; and the round goby, *Neogobius melanostomus*, which is native to central Eurasia (Caspian Sea) and can now be found in Europe and North America.

4.4.2 Ecosystem Impacts

The Chinese mitten crab is native to China and was later transferred to Europe. The crab is a freshwater species that can travel long distances on land, ruining fishing gear, undermining sediment and transporting parasites and diseases. A possible positive outcome is that the crab can be fished because it is considered a delicacy in some regions of the world.

The collapse of the fishing industry in the Black Sea has been correlated with the introduction of the comb jellyfish, *Mnemiopsis leidyi,* from the American east coast, which was introduced in the early 1980s. By 1989, the species had extensively spread and reproduced, i.e., up to 400 specimens per m^3 of water. It feeds not only on zooplankton but also on eggs and larvae of pelagic fish, causing a dramatic reduction in fish populations. The commercially attractive anchovy was also dramatically reduced due to competition for the same food sources. Since the mid-1990s, the Black Sea ecosystems have exhibited signs of recovery due to both reduced eutrophication and decreased comb jellyfish populations [124].

4.4.3 Estimated Costs and Societal Impacts

The zebra mussel is a bivalve native to western Asia, especially in the Aral, Black and Caspian Seas. In the 1700s, the zebra mussel spread through waterways in Europe, becoming established in lakes and rivers. In 1988, the zebra mussel was discovered in the Great Lakes (Canada and US), which has been estimated to be the most costly introduction in the US [125]. These mussels are small (maximum size of 5 cm), although they can appear in very high numbers, e.g., 700,000 m^2. The zebra mussel is known to clog intake pipes used in various industries and critical interiors of plant structures and wells, resulting in reduced flow or even complete shutdowns.

4.4.4 Human Health Impacts

4.4.4.1 Vibrio cholera

In 1991, thousands of people in Peru died due to an outbreak of cholera that was caused by *Vibrio cholera*, which was transported by ballast water from southern East Asia [126]. This epidemic rapidly spread throughout Latin America and Mexico, even reaching the US in July 1992. The Food and Drug Administration (FDA) determined that this outbreak originated from ship ballast water whose last port of call was in South America [127].

4.4.4.2 Toxins in Shellfish

People can become sick by eating shellfish (e.g., mussels, oysters, clams, and scallops) that filtered and fed on toxic microalgae. These microalgae can be found in resting forms within the sediment and hatch under appropriate environmental conditions. At the resting stages, they can be transferred with the sediments in ballast water tanks. Furthermore, microalgae can change state and become toxic under certain conditions. When humans consume shellfish containing microalga toxins, the transfer of the toxins can result in poisoning, potentially leading to fatal consequences. Shellfish toxins are classified into different forms, including amnesic shellfish poisoning, diarrhetic shellfish poisoning, neurotoxic shellfish poisoning and paralytic shellfish poisoning. The paralytic shellfish poisoning (PSP) form is known to be caused by toxic dinoflagellate species, which commonly occur in ship ballast waters as either viable forms or in resting cyst forms [128].

4.4.5 Regulations

The negative effects of aquatic introductions have been severe. As a result, IMO adopted the "*International Convention for the Control and Management of Ships Ballast Water and Sediments*" (BWMC) in 2004 to prevent, minimise and eliminate risks to the environment, human health, property and resources due to the transfer of harmful aquatic organisms via ship ballast water. A total of 15 guidelines were developed to support the implementation and provide technical guidance for the convention principles. Several regions have extra regulations on ballast water handling, including Australia, Canada, and the states of California, New York and Alaska in the USA. The convention sets a requirement that ships have an on-board ballast water treatment system, onshore pumping of the ballast water or other future solutions. The convention must be ratified by 30 states, constituting at least 35 % of the world tonnage, to enter into force. In September 2015, 44 states representing approximately 32.86 % of the world tonnage ratified the protocol. Thus, the number of states was fulfilled, although the required tonnage was not. When ratification occurs, the convention will be enforced 12 months afterward, although the regulations are dependent on the size and construction year of each vessel [123, 129].

To be approved by IMO, a ballast water management system must fulfil specific parameters. The system is not permitted to discharge more than 10 viable organisms per m^3 with sizes of at least 50 μm or 10 viable organisms per mL with sizes of 10–50 μm. In addition, the discharge of indicator microbes is not permitted to surpass specified concentrations. These limits are as follows: *Vibrio cholera* must be less than 1 colony forming unit (cfu) per 100 mL or less than 1 cfu per 1 g (wet weight) based on zooplankton samples, *Escherichia coli* must be less than 250 cfu per 100 mL, and intestinal *Enterococci* must be less than 100 cfu per 100 mL [129].

4.4.6 Ballast Water Exchange

Exchanging ballast water during oceanic voyages can be used as a ballast water management procedure and is one guideline in the Ballast Water Management convention (BWMC). Because oceanic water contains fewer plankton and marine organisms than coastal water, the exchange is intended to decrease the organism concentrations in ballast water tanks. Ballast water exchange is not a management method approved by the convention, although it can be used voluntarily. However, in some areas, ballast water exchange is required due to national regulations, e.g., for ships navigating the Great Lakes in Canada and the US and in Australia. Ballast water exchange under poor weather conditions can put ships at risk due to lost stability.

4.5 Marine Litter

Solid waste is also transferred to the marine environment from various sources. When solid waste enters the marine environment, it becomes “marine litter” or “marine debris”. In some cases, the term “floatables” is also used. All of these terms describe the same phenomenon and its adverse characteristics. Marine litter is defined as “any persistent, manufactured or processed solid material discarded, disposed of or abandoned in the marine and coastal environment. Marine litter consists of items that have been manufactured or used by people and deliberately discarded into the sea or rivers or on beaches; brought indirectly to the sea with rivers, sewage, storm water or winds; accidentally lost, including material lost at sea in bad weather (fishing gear, cargo); or deliberately left by people on beaches and shores" [130]. Waste material that is deposited in the ocean often becomes trapped in the centre of a gyre (circular ocean current), where material continually accumulates. For example, the Great Pacific Garbage Patch (also known as the Pacific Trash Vortex) refers to marine debris in the Pacific Ocean between Japan and North America [131].

Plastic is the most prevalent of floating marine debris [130]. Plastic litter is very durable and buoyant and adsorbs toxic compounds that are present at sea. Plastics do not usually biodegrade; instead, they fragment into smaller pieces that are both visible and invisible to the naked eye. UV light can damage synthetic materials; however, the deterioration of plastic litter under UV light requires long periods of time due to the haline environment and the cooling effect of the sea [132]. In addition, a living shield of marine organisms can protect the plastic from deterioration by UV light [132]. In the last thirty years, considerable attention has been given to very small plastic pieces (e.g., pellets and granules) with sizes of 5 mm or less. These pieces are called *microplastics* and have become more numerous in the natural environment [132]. Microplastics may have primary or secondary origins because they may be originally produced but also derived from the fragmentation of larger plastic items (Fig. 4.9) [133]. Microplastics can originate from cosmetics, airblast cleaning media, and fibres from fleece clothing [75].

Microplastics are an abundant component of marine debris, and they tend to accumulate in ocean gyres. However, coastal areas with industrial activities have also been identified as microplastic hotspots [133]. In the North Pacific Subtropical Gyre, a maximum number of 32.76 particles/m^3 and a maximum mass concentration of

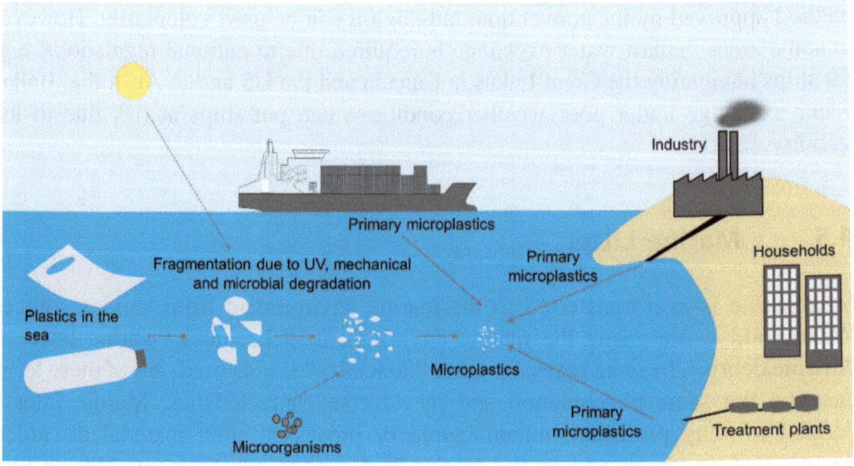

Fig. 4.9 Sources of microplastics in the marine environment, modified from Wright et al. [133]

250 mg/m^3 have been recorded [133], and in a Swedish harbour area adjacent to a polyethylene (PE) production plant, 100,000 particles/m^3 have been recorded in seawater samples [134]. Many forms of microplastics are commonly found in the marine environment. However, fibrous forms appear to be the most abundant [133].

The possible impact of microplastics on marine organisms and the marine food web remains a topic of research. These plastics can contain different additives (e.g., bisphenol A and phthalates) used during production. Plastic particles also tend to absorb hydrophobic persistent organic pollutants (POPs) after being discharged into seawater. Microplastics can be contaminated with waterborne POPs up to six orders of magnitude greater that the levels commonly found in ambient seawater [135].

Approximately 6.4 million tonnes of litter reach the oceans annually [130], whereas approximately 20 % of marine litter has a sea-based source, primarily from the shipping and fishing sectors. The main sea-based sources of marine litter are merchant ships, passenger ships, fishing vessels, military fleets, research vessels, pleasure crafts, offshore oil and gas platforms and fish farming installations [130]. The international regulations for the discharge of waste overboard vessels have become stricter. Currently, only food waste, non-harmful cargo residue, and non-harmful cleaning agents can be legally managed in this manner [139].

4.5.1 Impacts of Marine Litter

There are several adverse effects of marine litter (Fig. 4.10). Ship propellers may become entangled in litter, and the water intake pipe may become blocked, affecting navigational safety [136]. Coastal power stations can suffer from blocked cooling water intake screens. Beaches may lose their recreational and aesthetic value, in addition to the adverse safety effects, when marine litter washes up on land [136]. Human health may also be affected due to physical injuries during swimming or on beaches where marine litter is present. An additional disease risk exists when exposed to medical waste in the marine environment.

Another important impact of marine litter is related to marine wildlife, with both direct and indirect influences. Marine litter affects a minimum of 267 species

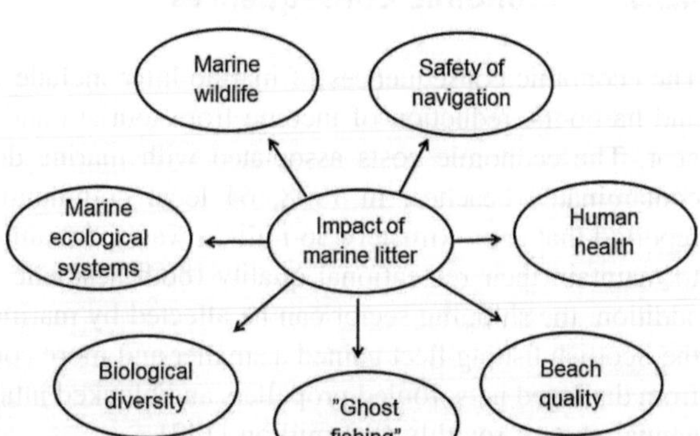

Fig. 4.10 Overview of the potential impacts of marine litter [136]

worldwide. Specifically, 86 % of all sea turtle species, 44 % of all seabird species and 43 % of all marine mammal species have experienced effects of marine litter [137]. Lost or abandoned fishing gear can move across coral reef systems, posing a threat and destroying the local flora and fauna [136]. The biodiversity in an ecosystem may be reduced because drifting plastic litter can act as transport platforms for the introduction of invasive species that can outcompete indigenous species [132]. The gas exchange between the overlying water and the pore waters of the sediment can be inhibited due to the accumulation of buoyant debris, which can harm the ecosystem by causing hypoxia [75]. Furthermore, marine debris can harm marine wildlife directly via entanglement and ingestion [137].

Food waste does not directly adhere to the common definition of marine litter because it is not persistent. However, it is a common garbage stream on vessels, is generated in large amounts, especially on passenger ships, and its discharge into the marine environment is still permitted. Discharge of food waste, which is an organic material, in significant amounts in sensitive areas can affect several characteristics of the sea, e.g., BOD, COD, TOC, turbidity and nutrient levels [70].

4.5.1.1 Entanglement

Seals, seabirds, whales and sea turtles are animals that often suffer from entanglement. Entangled animals can drown and have a reduced ability to catch food or to escape predators. Moreover, the attached debris can wound the animal by abrasion [75].

4.5.1.2 Ingestion

Marine litter may be mistaken for food or accidentally ingested by marine animals, in some cases blocking the digestive tract and causing internal injuries. The consumption of plastic by animals may limit their ability to lay down fat deposits (e.g., seabirds) or may block gastric enzyme secretion [75]. By consuming plastic litter, toxins potentially absorbed on their surface are also consumed, which can lead to bioaccumulation of toxins in the animal, e.g., ingested plastic particles can lead to PCB accumulation in bird tissues [138].

4.5.2 Economic Consequences

The economic consequences of marine litter include the cleaning costs of beaches and harbours, reduction of income from tourism and damage to ships and fishing gear. The economic costs associated with marine debris include the cleaning of contaminated beaches. In 1998, 64 local communities in the North Sea region reported that approximately $6 million was spent annually on cleaning the beaches to maintain their recreational quality (both aesthetic and safety aspects) [130]. In addition, the shipping sector can be affected by marine litter. It has been found that the Scottish fishing fleet gained a smaller and more contaminated catch and suffered from damaged nets, fouled propellers and blocked intake pipes, which resulted in an annual cost of roughly €12 million [142].

4.5.3 Regulations

The discharge of garbage generated on ships is governed by regulations in Annex V of MARPOL 73/78 for the prevention of garbage pollution from ships. The revised Annex V prohibits the discharge of all garbage into the sea except for food waste, animal carcasses, non-harmful cargo residues, cleaning agents and additives (Tables 4.5 and 4.6).

At international level a resolution MEPC.83(44) by IMO's marine environment protection committee states that facilities provided by ports must meet the needs of the ships that most often use the port and allow for the final disposal of ship waste to occur in an environmentally appropriate manner [140]. At the European regional level, EU Directive 2000/59/EC of the European Parliament and the Council was adopted on 27 November 2000; this legislation governs reception facilities for ship-generated waste and cargo residues. The directive states among others that waste management plans must be prepared for ports, that ships should notify ports regarding the waste delivery and that ports should implement fee systems that

Table 4.5 Overview of types of garbage allowed to be discharged at sea by ships operating outside of special areas, according to Annex V of MARPOL 73/78 [139]

No distance restriction although as far as possible is recommended	At minimum 3 NM from the nearest land, en route and as far as practicable	At minimum 12 NM from the nearest land, en route and as far as practicable
Cleaning agents and additives[a] contained in cargo hold wash water	Comminuted food waste	Food waste
Cleaning agents and additives[a] in deck and external surface wash water		Cargo residues[a] either contained or not contained in wash water
Carcasses of animals (carried as cargo that have died during the voyage)		

[a]These substances must not be harmful to the environment

Table 4.6 Overview of types of garbage allowed to be discharged by sea ships operating in special areas according to Annex V of MARPOL 73/78 [139]

No distance restriction	At minimum 12 NM from the nearest land, en route and as far as practicable
Cleaning agents and additives[a] in deck and external surface wash water	Comminuted food waste
	Cargo residues[a] contained in wash water[b]
	Cleaning agents and additives[a] contained in cargo hold wash water[b]

[a]These substances must not be harmful to the environment
[b]The discharge is only permitted if (a) both the port of departure and the next destination port are within the special area and the ship does not traverse beyond the special area between these ports and (b) if no adequate reception facilities are available at these ports

encourage ships to leave their waste at port reception facilities rather than discharge them at sea. According to the directive, some ships can be exempt from the requirements due to "regular traffic" or if categorized as "green ships" [141].

References

1. IMO. *The international convention for the prevention of pollution from Ships* (MARPOL), 1983, IMO: London, UK.
2. Townsin, R. L., *The ship hull fouling penalty.* Biofouling, 2003. **19**: p. 8–15.
3. Briski, E., Ghabooli, S., Bailey, S. A. & MacIsaac, H. J., *Invasions risk posed by macroinvertebrates transported in ship's ballast tanks.* Biological Invasions, 2012. **14**(9): p. 1843–1850.
4. Kingston, P. F., *Long-term Environmental Impact of Oil Spills.* Spill Science & Technology Bulletin, 2002. **7**(1–2): p. 53-61.
5. Sugiura, K., Ishihara, M., Shimauchi, T. & Harayama, S., *Physicochemical properties and biodegradability of crude oil.* Environmental Science & Technology, 1997(31): p. 45-51.
6. Teräväinen, M. J., Pakarinen, J. M. H., Wickström, K. & Vainiotalo, P., *Comparison of the Composition of Russian and North Sea Crude Oils and Their Eight Distillation Fractions Studied by Negative-Ion Electrospray Ionization Fourier Transform Ion Cyclotron Resonance Mass Spectrometry: The Effect of Suppression.* Energy & Fuels, 2007(21): p. 266-273.
7. Rodrigue, J., Comtois, C. & Slack, B., *The Geography of transport systems.* 2009, Oxon: Routledge. 352.
8. Rogowska, J. & Namiesnik, J., *Environmental Implications of Oil Spills from Shipping Accidents.* Reviews of Environmental Contamination and Toxicology. Vol. 206. 2010, New York: Springer. 95-114.
9. GESAMP. *Reports and studies 75: Estimates of oil entering the marine environment from sea-based activities.* 2007, London. p. 83.
10. Mitchell, R., MacDonald, I. R. & Kvenvolden, K. A., *Estimation of total hydrocarbon seepage into the Gulf of Mexico based on satellite remote sensing images.* Transactions, American Geophysical Union, 1999(80): p. 49-49.
11. Farrington, J. F. & McDowell, J. E., *Mixing and Water.* Oceanus, 2004. **43**(1): p. 46-49.
12. Zhao, H. T. *et al.*, *Size distribution and diffuse pollution impacts of PAHs in street dust in urban streams in the Yangtze River Delta.* J. Environ. Sci., 2009. **21**(2): p. 162-167.
13. Mascarelli, A., *After the oil.* Nature, 2010. **467**(7311): p. 22-24.
14. Burgherr, P., *In-depth analysis of accidental oil spills from tankers in the context of global spill trends from all sources.* Journal of Hazardous Materials, 2007. **140**(1-2): p. 245-256.
15. Feder, H. M. & Blachard, A., *The deep benthos of Prince William Sound, Alaska, 16 months after the Exxon Valdez oil spill.* Marine Pollution Bulletin, 1998. **36**: p. 118-130.
16. Vietes, D. R., Nieto-Roman, S., Palanca, A., Ferrer, X. & Vences, M., *European Atlantic: the hottest oil spill hotspot worldwide.* Naturwissenschaften, 2004(91): p. 535-538.
17. Hassler, B., *Accidental versus operational oil spills from shipping in the Baltic Sea: Risk Governance and Management Strategies.* Ambio, 2011. **40**: p. 170-178.
18. IMO. *MARPOL 73/78 (Consolidated edition 2002).* 1978, International Maritime Organization: London, UK.
19. NRC, *Oil in the Sea III:Inputs, Fates, and Effects.* 2003, Washington D.C., USA: The National Academies Press.
20. IMO. *ISM Code (International Safety Management Code) and revised guidelines on implementation of the ISM Code.* 2010, IMO: London, UK.
21. IMO. *A revised SOLAS chapter V (Safety of Navigation).* 2002, IMO: London, UK.

22. Farrington, J. W., *Oil Pollution in the Marine Environment Inputs, Big Spills, Small Spills, and Dribbles.* Environment, 2013. **55**(6): p. 3-13.
23. Lindgren, J. F., *Evaluating effects of low concentrations of oil in marine benthic communities.* Doctor of Philosophy thesis, Chalmers University of Technology, 2015.
24. ITOPF. *Fate of marine oil spills.* 2002.
25. Jernelov, A., *The Threats from Oil Spills: Now, Then, and in the Future.* Ambio, 2010. **39** (5-6): p. 353-366.
26. Mitchell, R. B., *International oil pollution at Sea: Environmental policy and treaty compliance.* 1994, Cambridge, USA: MIT press.
27. HELCOM. *Report on shipping accidents in the Baltic Sea area for the year 2008.* 2008: Helsinki: HELCOM.
28. EPA. *Understanding oil spills and oil spill response. Emergency response.* 1999.
29. Karakulski, K., Marawski, W. A. & Grzechulska, J., *Purification of bilge water by hybrid ultrafiltration and photocatalytic processes.* Separation and Purification Technology, 1998 (14): p. 163-173.
30. IMO. *MARPOL - How to do it, Manual of the Practical Implications of Ratifying and Implementing MARPOL 73/78.* 1993: London, UK.
31. US Legal, *Bevare of the Magic Pipe*, in *Ship Management International.* 2007.
32. Michel, J. *et al.*, *Potentially polluting wrecks in marine waters.* International Oil Spill Conference, 2005: p. 40.
33. SMA. *Miljörisker från fartygsvrak [Environmental risks from whipwrecks].* 2011. p. 82.
34. Hassellöv, I. M., *Förstudie om vraksanering [Prestudy on the remediation of shiwrecks].* 2007, Göteborg, Sweden: Alliance for Global Sustainability (AGS).
35. Landquist, H., Hassellöv, I. M., Rosén, L., Lindgren, J. F. & Dahllöf, I., *Evaluating the needs of risk assessment methods of potentially polluting shipwrecks.* Journal of environmental management, 2013. **119**(0): p. 85-92.
36. EPA. *Alkylatbensin i småbåtsmotorer- analys av miljöfördelar [Alkylate gasoline in leisureboat engines – analysis of advantages for the environment].* 2009.
37. STA. *Båtlivsundersökningen 2010 [The 2010 study of boating].* 2010, Swedish Transport Agency.
38. Alin, J. & Astnäs, T. *Jämförande studie av utombordsmotorers emissioner till vatten [Comparative study of emissions to water from outboard engines]* . 2001, Avdelningen för kemisk miljövetenskap, Chalmers Tekniska Högskola, Göteborg.
39. Gustavsson, T. *Update of gasoline consumption and emissions from leisure boats in Sweden 1990-2003 for international reporting.* 2005.
40. NRC. *Steering Committee for the Petroleum in the Marine Environment Update, Board on Ocean Science and Police, Ocean Sciences Board, Commission on Physical Sciences, Mathematics, and Resources. Oil in the Sea: Inputs, Fates, and Effects.* 1985, National Research Council: Washington DC. p. 273-274.
41. Fingas, M. & Fieldhouse, B., *What causes the formation of water-in-oil emulsions?* International Oil Spill Conference, 2001.
42. Fingas, M. & Fieldhouse, B., *Studies of the formation process of water in oil emulsions.* Marine Pollution Bulletin, 2003. **47**: p. 369-396.
43. Gong, Y. *et al.*, *A review of oil, dispersed oil and sediment interactions in the aquatic environment: Influence on the fate, transport and remediation of oil spills.* Marine Pollution Bulletin, 2014. **79**(1–2): p. 16-33.
44. Boehm, P. D. & Page, C. A., *Exposure elements in oil spill risk and natural resource damage assessments: A review.* Human and Ecological Risk Assessment, 2007. **13**: p. 418-448.
45. Burgess, R., M. *Evaluating ecological risk to invertebrate receptors from PAHs in sediments at hazardous waste sites.* 2009: Ecological Risk Assessment Support Center, Cincinnati, USA. p. 23.
46. Piatt, J. F. & Ford, R. G., *How many seabirds were killed by the Exxon Valdez oil spill.* American Fisheries Society Symposium, 1996. **18**: p. 712-719.

47. Page, C. A., Bonner, J. S., Summer, P. L. & Autenrieth, R. L., *Solubility of petroleum hydrocarbons in oil/water systems.* Mar Chem, 2000(70): p. 79-87.
48. Jewett, S. C., Dean, T. A., Smith, R. O. & Blanchard, A., *'Exxon Valdez' oil spill: impacts and recovery in the soft-bottom benthic community in and adjacent to eelgrass beds.* Mar. Ecol.-Prog. Ser., 1999. **185**: p. 59-83.
49. Xia, Y. & Boufadel, M. C., *Beach geomorphic factors for the persistence of subsurface oil from the Exxon Valdez spill in Alaska.* Environmental Monitoring Assessment, 2011. **183**: p. 5-21.
50. Cerniglia, C. E., *Biodegradation of polycyclic aromatic hydrocarbons.* Biodegrad, 1992(3): p. 351-368.
51. Sims, R. C. & Overcash, M. R., *Fate of polynuclear aromatic compounds (PNAs) in soil-plant systems.* Residue Rev, 1983(8): p. 1-68.
52. Fleeger, J. W., Carman, K. R. & Nisbet, R. M., *Indirect effects of contaminants in aquatic ecosystems.* Science of The Total Environment, 2003. **317**: p. 207-233.
53. Hack, L. A., Tremblay, L. A., Wratten, S. D., Lister, A. & Keesing, V., *Benthic meiofauna community composition at polluted and non-polluted sites in New Zealand intertidal environments.* Marine Pollution Bulletin, 2007. **54**(11): p. 1801-1812.
54. OECD. *Detailed review paper on aquatic arthropodes in life cycle toxicity test with an emphasis on developmental, reproductive and endocrine disruptive effects.* 2006: Paris, France. p. 125.
55. Juhasz, A. L. & Naidu, R., *Bioremediation of high molecular weight polycyclic aromatic hydrocarbons: a review of the microbial degradation of benzo a pyrene.* Int. Biodeterior. Biodegrad., 2000. **45**(1-2): p. 57-88.
56. Åkerblom, N., Goedkoop, W., Nilsson, T. & Kylin, H., *Particle-specific sorption/desorption properties determine test compound fate and bioavailability in toxicity tests with Chironomus riparius - high-resolution studies with Lindane.* Environmental Toxicology and Chemistry, 2010. **29**(7): p. 1520-1528.
57. Burns, K. A., Codi, S. & Duke, N. C., *Gladstone, Australia field studies: weathering and degradation of hydrocarbons in oiled mangrove and salt marsh sediments with and without the application of an experimental bioremediation protocol.* Marine Pollution Bulletin, 2000. **7**(41): p. 392-402.
58. Culbertson, J. B., Valiela, I., Pickart, M., Peacock, E. E. & Reddy, C. M., *Long-term consequences of residual petroleum on salt marsh grass.* Journal of a applied ecology, 2008 (45): p. 1284-1292.
59. Kirby, M. F. & Law, R. J., *Accidental spills at sea- Risk, impact, mitigation and the need for co-ordinated post-incident monitoring.* Marine Pollution Bulletin, 2010(60): p. 797-803.
60. Kasai, Y., Kishira, H. & Harayama, S., *Bacteria belonging to the genus Cycloclasticus play a primary role in the degradation of armomatic hydrocarbons released in marine environment.* Appl. Environ. Microb., 2002. **68**: p. 5625-5633.
61. Van Hamme, J. D., Singh, A. & Ward, O. P., *Recent advances in petroleum microbiology.* Microbiol. Mol. Biol. R., 2003. **67**: p. 503-549.
62. White, I. C. & Molloy, F. C. *Factors that determine the cost of oil spills*, in *International Oil Spill Conference.* 2003.
63. Etkin, D. S. *Estimating cleanup costs for oil spills*, in *International Oil Spill Conference.* 1999.
64. Etkin, D. S. *Worldwide analysis of marine oil spill cleanup cost factors*, in *Proc 23rd Artic and marine oilspill program technical seminar.* 2000.
65. Liu, X. & Wirtz, K. W., *Total oil spill costs and compensations.* Maritime Policy & Management, 2006. **33**(1): p. 49-60.
66. Loureiro, M. L., Ribas, A., Lopez, E. & Ojea, E., *Estimated costs and admissible claims linked to the Prestige oil spill.* Economical Economics, 2006. **59**: p. 48-63.
67. Kontovas, C. A. & Psaraftis, H. N. *Marine environment risk assessment: A survey on the disutility cost of oil spills*, in *2nd International Symposium on Ship Operations.* 2008.

68. HELCOM, *HELCOM recommendation 11/10. Guidelines for capacity calculation of sewage systems on board passenger ships.* 1990, Erweko Oy: Finland.
69. The Ocean Conservancy. *Cruise Control. A Report on How Cruise Ships Affects the Marine Environment,* 2002.
70. EPA. *Cruise Ship Discharge Assessment Report,* 2008.
71. Glassmeyer, S. T. *et al., Transport of Chemical and Microbial Compounds from Known Wastewater Discharges: Potential for Use as Indicators of Human Fecal Contamination.* Environmental Science & Technology, 2005. **39**(14): p. 5157-5169.
72. IMO. Resolution *MEPC.159(55). Revised guidelines on implementation of effluent standards and performance tests for sewage treatment plants, adopted on 13 October 2006.* 2006, Marine Environment Protection Committee, International Maritime Organisation: London.
73. Butt, N., *The impact of cruise ship generated waste on home ports and ports of call: A study of Southampton.* Marine Policy, 2007. **31**(5): p. 591-598.
74. Guilbaud, J., Massé, A., Andrès, Y., Combe, F. & Jaouen, P., *Influence of operating conditions on direct nanofiltration of greywaters: Application to laundry water recycling aboard ships.* Resources, Conservation and Recycling, 2012. **62**(0): p. 64-70.
75. Derraik, J. G. B., *The pollution of the marine environment by plastic debris; a review.* Marine Pollution Bulletin, 2002. **44**: p. 842-852.
76. Moore, C. J., *Synthetic polymers in the marine environment: A rapidly increasing, long-term threat.* Environmental Research, 2008. **108**(2): p. 131-139.
77. Thompson, R. C., Swan, S. H., Moore, C. J. & vom Saal, F. S., *Our plastic age.* Philosophical Transactions of the Royal Society B: Biological Sciences, 2009. **364**(1526): p. 1973-1976.
78. STA. *Toalettavfall från fritidsbåtar. Frågor och svar. Jämförelser med andra länder [Sewage from leisure boats. Q&A. Comparisons with other contries].* 2012, Swedish Transport Agency,http://www.transportstyrelsen.se/sv/Sjofart/Fritidsbatar/Batlivets-miljofragor/Toalettavfall-fran-fritidsbat/Fragor-och-svar/Jamforelser-med-andra-lander/,last accessed on 20 May 2015.
79. STA, *Föreskrifter om ändring i Transportstyrelsens föreskrifter och allmänna råd (TSFS 2010:96) om åtgärder mot förorening från fartyg [Regulation regarding change in STAs regulations and general advices regarding measures to reduce pollution from ships], TSFS 2012:13.* 2012, Swedish Transport Agency.
80. RIL. *Water supply and sewage, Part 1 (In Finnish: Vesihuolto I.* 2003, Association of Finish Civil Engineers: Liitto, Helsinki.
81. Hagy, J.D., Boynton, W.R., Keefe, C.W. & Wood, K.V., Hypoxia in Chesapeake Bay, 1950-2001: Long-term change in relation to nutrient loading and river flow. Estuaries, 2004. **27**(4): p. 634–658.
82. Bernes, C., *Change Beneath the Surface. An In-depth Look at Sweden's Marine Environment.* 2005: Swedish Environment Protection Agency.
83. IMO, Resolution *MEPC.227(64), Guidelines On Implementation Of Effluent Standards And Performance Tests For Sewage Treatment Plants, Adopted on 5 October 2012.* 2012, Marine Environment Protection Committee, International Maritime Organisation: London.
84. Office of the Federal Register National Archives and Records Administration, *33 CFR Part 159 - Discharge of Effluents in Certain Alaskan Waters by Cruise Vessel Operations.* 2001.
85. EPA, *Vessel general permit for discharges incidental to the normal operation of vessels (VGP).* 2013.
86. European Commission, *Commision directive 2012/49/EU of 10 December 2012 amending Annex II to Directive 2006/87/EC of the European Parliament and of the Council laying down technical requirements for inland waterway vessels.* 2012.
87. Dafforn, K. A., Lewis, J. A. & Johnston, E. L., *Antifouling strategies: History and regulation, ecological impacts and mitigation.* Marine Pollution Bulletin, 2011. **62**(3): p. 453-465.

88. Yebra, D. M., Kiil, S. & Dam-Johansen, K., *Antifouling technology - past, present and future steps towards efficient and environmentally friendly antifouling coatings.* Progress In Organic Coatings, 2004. **50**(2): p. 75-104.
89. Wahl, M., *Marine epibiosis. I. Fouling and antifouling: some basic aspects.* Mar. Ecol. Prog. Ser., 1989. **58**: p. 175-189.
90. Dayton, P. K., *Competition, disturbance and community organization: the provision and subsequent utalization of space in a rocky intertidal community.* Ecological Monographs, 1971. **41**: p. 351-389.
91. Marshall, C. K. *Biofouling - What's the problem.* in *Biofouling: problems and solutions.* 1994. University of New South Wales, Sydney, Australia.
92. Anderson, C. D. & Hunter, J. D. *NAV2000 Conference Proceedings.* 2000. Venice, Italy.
93. Lewis, A. J., *Fouling and fouling protection: A defense perspective.* Biofouling: Problems and solutions. 1994, The University of New South Wales, Sydney, pp 39-43.
94. Voulvoulis, N., Scrimshaw, M. D. & Lester, J. N., *Alternative Antifouling Biocides.* Applied organometal chemistry, 1999. **13**: p. 135-143.
95. Callow, M. E. & Callow, A. C., *Marine biofouling: a sticky problem.* Biologist, 2002. **49**.
96. Berntsson, K. & Jonsson, P., *Temporal and spatial patterns in recruitment and succession of a temperate marine fouling assemblage: A comparison of static panels and boat hulls during the boating season.* Biofouling, 2003(19): p. 187-195.
97. Blom, S.-E. & Nyholm, K.-G., *Settling times of Balanus balanoides (L.), Balanus crenatus Brug., and Balanus improvisus Darwin on the west coast of Sweden.* Zool. Bidrag Uppsala, 1961. **33**: p. 149-155.
98. Dahlström, M. *Pharmacological agents targeted against barnacles as lead molecules in new antifouling technologies.* Ph D thesis, Göteborg University, 2004.
99. Berntsson, K. J., PR. Lejhall, M. Gatenholm P, *Analysis of behavioural rejection of micro-textured surfaces and implications for recruitment by the barnacle Blanus improvisus.* J Exp Biol Ecol, 2000(251): p. 59-83.
100. Lagersson, N. C., *The ultrastructure of two types of muscle fibre cells in the cyprid of Balanus amphitrite (Crustacea: Cirripedia).* J. mar. biol. Ass. U.K., 2002. **82**: p. 573-578.
101. Dobretsov, S., Dahms, H. U. & Qian, P. Y., *Inhibition of biofouling by marine microorganisms and their metabolites.* Biofouling, 2006. **22**: p. 43-54.
102. Dreanno, C. *et al.*, *An [alpha]2-macroglobulin-like protein is the cue to gregarious settlement of the barnacle Balanus amphitrite.* Proceedings Of the National Academy Of Sciences Of the United States Of America, 2006. **103**: p. 14396-14401.
103. Glenner, H. & Høeg, J. T., *Scanning electron microscopy of metamorphosis in four species of barnacles (Cirripedia Thoracica Balanomorpha).* Mar. Biol., 1993. **117**: p. 431-439.
104. Gollasch, S., *The Importance of Ship Hull Fouling as a Vector of Species Introductions into the North Sea.* Biofouling, 2002. **18**(2): p. 105-121.
105. Roberts, J. & Tsamenyi, M., *International legal options for the control of biofouling on international vessels.* Marine Policy, 2008(32): p. 559-569.
106. IMO, *Addressing marine biosafety through global partnerships.* 2004, London, UK.
107. Müller- Steinhagen, H., *Heat exchanger fouling: Mitigation and cleaning techniques.* 2000, Julienstr. 40, D-45130 Essen, Germany: PUBLICO Publications. 382.
108. Qian, P. Y., Lau, S. C. K., dahms, H. U., Dobretsov, S. & Harder, T., *Marine Biofilms as mediators of Colonaziation by Marine Microorganisms: Implications for Antifouling and Aquaculture.* Marine Biotechnology, 2007.
109. Bernes, C., *Organiska Miljögifter [Organic pollutants].* 1995: Naturvårdsverkets förlag: Stockholm, Sweden.
110. Champ, A. M., *A Review of Organotin Regulatory Strategies: Pending Actions, Related Costs and Benefits.* Science and the Total Environment, 2000. **258**: p. 21-27.
111. Santillo, D., Johnston, P. & Langston, J. W. *Tributyltin (TBT) antifoulants: a tale of ships, snails and imposex.* 2001, EU Report.

112. IMO, *The International Convention on the control of Harmful anti-fouling Systems on Ships.* 2001, London, UK.
113. Fent, K., *Ecotoxicological effects at contaminated sites.* Toxicology, 2004. **205**: p. 223-240.
114. M., S. & W., S. G., *Managing the use of Copper-Based antifouling paints.* Environmental Management, 2007(39): p. 423-441.
115. Manzo, S., Buono, S. & Cremisini, C., *Predictability of copper, irgarol, and diuron combined effects on sea urchin Paracentrotus lividus.* Arch. Environ. Contam. Toxicol., 2008. **54**(1): p. 57-68.
116. Dahl, B. & Blanck, H., *Use of sand-living microalgal communities (episammon) in ecotoxicological testing.* Marine Ecology Progress Series, 1996. **144**: p. 163-173.
117. Gatidou, G., Thomaidis, N. S. & Zhou, J. L., *Fate of Irgarol 1051, diuron and their main metabolites in two UK marine systems after restrictions in antifouling paints.* Environ. Int., 2007. **33**(1): p. 70-77.
118. Karlsson, J., Breitholtz, M. & Eklund, B., *A practical ranking system to compare toxicity of anti-fouling paints.* Marine Pollution Bulletin, 2006(52): p. 1661-1667.
119. KEMI. *Kemiska ämnen i båtbottenfärger– en undersökning av koppar, zink och Irgarol 1051 runt Bullandö marina 2004 [Chemical compounds in antifoungpaints – an investigation of copper, zinc and Irgarol 1051 around Bullandö marina in 2004].* 2006.
120. Konstantinou, I. K. & Albanis, T. A., *Worldwide occurrence and effects of antifouling paint booster biocides in the aquatic environment: a review.* Environ. Int., 2004. **30**(2): p. 235-248.
121. Jakobson, A. H. & Willingham, G. L., *Sea-nine antifoulant: an environmentally acceptable alternative to organotin antifoulants.* Science and the Total Environment, 2000. **258**: p. 103-110.
122. Gollasch S. & Leppäkoski E., *Risk assessment and management scenarios for ballast water mediated species introductions into the Baltic Sea.* Aquatic Invasions, 2007. **2**(4): p. 313-340.
123. Gollasch S. *et al.*, *Critical review of the IMO international convention on the management of ships ballast water and sediments.* Harmful Algae, 2007. **6**(4): p. 585-600.
124. Kideys, A. E., *Fall and Rise of the Black Sea Ecosystem.* Science, 2002. **297**(5586): p. 1482-1484.
125. O'Neill C.R. Jr. & MacNeill D. B., *The Zebra Mussel (Dresissena polymorpha): An Unwelcome North American Invader.* 1991: New York Sea Grant Institute.
126. Ruiz, G. M. *et al.*, *Global spread of microorganisms by ships.* Nature, 2000. **408**(6808): p. 49-50.
127. Takahashi, C. K., Lourenco, N. G. G. S., Lopes, T. F., Rall, V. L. M. & Lopes, C. A. M., *Ballast water: A review of the impact on the world public health.* Venom. Anim. Toxins incl. Trop. Dis., 2008. **14**(3): p. 393-408.
128. Hallegraeff, G. M. & Bolch, C. J., *Transport of toxic dinoflagellate cysts via ships' ballast water.* Marine Pollution Bulletin, 1991. **22**(1): p. 27-30.
129. IMO, *Ballast water management convention and the guidelines for its implementation*. 2004, London, UK.
130. UNEP. *Marine litter. An analytical view.* 2005, Nairobi, Kenya.
131. National Geographic. *National Geographic Education* 2014, http://education.nationalgeographic.com/education/encyclopedia/great-pacific-garbage-patch/?ar_a=1, last accessed on 20 May 2015.
132. Barnes, D. K. A., Galgani, F., Thompson, R. C. & Barlaz, M., *Accumulation and fragmentation of plastic debris in global environments.* Philosophical Transactions of the Royal Society B-Biological Sciences, 2009. **364**(1526): p. 1985-1998.
133. Wright, S. L., Thompson, R. C. & Galloway, T. S., *The physical impacts of microplastics on marine organisms: A review.* Environmental Pollution, 2013. **178**(0): p. 483-492.
134. Noren, F. & Naustvoll, F. *Survey of Microscopic Anthropogenic Particles in Skagerrak.* 2010, N-Research, Commissioned by Klima- og Forurensiningsdirektoratet.

135. Mato, Y. et al., *Plastic Resin Pellets as a Transport Medium for Toxic Chemicals in the Marine Environment*. Environmental Science & Technology, 2001. 35(2): p. 318–324.
136. Chen, C.-L. & Liu, T.-K., *Fill the gap: Developing management strategies to control garbage pollution from fishing vessels.* Marine Policy, 2013. **40**(0): p. 34-40.
137. Laist, D., *Impacts of Marine Debris: Entanglement of Marine Life in Marine Debris Including a Comprehensive List of Species with Entanglement and Ingestion Records*, in *Marine Debris*, JamesM Coe & DonaldB Rogers, Editors. 1997, Springer New York. p. 99-139.
138. Ryan, P. G., Connell, A. D. & Gardner, B. D., *Plastic ingestion and PCBs in seabirds: Is there a relationship?* Marine Pollution Bulletin, 1988. **19**(4): p. 174-176.
139. IMO, *Simplified overview of the discharge provisions of the revised MARPOL Annex V.* 2013: London, UK.
140. IMO. Resolution MEPC.83(44). Adopted on 13 March 2000. Guidelines for ensuring the adequacy of port waste reception facilities. 2000: London, UK.
141. European Commission, *Directive 2000/59/EC of the European Parliament and of the Council of 27 November 2000 on port reception facilities for ship-generated waste and cargo residues.* Official Journal of the European Communities, 2000. **43**. 142. Mouat, J., Lozano, R.L., Bateson, H. Economic Impacts of Marine Litter. Report KIMO, 2010. Lerwick, UK
142. Mouat, J., Lozano, R.L., Bateson, H. Economic Impacts of Marine Litter. Report KIMO, 2010. Lerwick, UK

Emissions to the Air

5

Kent Salo, Maria Zetterdahl, Hannes Johnson, Erik Svensson, Mathias Magnusson, Cecilia Gabrielii and Selma Brynolf

Abstract

Seeing the black smoke coming out of the funnel of a manoeuvring ship makes it easy to understand that the ship's propulsion contributes to the emission of air pollutants. However, there is more than meets the eye going up in smoke. A vast majority of ships use fossil fuels, increasing a positive net contribution of carbon dioxide to the atmosphere when they are combusted. Because the fuels that are used are often of low quality and possess a high sulphur content, a number of other air pollutants are also emitted. Emissions to the air from ships include

K. Salo (✉) · M. Zetterdahl · H. Johnson · C. Gabrielii · S. Brynolf
Shipping and Marine Technology, Chalmers University of Technology, Gothenburg, Sweden
e-mail: kent.salo@chalmers.se

M. Zetterdahl
e-mail: maria.zetterdahl@chalmers.se

H. Johnson
e-mail: hannes.johnson@chalmers.se

C. Gabrielii
e-mail: cecilia.gabrielii@chalmers.se

S. Brynolf
e-mail: selma.brynolf@chalmers.se

E. Svensson
Researcher, Gothenburg, Sweden
e-mail: e-svensson@outlook.com

M. Magnusson
Volvo Group Trucks Technology, AB Volvo, Gothenburg, Sweden
e-mail: mathias.magnusson.2@volvo.com

K. Andersson et al. (eds.), *Shipping and the Environment*,
DOI 10.1007/978-3-662-49045-7_5

greenhouse gases (such as carbon dioxide, methane and nitrous oxide), sulphur and nitrogen oxides, with both acidifying and eutrophication effects, and different forms of particles, with impacts on health and climate. However, not all emissions to the atmosphere from ships originate from the combustion of fuels for propulsion and energy production. The handling of crude oil as cargo and compounds used in refrigeration systems cause emissions of volatile organic compounds and ozone-depleting substances. The sources of the most important emissions and relevant regulations are described in this chapter.

Keywords
Marine diesel engines · Emission formation · Greenhouse gases · GHGs · Carbon dioxide · CO_2 · Sulphur dioxide · SO_2 · Nitrogen oxides · No_x · Particles · Refrigerants · Ozone-depleting substances

Air pollution consists of a variety of substances, which range from visible particles of smoke and soot as seen in Fig. 5.1, to gaseous molecules of sulphur and nitrogen oxides that are not detectable by eye. Dating to preindustrial times, the problem of air pollution was described as being primarily caused by the combustion of wood and coal for heating and cooking purposes. After industrialisation and with the subsequent increase in the use of fossil fuels, the problem with air pollution increased. Currently, air pollution is a well-known problem on local, regional and global scales. The World Health Organisation reported that in 2012, nearly 7 million people died as a result of air pollution exposure (one in eight of the total global deaths) [1].

Emissions to the air from shipping are not only globally recognised as a source of air pollution but also as a source of air pollution that should be considered at both the regional and local scales. Emitted gases and particles can be transported in the atmosphere over long distances, i.e., from sea to land and over national borders and continents [2]. Given that approximately 70 % of the emissions from maritime transport are emitted within 400 km of land, shipping is a source of air pollution to consider in discussions of air quality in coastal areas and impacts on human health and the environment [3]. The emissions from shipping affect the environment, climate, and human health. Emissions include climate-related gases, such as carbon dioxide (CO_2), methane (CH_4), and nitrous oxide (N_2O), and halogenated hydrocarbons. Emissions of sulphur oxides (SO_x) and nitrogen oxides (NO_x) contribute to acidification of land and sea areas and to the formation of secondary particles. NO_x contributes to eutrophication and, together with volatile organic compounds (VOCs), induce the formation of ground-level ozone, which negatively impacts human health and the environment. SO_x, NO_x and emitted particles also have a negative impact on human health due to respiratory diseases and premature deaths from cardiopulmonary diseases. Different types of particles have both direct and indirect impacts on the climate. The International Maritime Organization (IMO) convention that regulates air pollution and GHG emissions is written in bold in Table 5.1.

Table 5.1 The various IMO conventions that control the discharge, handling and use of different hazardous products on ships

IMO convention	Regulates	Adoption (year)	Entered into force (year)
MARPOL 73/78 Annex I	Oil and oily mixtures	1973	1983
Annex II	Liquid chemicals in bulk	1973	1983
Annex III	Liquid chemicals in packaged form	1973	1992
Annex IV	Sewage	1973	2003
Annex V	Garbage	1973	1988
Annex VI	**Air pollution and GHG emissions**	**1997**	**2005**
International convention on the control of harmful anti-fouling systems on ships	Anti-fouling paints	2001	2008
International convention for the control and management of ships' ballast water and sediments	Invasive species	2004	Not yet in force
Hong Kong international convention for the safe and environmentally sound recycling of ships	Recycling of ships	2009	Not yet in force
Nairobi international convention on the removal of wrecks	Removal of wrecks	2007	2015

Convention discussed in this chapter is marked in bold

Oceangoing ships emitted 600–900 Tg[1] CO_2 in 2000; shipping accounted for nearly 15 % and 4–9 % of the global NO_x and SO_2 emissions. Future shipping scenarios in Europe for the year 2020 predict that certain maritime emissions will have a reduced impact on human health and the environment compared with land-based emissions because they are primarily emitted far from populated areas or sensitive ecosystems. These simulations also found that in port cities, emissions from shipping are the major source of urban air pollution in many cases and must be investigated further because the EU air quality limits for air pollutants are an issue, especially for fine particles. An increase in emissions from ships will counteract the benefits from controls on land-based emission sources in Europe [4].

The majority of the emissions from shipping are emitted in a coherent plume of exhaust gases. At sea, the exhaust emissions from shipping are released in relatively clean areas of the atmosphere and close to the sea surface. This area of the atmosphere is often referred to as the marine boundary layer (MBL) and is largely influenced by the sea (moisture and temperature). The emissions are locally released at high concentrations of different substances compared with background concentrations in the atmosphere; these emissions are diluted with the ambient air

[1]Tg = Teragram = 10^{12} g.

Fig. 5.1 The exhaust from a plume contains a mixture of different gases and both liquid and solid particles. The exhaust is diluted in the air, although it can travel long distances and have impacts on both human health and climate. *Photo* Kent Salo

via mixing. During the dilution process, species are chemically transformed, secondary species are formed and certain species are removed by wet or dry deposition. These processes depend on the background concentrations, concentrations of the primary emissions and the actual meteorological state of the atmosphere [4].

Emissions of particles and gases in ship plumes are diluted in the horizontal and the vertical directions in the atmosphere by the wind relative to the ship and by mixing throughout the depth of the boundary layer, respectively [5]. Expansion of a plume depends on its age. A young plume can expand in both the horizontal and vertical directions because the height of the plume is less than the height of the MBL. An older plume only expands in the horizontal direction because the top of the plume has already reached the top of the MBL. This shift in expansion direction is observed in plumes with ages of approximately 1000s, which is the earliest stage at which ship tracks can be formed because emitted particles have reached the top of the MBL [6].

To evaluate emissions to the air from ship operations, specific emission factors (EFs) are used. These factors, which are primarily collected using on-board measurements, are specific for each air pollutant or greenhouse gas and vary with engine type, fuel consumption, fuel type, operational mode and the use of abatement technologies [7]. Two common methods are used to calculate emissions:

top-down and bottom-up calculations. In a top-down calculation, the different emissions are quantified for all ships in a certain geographical area; however, in bottom-up calculations, the different emissions are calculated for each ship during specific routes. A top-down calculation is a simpler method and saves time, although the disadvantage is that it does not account for all of the ship details, different fuel types and various operational modes [4, 8]. In bottom-up calculations, it is possible to use the AIS (automatic identification system) to obtain data on routes and speeds, which can be related to specific ship data through IMO numbers [9]. The results of these calculations can be used to evaluate the emissions from shipping, compare with other sources or as inventories in dispersion modelling. The specific emission factors can be represented using different units depending on the goal of the calculations. For comparison with land-based transportation, the units of g/tonnes and kilometre (g/tonne-km) are useful. For a comparison of different fuel types, the units of g/kg fuel and g/kWh are useful. In studies of particle emissions, the number of particles (PN) are of interest, and the specific emission factors used include the number (#)/kg fuel or #/kWh. The basic formula for calculating a specific emission (E) is E (g) = EF (g/kWh) * engine load (kW) * time (h). An example of calculations of emissions is shown in Box 5.1.

Box 5.1 An example of calculations of emissions from ships

A certain amount of goods must be transported between Norrköping and Gothenburg in Sweden. In this case, this transport can be performed using either one large ship or 263 trucks. The distances between the two cities are 931 km by sea and 320 km by land. The EFs used for the lorries are taken from the Euro VI standard [10]. The emissions from the ship are calculated for three different scenarios: Ship 1 runs on HFO with 1 mass% S in SO_x emission control area (ECA) before 2015; Ship 2 runs on MGO with 0.1 mass % S in SO_x ECA after 2015 [11]; and Ship 3 runs on LNG [12]. For all three scenarios, the auxiliary engines (AEs) run on MGO with 0.1 mass% S [13]. The EFs are presented in Table 5.2.

The calculated total emissions for this transport are presented in Table 5.3, and the emissions are presented as the total emissions in grams (g), except for the particle number (PN), The PN is presented as number of particles (#), where PM stands for particle mass.

For comparison with land-based transportation, the units of g/tonnes cargo and kilometre (g/tonne-km) are useful. Table 5.4 presents the emissions calculated in g/tonne-km for all emissions except for PN, which is in #/tonne-km. Ship transport requires a distance that is three times longer, which will affect the total emission in the example. The calculated emissions in Tables 5.3 and 5.4 illustrate that there are advantages and disadvantages regarding the emissions to the air from the different methods of transport and with different fuel types.

Table 5.2 Emission factors used in the emission calculations

	CO	CO_2	NO_x	THC	SO_2	PM	PN
	g/kWh						#/kWh
Road	4	950*	0.4	0.16	0.12×10^{-2}*	0.01	6×10^{11}
Ship 1	3.91	584	11.9	0.04	3.66	0.09	2.35×10^{16}
Ship 2	2.6	561	11.2	0.04	1.8	0.05	1.93×10^{15}
Ship 3	1.4	402	1.0	1.1	Not defined	0.23×10^{-3}	3.47×10^{12}
Ship AE	2.26	613	12.0	0.18	1.89	0.08	1.27×10^{16}

Values marked with * have units of g/km

Table 5.3 Calculated total emissions due to the transport of goods by road and ship

	CO	CO_2	NO_x	THC	SO_2	PM	PN
	Total emission in grams (g)						#
Road	3.5×10^5	8.0×10^7	3.5×10^4	1.4×10^4	1.0×10^2	8.8×10^2	5.3×10^{16}
Ship 1	5.2×10^5	8.6×10^7	1.7×10^6	9.9×10^3	4.8×10^5	1.3×10^4	3.1×10^{21}
Ship 2	3.5×10^5	8.2×10^7	1.6×10^6	9.2×10^3	2.4×10^5	7.4×10^3	2.8×10^{20}
Ship 3	1.9×10^5	6.1×10^7	2.8×10^5	1.5×10^5	Not defined	9.0×10^2	2.4×10^{19}

Table 5.4 Calculated emissions due to the transport of goods by road and ship

	CO	CO_2	NO_x	THC	SO_2	PM	PN
	g/tonne-km						#/tonne-km
Road	0.1	28	0.01	0.005	3.5×10^{-5}	3.1×10^{-4}	1.9×10^{10}
Ship, scenario 1	0.06	10	0.2	0.001	0.06	0.002	3.7×10^{14}
Ship, scenario 2	0.04	10	0.2	0.001	0.03	0.001	3.3×10^{13}
Ship, scenario 3	0.02	7	0.03	0.02	Not defined	1.1×10^{-4}	2.9×10^{12}

5.1 Marine Diesel Engines and Emission Formation

5.1.1 Marine Diesel Engines

The most common engines or prime movers used to drive ships and other marine structures are diesel engines or compression ignition engines, the main types of reciprocating internal combustion engines [14, 15]. The power of such engines ranges from 1 MW for small high-speed engines up to 80 MW for large low-speed engines. More details about the performance parameters of different marine diesel engines are presented in Table 5.5. The advantages of diesel engines as prime movers are that this type of engine has high reliability, high efficiency, low costs

Table 5.5 Performance parameters of marine diesel engines, state of the art 2001 [15]

Specific data	Diesel engines		
	Low speed	Medium speed	High speed
Process	2-stroke	4-stroke	4-stroke
Output power range (kW)	80,000–8000	35,000–500	9000–500
Output speed range (rpm)	80–300	300–1000	1000–3500
Fuel type	Mainly HFO	HFO or MDO	MDO
Specific fuel consumption (SFC) (g/kWh)	160–180	170–210	200–220

(with respect to initial and operational costs) and high maintainability (due to a simple and established technology) and is load flexible (can operate efficiently at a low load) and rather insensitive to the quality of the fuel. The advantage with fuel flexibility leads to the possibility that low-quality fuels can be burned and in fact are usually used; however, at the same time, high amounts of air pollutants are emitted [15]. The diesel engine is named after the inventor and operates according to the Diesel cycle. Another type of internal combustion engines is spark ignition engines, which operate according to the Otto cycle. The main difference between these two types of internal combustion engines or operating cycles is how the fuel is ignited inside the cylinder during combustion. In a diesel engine, the increase in temperature and pressure during compression is high enough to cause a spontaneous ignition of the fuel, whereas for the spark ignition engine, a spark ignites the fuel [14]. The primary energy source for large low-speed marine diesel engines is heavy fuel oil (HFO). HFO can be used as an energy source for medium-speed engines as well, or marine diesel oil (MDO) can be used. The primary energy source for high-speed engines is MDO [15]. Another fuel type used for medium- and high-speed engines is marine gas oil (MGO) [16].

Marine diesel engines can be designed as either two- or four-stroke engines. Two-stroke engines are mainly low-speed engines, whereas four-stroke engines represent medium- and high-speed engines (Table 5.5). The four-stroke cycle of this type of engine requires four strokes of the piston, i.e., two revolutions of the crankshaft, to complete the events that produce one power stroke. The four-stroke cycle starts with the induction/intake stroke, which is followed by the compression stroke and the expansion/power/working stroke, ending with the exhaust stroke (Fig. 5.2).

Two-stroke engines require two strokes of the piston to complete the events that produce one power stroke. These engines were developed from the four-stroke engines to achieve a higher power output from a given engine design and a simpler valve design. The drawback with the two-stroke engine is ensuring that the two strokes occur efficiently and completely fill the displaced volume in the cylinder with a fresh charge of air. Some of the fresh charge of air will be diluted with the exhaust residuals, which then flows out of the cylinder during the scavenging process. The two strokes in the two-stroke cycle are the compression stroke and the power/expansion stroke (Fig. 5.3). Note that Fig. 5.3 only shows a schematic view over a cylinder in a two-stroke engine. There are different scavenging arrangements, i.e., the location of inlet and exhaust ports.

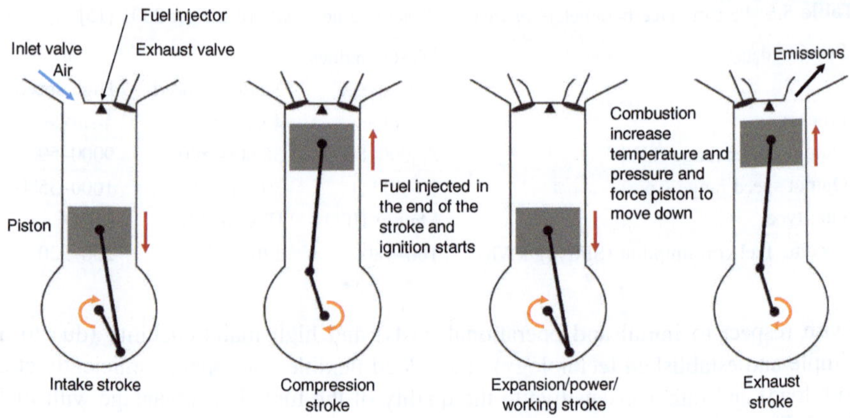

Fig. 5.2 Schematic picture of the four strokes in a four-stroke-cycle. *Red arrows* show the movement of the piston, *blue arrows* show the movement of air, and *black arrows* show the movement of emissions in the exhaust gases

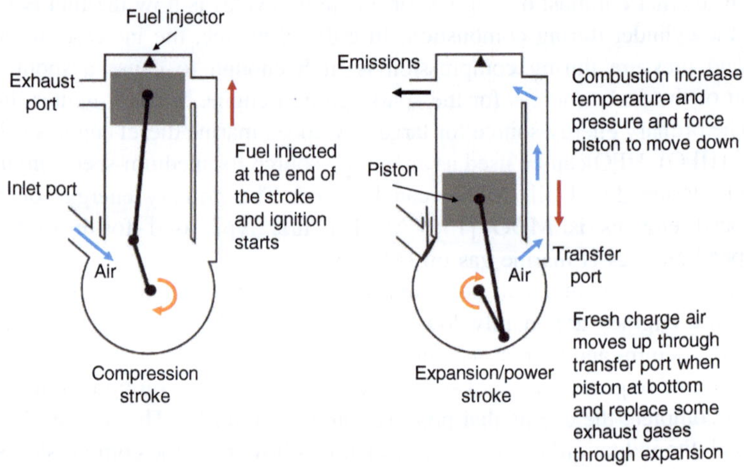

Fig. 5.3 Schematic picture of the two strokes in a two-stroke-cycle. *Red arrows* show the movement of the piston, *blue arrows* show the movement of air, and *black arrows* show the movement of emissions in the exhaust gases

5.1.2 Combustion Process in Diesel Engines

The combustion of the injected fuel into the cylinder starts at the end of the compression stroke and continues through the expansion stroke for both two- and four-stroke diesel engines. The fuel is injected through small orifices or nozzles as one or more jets with high velocity into the top of the cylinder. The injected fuel atomises into small droplets that can vaporise and rapidly mix with air of high temperature and pressure. This first phase of the combustion process, i.e., the time or crank angle interval between the start of injection to the initiation of combustion, is called the ignition delay. Chemical reactions may occur but very slow. The short period of ignition delay is followed by a phase called rapid or uncontrolled combustion. During this phase, the mixture of air and fuel, which occurs during the ignition delay, is ignited and causes a rapid increase in the pressure and temperature inside the cylinder. The rapid combustion is followed by diffusion or controlled combustion, which occurs with a so-called diffusion flame. During this phase, diffusion between the air and vaporised fuel occurs, and the rate of diffusion controls the combustion. The final phase of the combustion process is called final combustion and is a controlled combustion that is governed by diffusion; it occurs until all of the injected fuel and air are consumed [14, 17].

5.1.3 Thermochemistry Related to Combustion in Diesel Engines

Combustion of the mixture of fuel and air inside the cylinder is one of the processes that controls the efficiency, emissions and power of the engine. The mixing between fuel and air controls the combustion process of a diesel engine. The combustion process is a rapid exothermic, i.e., energy is released, gas-phase reaction. The reactions occurs in the reaction zone, also called the flame front, which is a zone that is very thin compared to the dimensions of the cylinder where the combustion occurs. The injected fuel is normally burned with air, which is 79 % nitrogen (N_2) and 21 % oxygen (O_2) on a molar basis; in the case of combustion, oxygen is the reactive component of air. The fuels used for diesel engines are a mixture of various compounds of hydrocarbons that are approximately 86 % carbon (C) and 14 % hydrogen (H) [17]. Marine fuels also contain considerable amounts of sulphur, and in HFO, many residual products such as ash, aromatics and minerals can also be found [8].

The general reaction for the combustion of fuel is that the reactants, here the fuel (hydrocarbons) and oxygen, react, producing the products carbon dioxide (CO_2) and water (H_2O) (Eq. 5.1).

$$\mathrm{Fuel} + O_2 \rightarrow CO_2 + H_2O \tag{5.1}$$

The chemical formula for a general fuel that contains hydrocarbons can be written as C_aH_b. Because air also contains nitrogen, this should be included in the reaction as well. An extended version of Eq. 5.1 is presented in Eq. 5.2, which shows the overall equation for the complete combustion of fuel.

$$C_aH_b + ((a+b)/4))(O_2 + 3.773N_2) \rightarrow aCO_2 + (b/2)H_2O + 3.773(a+b/4)N_2 \tag{5.2}$$

The stoichiometric (chemical correct or theoretical) proportions of fuel and air are defined in Eq. 5.2, i.e., the exact amount of oxygen needed to convert all of the fuel into completely oxidised products. Mixtures of fuel and air that contain more or less air than are required for stoichiometric conditions can also be burned. Excess air is called a fuel-lean mixture or lean conditions; the extra air added to the combustion is not involved in the reaction and can be found in unchanged form on the products' side in the reaction. Less air than required for stoichiometric conditions is called fuel-rich or simply a rich mixture or rich conditions, meaning that there is not enough oxygen to oxidise the carbon and hydrogen in the fuel to CO_2 and water. A mixture of CO_2 and H_2O with carbon monoxide (CO), hydrogen and nitrogen is found on the product side of the reaction. This means that the composition of the combustion products is different for fuel-rich compared to fuel-lean mixtures. Furthermore, the stoichiometric fuel/air ratio depends on the composition of the fuel. Therefore, to define the composition of the fuel-air mixture, the ratio of the actual fuel/air ratio to the stoichiometric fuel/air ratio can be used. The fuel/air equivalence ratio (Φ) is defined in Eq. 5.3; the relative air/fuel ratio (λ), i.e., the inverse of Φ, is defined in Eq. 5.4. For fuel-lean mixtures, $\Phi < 1$ and $\lambda > 1$; for fuel-rich mixtures, $\Phi > 1$ and $\lambda < 1$. For a stoichiometric mixture, both Φ and λ are equal to 1.

$$\phi = \frac{(Fuel/Air)actual}{(Fuel/Air)stoichiometric} \tag{5.3}$$

$$\lambda = \phi^{-1} = \frac{(Air/Fuel)actual}{(Air/Fuel)stoichiometric} \tag{5.4}$$

The formation of different emissions inside the cylinder depends on the distribution of the injected fuel and how this distribution changes with time in the cylinder due to mixing. During combustion, zones with lean- and rich-mixtures are found in the cylinder, and the fuel/air equivalence ratio (Φ) has an impact on the formation of various emissions in a diesel engine [17]. This will be used to describe the combustion conditions when the formation of primary nitrogen oxides (NO_x) and particles are discussed (see Sects. 5.4 and 5.5).

5.2 Greenhouse Gases (GHGs)

As described in Sect. 2.2, the greenhouse effect of Earth's atmosphere maintains the temperature at a reasonable level to sustain life. If the concentrations of greenhouse gases (GHGs) in the atmosphere are altered, these changes will influence the temperature. Globally, the most significant GHGs are CO_2 (primarily from energy use and deforestation), methane (CH_4; primarily from agriculture, waste and energy use), and nitrous oxide (N_2O; primarily from agriculture).

Over the last decade, several attempts have been made by individual researchers to assess total GHG emissions from shipping and by consortia that have produced reports for national, regional and international bodies. Several methods have been used to determine the GHG emissions from shipping and have provided differing results and associated uncertainties [18]. These methods can be divided into "bottom-up" and "top-down" methods. The former category refers to an analysis based on the activity of individual ships (e.g., using AIS data) coupled with technical assumptions. The latter method is based on international sales statistics of bunker fuel. Bottom-up methods are viewed as being more reliable in the literature [19, 20].

In the third GHG study commissioned by IMO [19], Smith et al. estimated that shipping contributed 972 million tonnes of CO_2, or 2.8 % of the global emissions in 2012. One of the reasons for this latest update was that previous studies did not anticipate the global economic crisis.[2] Indeed, one of the main results was that the economic crisis had a significant short-term effect. From 2007 to 2012, ships exhibited substantially reduced average speeds, resulting in a decrease in total emissions of approximately 13 %.

The problem with the shipping sector's impact on climate change has a longer time scale. Although forecasting the future is notoriously difficult, understanding the future contribution of shipping to climate change is used to assess the expected impact of policy changes [21]. A rough understanding of the current and future emissions from shipping can be gained from studying the main contributing factors as in Eq. 5.5:

$$\mathit{Total\ GHG\ emission} = \mathit{GDP} * \frac{\mathit{Transport\ work}}{\mathit{GDP}} * \frac{\mathit{Energy}}{\mathit{Transport\ work}} * \frac{\mathit{GHG\ emissions}}{\mathit{Energy}} \tag{5.5}$$

where GDP is the size of the global economy; the second term is the transport intensity of the economy, i.e., the amount of cargo transported over a certain distance (for example tonne * nautical mile) necessary to maintain the GDP; the third term is the energy required to perform that transport work (i.e., energy efficiency); and the final term is the amount of GHGs associated with each unit of energy that is used. Because the GDP is expected to continue to rise exponentially and the maritime transport work necessary for maintaining that economy has been nearly constant over the past few decades,[3] reducing total GHG emissions can be

[2] "The current estimate does not take account of the economic downturn experienced globally since 2009" (IMO, MEPC 64/5/5, Annex, p. 1).

[3] The use, extraction and transport of materials is inherent to our economic system [22]. Globally, material use increased by a factor of 8 during the last century, slightly slower than economic development, although faster than population growth [23]. Although developed regions have adopted strategies for dematerialisation—using less materials for the same economic output—developing regions of the world still have a need for raw materials to construct roads, buildings, ports, railways and other forms of infrastructure, and a growing number of middle-class citizens will buy more goods. Per capita material and energy use is 5–10 times higher in developing regions than in developed regions. It has been forecasted that energy and material use will grow by a factor of 2–3 in the coming decades as agrarian regions become industrial [24].

viewed foremost as a matter of improving energy efficiency (the third factor) and switching fuels to less GHG-intensive fuels (the fourth factor).

Projections have demonstrated that it is very difficult to find scenarios in which total emissions from shipping are reduced. Even assuming a 60 % increase in energy efficiency, low economic growth and increased use of LNG in shipping (which is likely to have lower GHG emissions from a life-cycle perspective; see Chap. 10), the third GHG study did not find that total emissions from shipping would decrease from today's levels by 2050 [19].

Future scenarios for total global emissions vary widely depending on the required certainty of the projection with respect to increased temperatures and based on the assumed point in time at which the world reaches a peak in emissions. The concept of a carbon budget has proven useful for such a discussion; to reach a given temperature increase with a given certainty, the total amount of anthropogenic carbon emitted to the atmosphere can be calculated. This concept means that the longer it takes for global emissions to peak (and they are still exponentially increasing), the more sharply emissions will need to be reduced after the peak. In scenarios with a high probability of not exceeding a 2 °C warming, shipping could constitute 15–25 % of the available global budget in 2050.[4]

5.2.1 Sources

5.2.1.1 Carbon Dioxide

The main contributor to greenhouse gas emissions from shipping is CO_2, which is formed from the combustion of the carbon in the fuel used for propulsion and from energy and heat production on ships. Therefore, CO_2 emissions are directly connected to a ship's fuel consumption. Reducing CO_2 emissions is foremost a matter of improving energy efficiency or changing to renewable fuels.

5.2.1.2 Methane

As described in Sect. 10.4.2.2, the interest in the use of liquefied natural gas (LNG) as a shipping fuel has increased with existing and upcoming regulations on the quality of marine fuels, which are designed to reduce SO_2 emissions. Produced from natural (fossil) gas, LNG and its composition depend on the origin of the gas, although a major component (90–95 %) is methane. The lower carbon content of LNG compared with common marine fuels, such as HFO, MGO and MDO, leads to reduced carbon dioxide (CO_2) emissions by 20–25 %. However, possible spills and slippage of methane (CH_4) during the handling and combustion of LNG could increase the shipping-related contribution of GHGs because CH_4 is a potent GHG (see Table 2.5) with a GWP that is more than 20 times higher than that of CO_2. In

[4]Meinshausen et al. [25] showed how the probability of not exceeding a 2-degree warming ranged from 85 to 36 % when emissions ranged from the equivalent of 10 Gt CO_2 per year to 36 Gt CO_2 eq per year by 2050.

addition, the processing of natural gas, which is necessary to produce LNG that is suitable for marine fuel, is complicated and consumes large quantities of energy.

5.2.1.3 Nitrous Oxide

Nitrous oxide (N_2O) (also known as "laughing gas") is a powerful greenhouse gas that has a long residence time in the atmosphere. N_2O is formed during the combustion of fuel under certain conditions. This topic is described in detail in Box 5.3.

5.2.1.4 Halocarbons

Halocarbons i.e., halogenated hydrocarbons, which are commonly used as refrigerants, are notably strong greenhouse gases with GWPs of up to 10,000; the concept of GWP is explained in Sect. 2.7.4. The contribution due to shipping of halocarbons acting as GHGs is small, although halocarbons have potential ozone-depleting capabilities; this topic is further described in Sect. 5.7.

5.2.2 Human and Environmental Implications

In 1988, the United Nations Environment Programme (UNEP) and the World Meteorological Organization (WMO) created the International Panel for Climate Change (IPCC) to regularly review climate change science and produce review reports [26]. In their fifth and latest review, the IPCC asserted an "extreme likelihood" (>95 %) that humans have "exerted a substantial warming influence on climate" since 1750. Moreover, since 1950, it is "exceptionally unlikely" (<1 %) that natural processes have had an influence comparable to that attributable to processes that are anthropogenic [27].

Large differences once existed between what are often denoted as developing countries and developed countries in terms of energy use and CO_2 emissions. Per capita emissions are still highly skewed [28] (developed countries typically have larger emissions per capita), although developing countries have now overtaken developed countries in terms of *total* emissions [29]. Despite the economic downturn of 2009, the increase in CO_2 emissions from the combustion of fossil fuels was the largest ever recorded. Emissions decreased in developed countries and increased in developing countries.

Compared to other forms of adverse impacts that humans have on the environment, such as acidification and eutrophication, the effects of CO_2 and other GHGs persist over long periods of time. Even if societies were to rapidly decrease or even halt CO_2 emissions, the concentration in the atmosphere decreases much slowly. Increases in temperature, changes in precipitation, and rising sea levels are expected to be irreversible, at least in the coming 1000 years [30].

Currently, the global mean temperature is nearly 1 °C warmer than pre-industrial levels. The effects of climate change are visible, e.g., in the form of shrinking glaciers, melting permafrost and costal erosion and in changes to biological systems [31]. Certain processes are occurring faster than earlier predictions; the Arctic is melting faster than forecasted in the latest report of the IPCC [32] and might be ice free during summertime in 30 years [33]. In many business-as-usual scenarios, the global mean temperature might rise by 5 °C, which is comparable to the difference between present day and the last ice age. An equally large additional temperature increase is expected to be devastating [34]. Even with a 2 °C warming, the goals set by nations in the Copenhagen Accord[5] will have a profound effect. For example, the temperature increase required to preserve more than 10 % of global coral reefs is approximately 1.5 °C.

Effects have already been observed in the maritime environment. Decreased ocean productivity, altered food web dynamics and a greater incidence of disease have been noted [35]. Larger effects are expected in oceans located in polar regions and in the tropics due to decreased ice levels and perturbations in coral ecosystems [36]. Increased acidity due to the deposition of atmospheric CO_2 in the ocean combined with higher temperatures [37] will force marine ecosystems into conditions that have not appeared for hundreds of millions of years (if ever). Such changes were previously associated with mass extinction events [38]. Sea level rise will have profound effects on many communities [39], especially small island states, which are likely to be submerged [40].

Climate change is expected to also directly affect shipping through more severe weather and new trade routes. As the Arctic ice grows smaller, especially in the summer, shorter routes from China to Europe will become available through the Arctic [41] (see Box 1.1 in Chap. 1 and Sect. 12.2.4).

5.2.3 Regulations

Together with civil aviation, the shipping sector is a globally special case with respect to GHG emissions. Instead of being handled in global diplomatic discussions through the UN Framework Convention for Climate Change (UNFCCC), this task was handled through the Kyoto Protocol. The role of shipping in the mitigation of GHG emissions depends on many factors, e.g., the timing of global regulations for the rest of the world, the resulting contribution of other sectors in magnitude and time, the emissions from the current shipping sector and its projection into the future based on the likely demand for transportation, and the available potential for improvement and associated costs.[6] However, this is ultimately a political question that can only be answered via negotiations in different international, regional and

[5]The Copenhagen Accord is a non-legally binding document presented and taken note of at the 15th meeting of the Conference of The Parties (COP) to the Framework Convention on Climate Change (FCCC) in Copenhagen, 2006.

[6]In certain contexts, the word "reduction" is used to indicate reduction vis-á-vis a business-as-usual baseline and not a reduction of total emissions. It is important to note that the issue of climate change concerns not efficiency per se, but whether efficiency can be improved at a sufficiently fast rate, i.e., faster than the growth of the demand for transport work by sea.

national fora. In the shipping sector, discussions on policy instruments that could mitigate the climate impacts of shipping that occur primarily in IMO and the EU.

GHG emissions first appeared on the IMO's table in 1991. It was argued by the UK delegation that "consideration should be given to accurate monitoring and forecasting of carbon dioxide emissions from shipping and the difficulty of allocating responsibilities to individual states".[7] However, no action was taken until much later, until the Kyoto Protocol was linked to the UNFCCC, wherein countries were assigned the task of "reducing or limiting" GHG emissions from shipping. This agreement was a compromise solution because countries could not agree on how to apportion emissions from either aviation or shipping to individual countries [26] and was applied through the following text:

> The Parties included in Annex I shall pursue limitation or reduction of emissions of greenhouse gases not controlled by the Montreal Protocol from aviation and marine bunker fuels, working through the International Civil Aviation Organization and the International Maritime Organization, respectively.[8]

The implications of "pursue" and "work through" have been extensively discussed in IMO meetings. Diplomatic discussions between countries have been reported to be in a "deadlock" due to disagreement between the interpretation of two basic concepts underlying IMO and the Kyoto protocol [44]. First, the concept of common but differentiated responsibilities (CBDRs) is a fundamental component of the UNFCCC:

> The Parties should protect the climate system for the benefit of present and future generations of humankind, on the basis of equity and *in accordance with their common but differentiated responsibilities* and respective capabilities. Accordingly, the developed country Parties *should take the lead in* combating climate change and the adverse effects thereof.[9]

In the IMO regulation, the concept of no more favourable treatment (NMFT) is universally used. A port state can apply IMO legislation to any ship entering its waters or ports. The IMO secretariat has made it clear that they have "not identified any potential treaty law conflict between the Kyoto Protocol and the provisions that may be developed by the Committee on GHG emissions from the combustion of

[7]Report from the BCH subcommittee, MEPC 32/12, quoted in Strong [42]. This issue has still not been resolved. Any apportionment scheme for a nation or region must be sufficiently accurate to capture the implementation of abatement measures and must also be directed to those aspects that are within the sphere of influence of the same entity. Heitmann 43 reviewed apportioning regimes and concluded that allocation based on the nationality of the commercial operator is most fair because operators "have the most control over the emission levels of their ships by regulating speeds and routes", although no single option can be thought of as "fair" [43]. The Kyoto protocol does not in any way hinder countries from including marine bunker fuel in their national emission inventories and might even provide countries with further motivation to take action themselves and increase pressure on the IMO process [26].

[8]Article 2.2, Kyoto Protocol.

[9]Article 2, Framework Convention on Climate Change, our emphasis.

marine bunker fuels".[10] Among other reasons, they stated that "pursue limitation… is not the same as limiting the outcome of IMO's decision-making process to application to Annex I countries exclusively". Finally, it has been concluded that shipping regulation must be global in scope for both principle and in practical matters:

> It is due to the complexities of the international shipping trade (i.e., the interaction of private and public law in connection with registration; the right and obligation to fly a flag; and the further interaction between flag, port and coastal State jurisdiction) that IMO shipping regulations are, as a matter of principle, and must be, as a practical matter, global in nature and applicable to all commercial ships, with appropriate differences, if any, to be based on factors such as their type, structure, manning and operational features, irrespective of the flag they are flying or the degree of industrial development of the flag State or the State of nationality of the owner or the operator.[11]

A compromise solution has been argued for because only applying the NMFT principle might weaken the negotiating position of developing countries in global climate discussions [45]. Several approaches have been proposed to reconcile these principles. As an example, an instrument could be globally applied with economic compensation to non-Annex I countries. Alternatively, the application could be limited to passages to Annex I countries with or without compensation to other parties [44].

In 2008, countries in IMO agreed to introduce two measures to address the sector's GHG emissions: the Ship Energy Efficiency Management Plan (SEEMP) and the Energy Efficiency Design Index (EEDI). The first measure is directed towards encouraging better management practices in energy use on ships, whereas the latter is a design standard for new ships. The EEDI is described in Box 5.2, and the SEEMP is extensively discussed in Chap. 10. A report to IMO demonstrated that these measures will not be sufficient to reduce total emissions from the sector; instead, only slow growth is expected. Further measures, primarily market-based tactics in the form of a tax on bunker or an emissions trading scheme (as examples), have been stalled due to the CBDR conflict [44].

The perceived lack of progress in IMO has spurred discussions for at least a decade on the possibilities for regional and even national policies. For example, the EU has pushed forward with its own regional monitoring, verification and reporting (MRV) scheme. The European Commission (EC) also expects a short-term improvement in energy efficiency of approximately 2 % as valid and reliable data sets on ship-based energy consumption become available in shipping markets and within shipping organisations [46]. For the longer term, the EC has expressed the

---

[10]In "Review of proposed market-based measures: relation to relevant conventions and rules". The organization's work on GHG emissions and the United Nations Framework Convention on Climate Change and its Kyoto Protocol. Submitted by the IMO secretariat to the Third Intersessional Meeting of the Working Group on GHG Emissions from Ships, 2011. Document number GHG-WG 3/3/9.

[11]GHG-WG 3/3/9.

intent to combine the scheme with a market-based measure, such as an emissions trading scheme. A concurrent discussion on reporting emissions from shipping is taking place in IMO, and the EU has emphasised that they will merge their plans with those of IMO if an international solution is brought forward.

In summary, much remains to be achieved in terms of regulation if shipping is to reduce its total emissions. The EU has set a goal of at least 40 % reduction from 2005 levels by 2050 [47], and researchers have argued for the need of even larger cuts of up to 85 % by 2050 [48].

Box 5.2. Instruments used to mitigate GHG emissions from shipping: EEDI and SEEMP

The Energy Efficiency Design Index (EEDI) sets a limit on the energy efficiency of new vessels based on a measure of the transport work (e.g., tonnes-miles) it can produce compared with the energy it requires. All new ships exceeding 400 GT must carry a document detailing its *attained* EEDI, which must be lower than a pre-set *required* EEDI. These required levels are subsequently decreased over time. The attained EEDI is calculated according to the following formula:

$$EEDI = \left(\prod_{j=1}^{n} f_j \sum_{i=1}^{n_{ME}} P_{ME(i)} \cdot C_{FME(i)} \cdot SFC_{ME(i)} + (P_{AE} \cdot C_{FAE} \cdot SFC_{AE}) \right.$$
$$+ C_{FAE} \cdot SFC_{AE} \prod_{j=1}^{n} f_j \sum_{i=1}^{n_{PTI}} P_{PTI(i)} - \sum f_{eff(i)} \cdot P_{AEeff(i)}$$
$$\left. - \sum_{i=1}^{n_{eff}} f_{eff(i)} \cdot P_{eff(i)} \cdot C_{FME} \cdot SFC_{ME} \right)$$
$$\cdot \frac{1}{f_i \cdot f_l \cdot f_w \cdot f_c \cdot capacity \cdot v_{ref}}$$

or EEDI = (CO_2 emissions from main engine(s) + CO_2 emissions from auxiliary engines—CO_2 emission reduction due to innovative technology(s))/ transport work.

The required EEDI reference levels have been calculated based on historical data on ships divided into different categories (e.g., container and bulk). For most ship types, the required levels were decreased from the initial reference level by 10 % in 2015, 20 % in 2020 and 30 % in 2025. A report commissioned by IMO showed that the combination of the EEDI and the SEEMP should not be expected to be sufficient to stop the growth of emissions from the shipping sector [21]. Both strengthening the requirements of the SEEMP and sharpening the EEDI limits have been proposed. However, only a limited discussion on this topic in the scientific literature has occurred.

It has been shown that the final SEEMP was quickly watered down into its current form, which lacks many typical management system features, although initial proposals to IMO for the SEEMP were more ambitious and connected to the International Safety Management (ISM) Code that regulates safety management systems in shipping companies [49]. Suggestions to strengthen the SEEMP have appeared as a submission to IMO [50] and in reports to the European Commission [51]. It has been argued that the EEDI limits were set based on a data set composed of ships that had been optimised for (high) speeds and carrying capacities rather than energy efficiency. Many recently constructed ships were argued to be less efficient than those built in the 1990s [52]. As a result, new ships have been shown to meet EEDI targets with a large margin, even those set for 2030 [53].

5.3 Sulphur Oxides

Since the 1970s, much attention has been focused on reducing SO_x emissions, primarily in Europe and North America. For many years, this attention was directed at an environmental problem known as "acid rain", i.e., the *acidification* of lakes, watercourses and soils caused by the deposition of air pollutants [54]. The abbreviation SO_x is often used for sulphur dioxide (SO_2) and sulphur trioxide (SO_3), although nearly all sulphur is emitted as SO_2. In many areas, SO_2 is the primary air pollutant that causes acidification. Other pollutants include nitrogen oxides (NO_x) and ammonia (NH_3). Recently, additional emphasis has been placed on problems associated with the atmospheric formation of particles from SO_x emissions. These sulphate particles have impacts on human health, visibility and climate, although the latter is due to a cooling effect (see further information in Sect. 5.5 [54]. Other impacts of SO_x emissions include significant damage to buildings and structures, which incur economic costs [56].

By examining the developments of global SO_2 emissions from all types of anthropogenic sources dating to 1850, studies have found that global emissions peaked in the 1970s and subsequently decreased significantly. This decrease was followed by a global increase during the period 2000–2005; however, global emissions subsequently decreased up to 2011 [55, 57]. A substantial portion of the decrease (up to 2005) was due to considerable emission reductions from land-based sources, primarily in Europe and North America [55]. These reductions resulted from abatement measures enabled by the 1979 Convention on Long-range Transboundary Air Pollution (CLRTAP) and the air pollution policies of the US and the EU [54, 58, 59]. In the EU-27[12], SO_x emissions decreased by 82 % between 1990 and 2010 [60]. The corresponding figure for the US was approximately 61 % [61].

[12]EU-27 represents the Member States in the European Union from 1 January 2007 to 30 June 2013

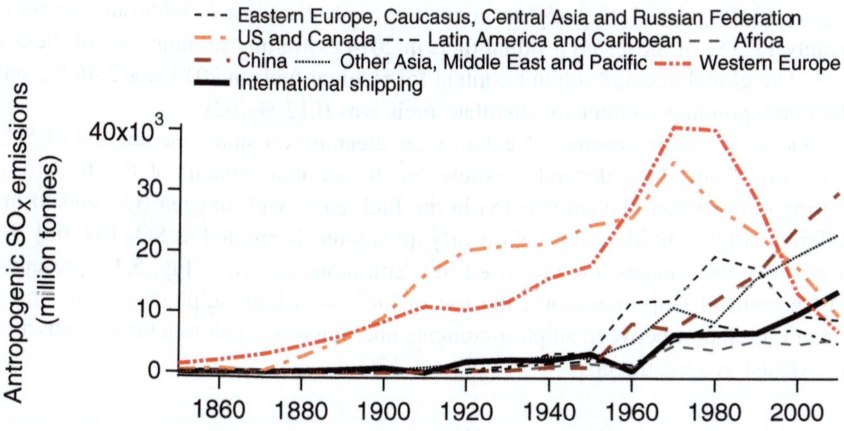

Fig. 5.4 Global anthropogenic SO_2 emissions from 1850 to 2010 from world regions and international shipping (data for 1850–1980: Smith et al. [55], data for 1990–2010: Klimont et al. [57]). It is acknowledged that these data are estimations and that these two studies use different estimation methods for the years 1990–2005

Although China (and the developing countries in general) represented a major portion of the increase between 2002 and 2005, emissions from international shipping accounted for a significant and increasing contribution [55]. The black line in Fig. 5.4, which shows estimates of global anthropogenic SO_2 emissions from 1850 to 2010, illustrates this contribution. Notably, Fig. 5.4 shows a sharp increase in emissions from international shipping from 1990 to 2010. Based on these data, global emissions from international shipping in 2010 were higher than the European and North American emissions. However, it should be noted that the figures for international shipping are estimated and contain large uncertainties [57].

5.3.1 Sources

In general, SO_x emissions are primarily associated with relatively high sulphur contents in fossil fuels [54, 55]. Crude oil naturally contains sulphur due to its origin from plant and animal material. The sulphur contents are even higher in the heavy fuel oils (HFOs or residual fuels) used by ships than in the original crude oil because the sulphur is enriched in the heaviest fractions during the refining process (see Box 2.4 in Chap. 2). In 2007, nearly 70 % of the fuels used by ships were HFOs, and the remaining were distillate fuels. [20] The sulphur content of HFOs depends on the sulphur content of the crude oil, which varies in different regions of the world. In general, the sulphur content exceeds 1 mass%,[13] with only approximately 12.3 % of the tested residual fuels in 2014 exhibiting sulphur contents below 1 %. The global average sulphur content for residual fuels in 2014 was 2.46 %, and the corresponding content for distillate fuels was 0.12 % [62].

Due to a historic absence of exhaust gas cleaning on ships, the amount of SO_x emissions from ships depended solely on the sulphur content of the fuel [63]. During combustion, the sulphur (S) in the fuel reacts with oxygen (O_2) and forms different sulphur oxides, although nearly all sulphur is emitted as SO_2 [64, 66]. To summarise the sources, the increased SO_x emissions shown in Fig. 5.4 represent a vast growth of ship operations, the use of fuels with high sulphur contents and a direct correlation between sulphur contents and emissions due to a historic absence of exhaust gas treatments.

5.3.2 Transboundary Impacts

Most gases that enter the atmosphere from both natural and anthropogenic sources undergo oxidation because they exist in a chemically reduced state. The chemical conversions that occur in the atmosphere change the properties of the emitted substances, affecting their lifetimes, transport distances and deposition rates. Oxidation of SO_2 emissions occurs in the gas phase, in cloud droplets, or on the surface of aerosol particles. In the gas phase, SO_2 is oxidised by hydroxyl (OH) radicals. The result is an adduct, i.e., $HOSO_2$, that is oxidised into SO_3, which reacts with water (H_2O) in the atmosphere to form sulphuric acid (H_2SO_4) [66]. Acidic gases and particles formed from SO_2 emissions can be transported to the surface and become absorbed or adsorbed by land surfaces, materials or water surfaces in a process known as *dry deposition.* When dissolved in clouds, fog, rain or snow, these substances are deposited on land or water surfaces via *wet deposition.* In water, i.e., in rain or snow, sulphuric acid is dissolved into two hydrogen ions ($2H^+$) and one sulphate ion (SO_4^{2-}) [66].

As mentioned previously, atmospheric conversions affect the distance over which a pollutant can be transported; in the case of SO_2 emissions, this process enables transport at long distances, at least the portion that is deposited by wet deposition. In general, the importance of dry deposition decreases with distance from the source, and the importance of wet deposition increases [67]. In this regard, wet deposition of sulphur can be viewed as the transboundary component of SO_x emissions. The term *transboundary* indicates the crossing of national or other political borders, and SO_2 and its atmospheric oxidants can be transported over long distances, i.e., from hundreds of km to more than 1000 km before deposition [63, 65, 67, 68]. Therefore, SO_x belong to the category of *long-range*

---

[13] Percentage by mass, i.e., $m_{sulphur}/m_{fuel}$.

transboundary air pollution, which is defined in Article 1b of the 1979 Convention on Long-Range Transboundary Air Pollution (CLRTAP) [69]:

> air pollution whose physical origin is situated wholly or in part within the area under the national jurisdiction of one State and which has adverse effects in the area under the jurisdiction of another State at such a distance that it is not generally possible to distinguish the contribution of individual emission sources or groups of sources (Art. 1b) [69].

With respect to land-based emission sources, transboundary air pollution can be understood in the way that countries send out emissions from their territories via direction- and distance-dependent factors, such as wind currents. Certain countries could be net polluters and others net recipients of these pollutants [70] (see further information on acidification in Sect. 2.7.8.2). This is also an important aspect for understanding the impacts caused by SO_x emissions from ships and the need for measures beyond national borders. Approximately 70 % of the emissions from ships occur within 400 km of land. Due to the transboundary nature of SO_x, ships are capable of contributing to pollution hundreds of km inland [4, 63].

To understand the contribution of ship emissions to acidification on land, it is important to separate the *contribution to total emissions* (as shown in Fig. 5.4) and the *contribution to the deposition of sulphur* (which causes acidification). The contribution of ships to the wet deposition of sulphur in northwestern North America and Scandinavia has been estimated to be 15–25 %. For southwestern Europe and northwestern Africa, the corresponding contribution has been estimated to be 15–20 % [71]. Scenarios have suggested that by 2030, the benefits from reduced land-based SO_2 emissions would be significantly counteracted by increasing ship emissions in several areas, such as Europe [4].

As highlighted in Chap. 2, the extent of acidification impacts caused by sulphur depositions depends on natural buffering capacities. Depending on geological characteristics, certain environments are more sensitive than others. Scandinavia is often highlighted as a typical example of a sensitive environment for acidification. Taking Sweden as an example, it is one of the European countries where the ecological damage caused by acidification has been most evident and severe. [56] Over 90 % of the sulphur deposited over Sweden in 2012 originated from foreign emission sources, and shipping has been identified as the largest single source of these depositions [72].

The political importance of reducing SO_x emissions from ships has thus been particularly high in States with sensitive environments and where considerable measures have been taken to reduce land-based sources via cooperation with other States. When viewed in terms of the more recent awareness of the health impacts, the importance of reducing SO_x emissions from ships has shifted from a regional dimension toward a more global dimension in which ship emissions are represented globally along the coastlines of the major international trade routes [73].

5.3.3 Regulations

SO_x emissions from ships are addressed in Regulation 14 in MARPOL 73/78 Annex VI (see Sect. 3.3.4). Regulation 14 sets a global limit on the sulphur content of bunker fuels and stricter limits in certain areas that are referred to as SO_x Emission Control Areas (ECAs). This division of a global limit and a stricter regional limit should be viewed in the context of concerns over acidification in countries with particularly sensitive environments, which was highlighted previously. Most importantly, however, it should be viewed as a compromise after negotiations to reach a global solution in IMO. The negotiations were characterised by a strong focus on high costs to the oil industry of a strict global regulation [73].

Drafted in this spirit of compromise, Regulation 14 initially limited the sulphur content to 4.5 %[14] globally and 1.5 % in SO_x ECAs. An alternative to reduce SO_x emissions in SO_x ECAs is the use of an exhaust gas cleaning system or other onboard abatement technologies [88] (see Sect. 11.8). During the period from the adoption of Annex VI in 1997 to its entry into force in 2005, SO_x emissions from ships increased due to increased shipping activity. Moreover, awareness of the health impacts associated with particles formed from SO_x emissions also increased, primarily in the EU and the US [73]. In 2005, the Marine Environment Protection Committee (MEPC) of IMO (see Sect. 3.4.1) consequently decided to start a revision process. The revised Annex VI was adopted in 2008 and entered into force on the 1st of July 2010 [75].

With the revised Annex VI, the sulphur content of bunker fuels was to be reduced progressively from 2010 to 2020, as illustrated in Fig. 5.5. The initial 4.5 % global sulphur limit was first reduced to 3.5 % on 1 January 2012, and this target is to be followed by another reduction to 0.5 % on 1 January 2020. However, the latter date depends on the results of a review in 2018 on the availability of compliant fuel oil for 2020. If the parties to Annex VI conclude from this review that it is not possible for ships to comply by 2020, it can be decided to extend the date to 1 January 2025. For SO_x ECAs, the sulphur limit was first reduced to 1.0 % on 1 July 2010 and subsequently to 0.1 % on 1 January 2015 [88]. Additional environmental aspects relating to the use of exhaust gas cleaning systems, such as scrubbers, are addressed in Sect. 11.8.

Since the entry into force of the revised Annex VI, the SO_x ECA concept has been expanded to include NO_x, SO_x and PM[15] under the term ECA [76]. Currently, ECAs exist in the Baltic Sea, the North Sea, the English Channel, around the coastlines of North America and in a small area in the Caribbean Sea that are controlled by the US [77]. Table 5.6 shows the existing ECAs with adoption, entry into force and effective dates.

---

[14]Percentage by mass.

[15]No specific regulation for PM emissions has been adopted in MARPOL Annex VI, although PM emissions became indirectly regulated by the adoption of Regulation 14 [73].

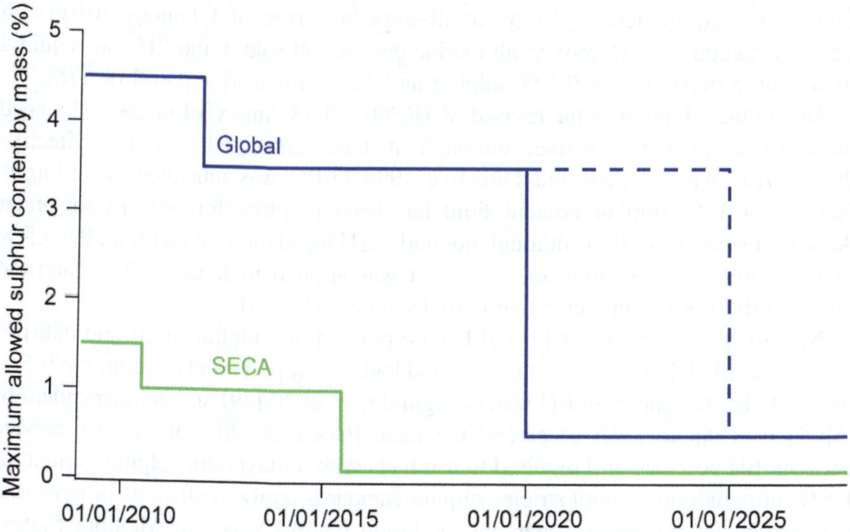

Fig. 5.5 Maximum permitted sulphur contents in bunker fuels (Regulation 14 of the Revised MARPOL 73/78 Annex VI, 2008). The global limit of 0.5 % sulphur by mass might be postponed to 1 January 2025, depending on a review in 2018; this is represented by the *dashed line*

Table 5.6 Key dates and emissions for ECAs [77]

ECA	Included emissions	Adopted	Entry into force	Effective
Baltic Sea	SO_x	26 Sept 1997	19 May 2005	19 May 2006
North Sea	SO_x	22 Jul 2005	22 Nov 2006	22 Nov 2007
North American	SO_x, NO_x and PM	26 Mar 2010	1 Aug 2011	1 Aug 2012
United States Caribbean Sea ECA	SO_x, NO_x and PM	26 Jul 2011	1 Jan 2013	1 Jan 2014

In addition to MARPOL 73/78, the European Union Directive 2012/33/EU deserves particular attention. This directive incorporated Regulation 14 of the revised MARPOL 73/78 Annex VI into EU legislation regarding the sulphur content in marine fuels. The EU's legislation on the sulphur content of certain liquid fuels dates back to 1993 with Directive 93/12/EC, although Directive 1999/32/EC replaced the former and provides the foundation for the later amendments in 2005 and 2012. Initially, Directive 1999/32/EC included limits on the sulphur content in MDO and MGO used by ships in EU territorial waters and inland waterways. The 2005 amendment adopted by Directive 2005/33/EC implemented the sulphur regulations in MARPOL 73/78 Annex VI, which entered into force in 2005 and included a 1.5 % sulphur content limit in SO_x ECAs. This policy further added a limit of 1.5 % sulphur content for passenger ships in regular service between EU ports and a requirement of 0.1 % for all ships in port as of 1 January 2010 (with selected exceptions). Moreover, all marine gas and oil sold in the EU was required to contain a maximum of 0.1 % sulphur and 1.5 % for marine diesel oil [78].

Due to the adoption of the revised MARPOL 73/78 Annex VI in 2008, the need arose to adapt the directive, although it took until 2012 before Directive 2012/33/EU was adopted and Directive 1999/32/EC was amended accordingly. Notably, a 1 % sulphur content limit has been in place for SO_x ECAs under Annex VI since July 2010, although not under EU legislation. In addition, the 3.5 % global sulphur content limit of Annex VI was applied in January 2012, and the directive did not become effective until 18 June 2014 [78].

Nonetheless, Directive 2012/33/EU incorporated the sulphur limits and dates of the revised MARPOL 73/78 Annex VI and had one important exception: the 0.5 % limit is to be mandatory in EU waters beginning in 2020 [79] and is independent of IMO's decision to keep or extend the date. Passenger ships in regular service between EU ports are still required to use fuels with a maximum sulphur content of 1.5 %, although only "until stricter sulphur standards apply to all ships in territorial seas, exclusive economic zones and pollution control zones of Member States" (Para. 14, preamble) [79]. In addition, the directive includes a general ban on the use of marine fuels with sulphur contents exceeding 3.5 % within a member State's territory "except for fuels supplied to ships using emission abatement methods subject to Article 4c operating in closed mode" (Art. 3(a)) [79] For further information on closed loop scrubbers and other abatement methods see Chap. 11. Additionally, the 0.1 % sulphur content limit for fuels used by ships at berth in EU ports still applies [79].

Furthermore in Directive 2009/30/EC the sulphur content of fuels used by ships operating in inland waterways was limited to 10 ppm on January 1, 2011 [174].

5.4 Nitrogen Oxides

Nitrogen oxides (NO_x) are generally defined as the sum of nitrogen monoxide (NO) and nitrogen dioxide (NO_2). For combustion engines, such as marine diesel engines, the formation of NO_x is predominately related to the oxidation reaction between atmospheric nitrogen (N_2) and oxygen (O_2), which forms NO_x. However, this reaction requires high combustion temperatures, e.g., above 1500 °C, and is hence usually denoted as thermal NO_x. If the fuel that is combusted contains nitrogen, this oxygen can be oxidised and form NO_x.

When NO_x is released to the atmosphere, it contributes to a variety of different environmental impacts, such as eutrophication (increased nitrogen nutrients to biomass), acidification and also as a precursor to the formation of ground-level ozone and secondary particulate matter. Hence, the anthropogenic NO_x emissions have been given increasing attention; globally, NO_x emissions (Tg/year) have decreased in the US from 25.5 to 17.7 and in the EU-27 from 17.1 to 11.3 between

1990 and 2006 [65]. In contrast, these reductions are thought to be comparable in magnitude to the increased NO_x emissions from international shipping, which increased from 12 to 20 Tg/year between 1990 and 2006 [20]. Additionally, the registered fleet of operating ships in 2000 gave rise to approximately 15 % of all anthropogenic NO_x emissions at a global level [4]. Based on this information, in 2014, IMO decided to postpone its most stringent international regulation for decreased NO_x emissions from marine engines (Tier III) from 2016 to a later date.

On a European level, the European Environment Agency [169] addressed the effects of international shipping on European air quality and climate forcing. Their report highlighted that fact that NO_x emissions from international shipping in European waters are projected to increase and could equal all land-based sources (EU-27) by 2020. Furthermore, the European Commission [84, 170] stated the following: "NO_x emissions from shipping are a direct contribution to eutrophication of inland and marine waters and terrestrial habitats, and to the formation of (secondary) particulate matter affecting health."

On a regional level, during 2011, the total NO_x emissions from ships operating in the Baltic Sea were approximately two and half times higher than the NO_x emissions from Sweden during the same period. The countries near or close to the Baltic Sea and the North Sea are also exposed to the deposition of atmospherically oxidised nitrogen from activities in these regions. In Sweden in 2011, 11 % of the deposition originated from the Baltic Sea and 13 % came from the North Sea, respectively [85] see also Box 2.1.

In summary, the contribution of NO_x emissions from vessels might not be negligible on a global scale, on a European scale or in the Baltic and North Sea areas. Instead, these emissions represent a substantial proportion of all anthropogenic NO_x emissions. Hence, the contribution of NO_x from ship operations should be considered when striving for an overall reduction in anthropogenic NO_x emissions.

5.4.1 Formation

The formation of NO_x in a diesel engine depends on the combination of high temperatures, the availability of oxygen and nitrogen, and the duration of the combustion [14]. Because the formation of NO_x is highly dependent on the combustion temperature, the formation rate increases at higher temperatures [86]. Therefore, the highest concentrations of NO_x should be found in slightly rich mixtures (excess of fuel) because the maximum temperatures occur under slightly rich conditions [87]. However, the formation of NO_x also depends on the presence of oxygen. Therefore, in total, the highest NO_x formation rate occurs on the lean side of stoichiometric conditions [87]. The formation of NO_x is also influenced by the combustion duration. For a diesel engine, the burn rate during the mixing-controlled combustion phase is proportional to the engine speed (rpm), although it remains relatively constant on a crank angle basis, which results in decreased burn rates at lower rpm [17]. Hence, at reduced engine speeds (rpm), the available time for NO_x formation is prolonged, and the NO_x emissions increase [87]. The relation between rpm and NO_x emissions (g NO_x/kWh) can also be found in the NO_x emissions regulation for ship operations [88].

The formation of NO in combustion flames can be explained by three mechanisms: *thermal*, *prompt* and *nitrous oxide*. For marine diesel engines, the majority of the NO_x formation is due to the thermal mechanism, whereas the prompt mechanism and the nitrous oxide mechanism account for a smaller portion and are further discussed in Box 5.3.[16] The generally accepted governing reactions for the formation of NO from atmospheric nitrogen (N_2) at near-stoichiometric fuel-air conditions are described as follows [17, 87]:

$$O^{\cdot} + N_2 \leftrightharpoons NO + N \tag{5.6}$$

$$N + O_2 \leftrightharpoons NO + O \tag{5.7}$$

$$N + OH^{\cdot} \leftrightharpoons NO + H \tag{5.8}$$

The first two reactions (5.6 and 5.7) form the Zeldovich mechanism. The third reaction (5.8) was added by Lavoie et al. [90] and substantially contributes to NO formation [87]. Thermal NO formation is assumed to occur in high-temperature combustion gases in which all other species are in equilibrium except for NO [87]. Thus, the thermal formation of NO occurs both at the flame front and in post-flame gases [17]. However, because the pressure in the cylinders increases throughout the combustion period, the temperature of early-burned gases increases above their immediate temperatures after combustion and results in increased NO formation. The thermal NO formation in post-flame gases is generally more pronounced than NO formation at the flame front [17]. The overall NO formation rate can be described by combining reactions 5.6–5.8 with the individual species concentrations in moles per cubic centimetre (denoted by [ ]) and their individual reaction rate constants. The NO formation rate can be further simplified to Eq. 5.9, which illustrates the strong dependency of the NO formation rate on temperature (T) and that high temperatures and oxygen concentrations result in high NO formation rates [17]:

$$\frac{d[NO]}{dt} = \frac{6 \times 10^{16}}{T^{1/2}} \exp\left(\frac{-69,090}{T}\right) [O_2]_e^{1/2} [N_2]_e \quad \frac{mol}{cm^3 * s} \tag{5.9}$$

[16] The maximum combustion temperature occurs under slightly rich conditions, e.g., approximately $\Phi = 1.05$. This is due to that the heat release from the combustion process and the heat capacity of the formed combustion products ($N_{products} * C_{p,\ products}$) both decrease for an equivalence ratio greater than one ($\Phi > 1$), while between $\Phi = 1$ and Φ (T_{max}), the heat capacity of the combustion products decreases more rapidly than the heat release from the combustion (ΔH_c). Hence, the combustion temperature increases. However, beyond Φ (T_{max}), the heat capacity of the products starts to decreases slower than ΔH_c whereby the combustion temperature starts to decrease. Therefore, the highest combustion temperature occurs at approximately $\Phi = 1.05$ [89].

The thermal formation of NO in a diesel engine typically begins when the temperature in the cylinders during the combustion process exceeds 1200 °C, although the formation will not become significant until the temperature reaches 1500 °C [87]. Hence, as previously noted, an important period in the combustion process occurs between the initiation of combustion and shortly after reaching the peak cylinder pressure. During this period, gas elements that were burned early in the combustion process are further compressed, resulting in an increased temperature and NO formation rate. After reaching the peak cylinder pressure, the burned gas elements begin to expand, which is followed by a decrease in temperature. This expansion, combined with the mixing of high- and low-temperature gas elements, freezes the NO chemistry, i.e., the formation and decomposition of NO stops [17]. Hence, the chemical equilibrium NO/NO_2 ratio at typical flame temperatures should result in small amounts of NO_2, whereas experimental data show that 10–30 % of the total NO_x in diesel exhaust can consist of NO_2 [17]. For large two-stroke engines, approximately 5–7 % of NO is converted to NO_2, which depends on the oxidation during the expansion stroke and in the exhaust system [91]. It might be possible to explain the NO_2 formation in the combustion process by the rapid conversion of the formed NO into NO_2 via reaction 5.10, whereas NO_2 can be reduced to NO again via reaction 5.11. However, if the NO_2 formed at the flame front is rapidly cooled by mixing with low-temperature fluid elements, the conversion of NO_2 into NO can be quenched. This process is consistent with the high NO_2/NO ratios that occur at low load in diesel engines, in which regions with cooler fluid elements are more pronounced [17].

$$NO + HO_2 \leftrightharpoons NO_2 + OH^{\cdot} \tag{5.10}$$

$$NO_2 + O^{\cdot} \leftrightharpoons NO + O_2 \tag{5.11}$$

Box 5.3 The prompt mechanism and the nitrous oxide mechanism.
NO is generally formed in post-flame gases because the time scale for the fuel oxidation process is shorter than that of the thermal NO formation mechanism [89]. However, measurements of NO in combustion gases cannot be extrapolated to zero at the flame front, where the fuel is oxidised and not all gases have reached equilibrium. Therefore, some NO is formed in the flame zone, i.e., by the so-called prompt mechanism. The prompt mechanism is important when fuel-bound nitrogen is present or when the combustion temperature is sufficiently low for the thermal mechanism to become negligible [87]. It is suggested that this mechanism is governed by the following reactions (reactions 5.12 and 5.13):

$$CH + N_2 \leftrightharpoons HCN + N \tag{5.12}$$

$$C_2 + N_2 \leftrightharpoons 2CN \tag{5.13}$$

The general scheme of the prompt mechanism includes the reaction between fuel-originating hydrocarbon radicals and molecular nitrogen, which form compounds that are further converted into NO. Reaction 5.12 is considered to the main path and the rate-limiting step [89].

The final NO formation mechanism is the *nitrous oxide* mechanism with subsequent decomposition into NO. The nitrous oxide mechanism is important at low temperatures and under lean-fuel conditions (excess of air, $\Phi < 0.8$) [87, 89]:

$$N_2 + O + M(\text{any molecule}) \leftrightharpoons N_2O + M \tag{5.14}$$

5.4.2 Human and Environmental Implications

Emissions of NO_x have a variety of environmental implications, such as eutrophication and acidification [92]. NO_x is also a precursor to the formation of ground-level ozone and secondary particulate matter [93]. The formation of NO_x originates from both anthropogenic and natural sources, and combustion processes are the largest anthropogenic source. Natural sources of NO formation consist of burning biomass, biogenic sources (such as nitrification, denitrification and decomposition of nitrite [NO_2^-]) and NO_x produced by lightning [66].

NO_x emissions from anthropogenic sources contribute to an increased concentration of atmospheric nitrate (NO_3^-), which causes eutrophication via dry and wet deposition of nitrate to the ecosystem [65] High inputs of nutrient nitrogen might also cause a substantial reduction in water quality, resulting in decreased biodiversity and changes in toxicity effects. The addition of nitrogen nutrients stimulates the primary production and growth of phytoplankton, which could result in an increase in the presence of zooplankton and benthic suspension feeders that feed on the phytoplankton. If the addition of nitrogen nutrient results in an excess of primary production, these organisms will eventually fall to the bottom and be degraded by aerobic bacteria that consume oxygen. This process might lead to reduced dissolved oxygen (hypoxia) or even to the depletion of dissolved oxygen (anoxia) in the water [94].

Furthermore, the effects of NO_x on eutrophication, acidification, the formation of ground-level ozone and the formation of secondary particulate matter primarily involve NO_2 as the source of NO_x. Because the majority (>90 %) of NO_x formed in combustion processes is emitted as NO [66], the oxidation of NO to NO_2 in the atmosphere is an important process. The daytime conversion of NO to NO_2 can be explained by several reactions that include $OH^\cdot$, organics and $HO_2^\cdot$ and $RO_2^\cdot$ radicals. Peroxy radicals ($HO_2^\cdot$ and $RO_2^\cdot$) are the specific species that convert NO to NO_2 at ambient concentrations [66].

Considering the effects of NO_x on acid deposition, this process might occur through the formation of nitric acid (HNO_3) [66], where M denotes a random molecule:

$$NO_2 + OH^{\cdot} + M \rightarrow HNO_3 + M \quad (5.15)$$

NO_2 is also considered to have human health effects, such as inflammation of the airways when high concentrations of NO_2 are present. Moreover, as previously mentioned, NO_2 is also associated with the formation of ground-level ozone and secondary particles (see Sect. 2.7.9). The formation of secondary particulate matter (in either mass or number) in the atmosphere can also occur via several different mechanisms [66]. One of these mechanisms is the reaction of gases on the surface of existing particles to form condensed products. The reaction between HNO_3 and sea salt is an example.

NO_x emissions also have an impact on the climate. The radiative forcing (RF) related to NO_x emissions from shipping is associated with an increase in ozone and a decrease in CH_4. The overall effect of NO_x emissions from ships is thought to result in a net negative RF, i.e., cooling of the climate [4]. To summarise, NO_x might give rise to several different human and environmental implications that need to be considered when evaluating the potential effects of anthropogenic NO_x emissions.

5.4.3 Regulations

At an international level, NO_x emissions from operating ships are regulated by the IMO's Revised MARPOL 73/78 Annex VI Regulation 13 [88]. This regulation applies to each marine diesel engine with a power output exceeding 130 kW that is to be installed on a ship and each marine diesel engine (already installed on a ship) with a power output of more than 130 kW that undergoes a major conversion on or after 1 January 2000. A major conversion is defined as a modification that has been performed on or after 1 January 2000 that has not already been certified according to the emission standards set by Regulation 13. There are three general modifications that imply a *major conversion*: the engine is replaced by a marine diesel engine or an additional marine diesel engine is installed, a substantial modification (as defined by NO_x Technical Code 2008) is made to the engine, or the maximum continuous rating (MCR) of the engine is increased by more than 10 % of the original MCR [88].

The NO_x regulations in Regulation 13 of MARPOL 73/78 Annex VI are defined by three separate NO_x emission levels, i.e., Tier I, Tier II and Tier III, which are illustrated in Fig. 5.6

Tier I applies globally to the operation of marine diesel engines installed on ships constructed on or after 1 January 2000 and prior to 1 January 2011, and Tier II applies globally to the operation of marine diesel engines installed on ships constructed on or after 1 January 2011. Moreover, Tier III represents a NO_x reduction

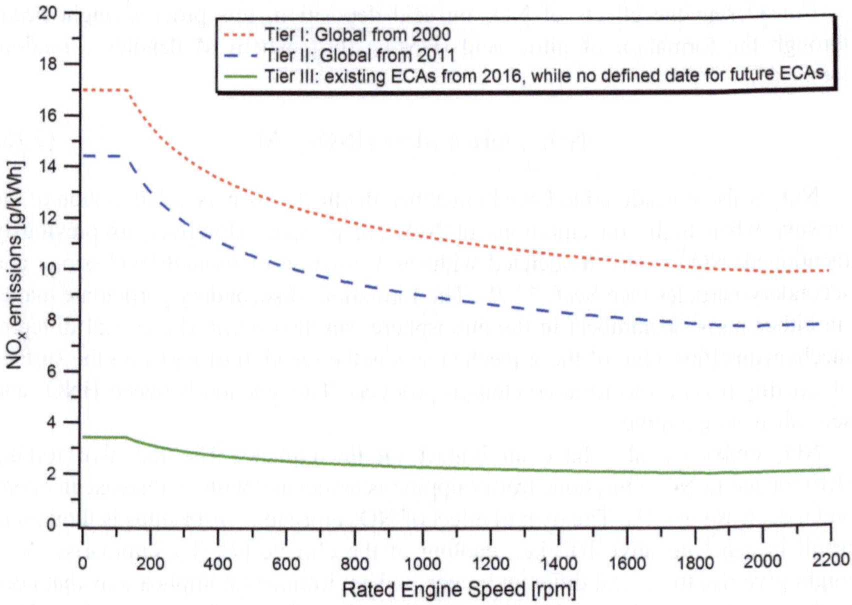

Fig. 5.6 IMO Revised MARPOL 73/78 Annex VI Regulation 13 for Nitrogen Oxides (NO_x)

of approximately 80 % compared with Tier I engines, although it is only applicable in specific ECAs that are designated by IMO [88, 95]. Tier III will begin being enforced on 1 January 2016 for the existing ECAs, whereas no date has been defined for future ECAs [96]. This topic is further discussed in Box 5.4.

Regardless of the enforcement date for the Tier III regulation, it is important to establish ECAs to achieve practical effects from this stricter regulation because Tier III will only be applicable in ECAs. Currently, according to Appendix VII of MARPOL 73/78 Annex VI [88], the only ECAs in the world (except for the SECAs in the Baltic Sea and the North Sea) are the North American area and the United States Caribbean Sea area. However, an ongoing discussion is occurring within HELCOM to designate the Baltic Sea area as a NO_x ECA. A complete written proposal is ready for submission to the MEPC when the time for submission has been decided [97, 98]. Additionally, the environmental impact, health benefits and economic impact of a NO_x ECA in the North Sea have been assessed [99, 100]. Moreover, due to the lack of NO_x ECAs and the fact that the IMO's NO_x limits only refer to new ships, the effects of the MARPOL 73/78 restrictions are currently limited [101].

Box 5.4 IMO review of the status of technological development required to implement the Tier III NO_x emission standards.
According to IMO [88] Regulation 13 §10, the organisation shall review the status of the technological development to implement the Tier III NO_x emission standards no later than 2013. Accordingly, at the 62nd IMO MEPC meeting in 2011, the committee agreed to establish a correspondence group under the coordination of the United States [102], which provided an interim report to MEPC 64 [103, 171] and a final report to MEPC 65 [104, 172]. The final report was submitted to MEPC 65 [172] and suggested that the effective date of Tier III should be retained, i.e., 1 January 2016; however, this report also suggested that marine engines fuelled solely by gaseous fuels, e.g., LNG, could be considered for Tier III compliance at a future date. However, in a submitted document to MEPC 65 [105], the Russian Federation questioned the appropriateness of retaining the effective date of Tier III and proposed that it should be shifted to 2021. This proposal was further discussed during MEPC 65, and the committee agreed to the proposal for the effective date to be amended and postponed to 1 January 2021 [106]. Accordingly, IMO [107] sent the text of draft amendments to MARPOL 73/78 Annex VI and the NO_x Technical Code 2008, which were circulated under cover of Circular Letter No. 3370 [108]. Before MEPC 66, additional information on the technological status and certain critical aspects of Tier III technologies was provided by European Association of Internal Combustion Engine Manufacturers (EUROMOT), with the conclusion that there are mature Tier III-compliant technologies [109]. At the 66th IMO MEPC meeting, it was decided to retain the effective date of 1 January 2016 for the existing ECAs, i.e., the North American ECA and the US Caribbean Sea ECA, and that the effective date for Tier III in future ECAs should be the date of adoption of such an ECA or a later date that can be specified in the amendment designating the NO_x Tier III ECA, whichever is later [96].

At the European level, there are currently no enforced direct NO_x emission regulations for international shipping in European waters. A document from the European Commission brings attention to the EU policy and legislation on air pollution from shipping [84], specifically concerning the possible postponement of IMO Tier III to 2021. The European Commission [84] identified the current European policies that attempt to improve air and water quality, e.g., by preserving, protecting and improving the environment and protecting human health, and stated that a postponement of IMO Tier III would adversely affect the objectives of the EU policies. Furthermore, the European Commission [84] noted that the current EU legislation, i.e., the National Emission Ceilings (NEC) Directive [81] and the Ambient Air Quality Directive [83], provide limits for the NO_x concentration in the air. The Ambient Air Quality Directive requires that the European Commission and the Member states should strive for international cooperation to reduce atmospheric pollution,[17] e.g., NO_x emissions, whereas the NEC Directive introduces specific national emission ceilings. However, the NEC Directive only includes NO_x emissions from national maritime shipping and international shipping within inland waterways. Therefore, emissions from international maritime shipping other than within inland waterways are not included in the current regulations. The mandate to regulate emissions from international shipping is preferably placed on IMO, which is expressed in the NEC directive,[18] although the mandate has a substantial effect on air quality in the EU [81]. Additionally, the European Commission [84] recognises that the Commission adopted the "Clean Air Quality Package" on 18 December 2013, which highlights the effects of shipping emissions on land-based air quality and the cost-effectiveness of further measures in this sector. Hence, the European Commission [84] suggested that the postponement of IMO Tier III to 2021 should be opposed and stated the following: "NO_x emissions from shipping are a direct contribution to eutrophication of inland and marine waters and terrestrial habitats, and to the formation of (secondary) particulate matter affecting health."

However, NO_x emissions from inland waterway vessels are directly regulated not only by the NEC Directive. NO_x emissions are not separately regulated; instead, they are regulated together with the emissions of hydrocarbons (HC), i.e., x g of (NO_x + HC). The emission limits range between 7.2 and 11.0 g HC + NO_x per kWh [80, 82, 110].[19]

As stated previously, between 1990 and 2006, NO_x emissions (Tg/year) decreased in the EU-27 from 17.1 to 11.3 (−5.8) [92]. The reduced NO_x emissions in the EU can be attributed to three sectors: road transport, energy industries and energy generation in industry. Emission reductions in the road transport sector were achieved, whereas transport demand increased during the same period due to the installation of catalytic converters in vehicles to meet the European NO_x emission standards, also known as the 'Euro' standards [92]. The EU emission standards for heavy-duty diesel engines were introduced in Directive 88/77/EC and further developed in several steps, i.e., Euro levels [111]. The current regulation, i.e., Euro VI, is the most stringent and requires a NO_x reduction of approximately 95 % compared with Euro I, i.e., from 8.0 to 0.4 g NO_x/kWh, while simultaneously

[17]Dir. 2008/50/EC Preamble at (2) reads as follows: "In order to protect human health and the environment as a whole, it is particularly important to combat emissions of pollutants at source and to identify and implement the most effective emission reduction measures at local, national and Community level. Therefore, emissions of harmful air pollutants should be avoided, prevented or reduced and appropriate objectives set for ambient air quality taking into account relevant World Health Organisation standards, guidelines and programmes.".

[18]For further information, see Dir. 2001/81/EC Art. 1, Art. 2(a) and Art. 11.

[19]The NO_x emissions for inland waterway vessels are regulated via Dir. 97/68/EC, which was amended by Dir. 2004/26/EC and later amended (testing and approval) by Dir. 2010/26/EU. The specific dates and emission limits can be found in Dir. 2004/26/EC Art. 1(6.)(d) and Annex I(4.)(b).

introducing an upper limit for ammonia (NH_3) emissions of 10 ppm [10, 112].[20] The sum of the emission limits for HC and NO_x under Euro VI can be compared to the emission standards for inland waterway vessels and there is a notable difference. The most stringent regulation for inland waterway vessels has a limit of 7.2 g (NO_x + HC)/kWh, whereas Euro VI (transient testing) has a limit of 0.62 g (NO_x + HC)/kWh.

The development of the Euro standard for diesel engines used for land-based transport can serve as an example, in which the absolute NO_x emission limits are below the Tier III standards. This example also illustrates a development in emission standards that include more than three steps and include an absolute level for ammonia slip. However, if the emissions standards for land-based diesel engines are to be compared with marine diesel engines, the higher volume and/or weight of the transported goods per individual vessel must be considered when comparing with a truck or equivalent vehicle. Hence, NO_x emissions from shipping activities in European waters might be considered when striving for overall reduced NO_x emissions in Europe.

At the national level, some nations have introduced economic incentives to further encourage the installation of NO_x abatement technologies on merchant vessels independent of a vessel's age. Two examples of such incentives are the Norwegian NO_x fund and the Swedish environmentally differentiated fairway dues. The Norwegian NO_x fund is a cooperative effort in which the participant enterprises, i.e., primarily ship owners and other enterprises that are obligated to pay the NO_x tax [113–116] are exempted from the Norwegian NO_x tax[21] and are instead obligated to pay a NO_x fee to the fund. The participants are later allowed to apply for economic support from the fund to install and operate NO_x abatement technologies. However, the support is not granted for NO_x-reducing measures that only reach mandatory IMO regulations [119].

The Swedish environmentally differentiated fairway dues were introduced by the Swedish Maritime Administration in 1998. The fairway dues are a financial incentive intended to reduce NO_x emissions from operating ships and are designed using a stepwise reduction in fairway dues as a function of the reduced NO_x emissions from a particular ship [173, 74]. To be granted reduced fairway dues, the NO_x emissions from an operating ship must be measured both on the main engines (at 75 % engine load) and on the auxiliary engines (at 50 % engine load), where after an average X g NOx/kWh is calculated for the entire ship. The average NOx emission value is later used to apply for a certain fairway rebate. The measurements must also be conducted according to ISO 8178 and IMO Technical Code on Control of Emission of Nitrogen Oxides from Marine Diesel Engines [120]. If a ship is equipped with a selective catalytic reduction (SCR) system (further

[20]For steady-state conditions, the Euro VI limit is 0.4 g NO_x per kWh, whereas at transient conditions, the limit is 0.46 g/kWh. For further information on the Euro VI emission limits, including NH_3, see Com. Reg. (EU) No 582/2011 Annex xV. I.

[21]The NO_x tax was established in 2006 by the Norwegian parliament [117] as a consequence of Norway signing the Gothenburg Protocol and agreeing to reduce NO_x emissions [118].

described in Chap. 11), the ship owner must prove that the SCR is in operational use, i.e., the urea consumption and provision of urea purchase must be reported and the NH_3 emissions must be measured [121]. The reduced fairway dues for an operating ship are valid for three years; thereafter, the measurements must be conducted again.

5.5 Particles

Interest in atmospheric particles has increased. Information on their impact on human health and climate and a deeper knowledge of sources, formation, fates and how the chemical and physical properties impact health and visibility are needed. Particles from combustion processes in engines are formed due to the incomplete combustion of hydrocarbons in the fuel particle emissions. Emissions of primary particles from ship operations are of the same magnitude as that of road traffic, i.e., 1.7 Tg and 2.1 Tg annually, respectively [122]. Modelling emphasises that particle emissions from ship operations might be responsible for approximately 60 000 cardiopulmonary and lung cancer deaths per year globally [3]. Studies of ice cores in the Arctic region have shown that sulphate concentrations have increased by a factor of 4 from pre-industrials levels to 1980, and nitrate concentrations have increased by a factor of 2–3 from the 1950s to 1980s. Ice cores from the Alps have shown an even larger increase due to their close proximity to anthropogenic sources in Europe. The current levels of BC are 2 times higher than in the 1900s [123]. This information highlights the notion that particle emissions from ship operations and from other combustion sources cause air pollution that should be considered; for operating ships, it is important to consider these emissions in densely populated coastal areas.

The nomenclature used for particle emissions can be confusing. In general, the emissions of particles are referred to as emissions of particulate matter. The abbreviation that is typically used for particulate matter is PM, and the known expressions are PM10 and PM2.5, which refer to the masses of particles with diameters less than 10 and 2.5 μm, respectively. In this work, the abbreviation PM is used for particle mass, i.e., the mass of particles emitted. An earlier focus was on the mass of particles emitted (PM); however, in recent years, there has been an increase in the number of measurements that consider the number of particles emitted. Therefore, particle emissions are also discussed with respect to the number of particles; the abbreviation used in this case is PN.

Regarding particle emissions from combustion sources, the terms black carbon (BC), organic carbon (OC), elemental carbon (EC) and soot are also frequently used and can be confusing. In this section, a summary of definitions is presented to provide the reader an opportunity to understand that differences exist between these types of carbonaceous particles. However, the reader should also note that the definitions are not always clear, and further research is underway to improve them because particles composed of various carbonaceous species are the least understood and most difficult to characterise out of all chemical compounds found in aerosols. Black carbon (BC) particles are light-absorbing particles that can absorb visible light with a wavelength exceeding 550 nm. These particles are refractory, and a temperature above 4000 K is needed to volatilise these particles. Elemental carbon (EC) particles are also thermally stable, can be viewed as solid particles under atmospheric conditions and are insoluble in any solvent. Earlier definitions of EC referred to carbonaceous particles composed of pure carbon and not combined with oxygen and/or hydrogen, which could be the case for organic carbon (OC). OC particles refer to particles composed of carbon that is chemically combined with hydrogen, oxygen, sulphur or nitrogen, as examples [124]. Soot particles are formed in the cylinder during the combustion of the fuel, are primarily composed of carbon and can be viewed as agglomerates of spherules composed of graphite-like micro-crystallites [17, 124].

5.5.1 Formation

Particles emitted from anthropogenic sources are dominated by fine particles, i.e., 0.002–2.5 μm in diameter, whereas particles from natural sources are primarily larger. Examples of natural particles in the atmosphere are sea spray/salt and dust. The main anthropogenic sources of particle emissions are different types of combustion for transportation, energy production and industry [65].

5.5.1.1 In the Engine

Within the shipping sector, particles are formed during the combustion of marine fuels in engines. The formation and growth processes occur during the expansion and exhaust stroke in a four-stroke engine and during the power stroke of a two-stroke engine. Primary particles are formed during the expansion stroke (four-stroke) or power stroke (two-stroke) and are the result of the incomplete combustion of hydrocarbons in the fuel and lubrication oil. The formation of primary particles, also known as soot, begins immediately after the fuel is injected and occurs under high-temperature and fuel-rich conditions ($\Phi > 1$) see Sect. 5.1. The available time for particle formation is several milliseconds. The soot particles are small residual carbon particles/spherules with diameters of 1–10 nm that are formed from the thermal decomposition of large hydrocarbon molecules in the fuel. After formation, these newly formed particle nuclei begin to grow. The particle growth processes include surface growth, aggregation and coagulation. Surface growth involves the attachment of organic (unburned hydrocarbons and oxygenated carbons) and inorganic species, such as SO_2, H_2SO_4, and NO_x, in the gaseous phase to the existing particles with an unchanged number of particles (PN), whereas coagulation reduces the PN as particles collide to form new particles. Aggregation results in chains and clusters of the primary soot particles and an increase in size (to

10–100 nm in diameter) [5, 14, 17]. The newly formed primary particles might undergo oxidation, i.e., burning of soot, which results in the formation of gaseous species, such as CO and CO_2, and occurs in the lean zones ($\Phi < 1$) of the cylinder during the expansion stroke [14, 17].

The soot particles formed during the expansion stroke undergo further condensation, adsorption and agglomeration when the exhaust valves are opened, which begins a rapid cooling of the exhaust gases, and this process occurs during the exhaust stroke [5]. The valves remain open throughout the exhaust stroke, and the exhaust gases leave the engine and enter the exhaust gas channel/funnel. The amount of emitted soot depends on the differences between the rate of formation and the rate of oxidation in the cylinders during combustion [14]. The transformation of diesel particles within the funnel depends on the residence time, initial particle concentration, concentration of volatile compounds and temperature. The transformation increases the mass concentration, volatile fraction of the particles and average diameter of the particles and reduces the number concentration and hydrocarbon concentration [125].

The combustion conditions, i.e., engine type and load, impact the number concentration and sizes of the emitted particles, whereas the fuel type and lubrication oil impact the chemical content of the particles [4]. High temperature and pressure favour the production of particles and result in higher particle number concentrations, i.e., a slow-speed marine diesel engine (SSD) emits more particles than a medium-speed marine diesel engine (MSD) [126]. The timing and rate of the fuel injection impact the emissions of particles from the engine. The fuel injection timing can be either retarded or early. Early injection of fuel will reduce the particle emissions because it causes a higher temperature during expansion and provides additional time for oxidation of particles. However, this situation will cause an increase in NO_x emissions [14, 17] (Box 5.5). Furthermore, the formation of particles is related to an increased engine load, which means that a higher amount of fuel is injected [14]. The particle emissions will also be affected by engine design. For example, turbocharging an engine reduces the particle emissions due to higher-temperature air in the combustion process, which favours the oxidation of soot [14]. Marine diesel engines are generally turbocharged [17].

Box 5.5 The diesel dilemma

High efficiency and low emissions are the desired goals of any diesel engine designer. By running an engine at high temperature and pressure, the following achievements are possible:

Maximise the power output: The engine-delivered power strongly depends on the average pressure in the cycle.

Maximise the efficiency: Efficiency depends on the cycle's maximum temperature.

Minimise emissions of PM, HC and CO: These emissions are a consequence of incomplete combustion, which in turn is generated by the imperfect mixing of the fuel with air and by low combustion temperatures and duration, see Sect. 5.1.

However, as discussed in Sect. 5.4, measures for reducing NOx emissions intend to reduce the combustion temperature, excess air, and combustion duration. Therefore, it is clear that a high engine efficiency; low particle, HC and CO emissions; and low NOx emissions are competitive goals.

One example of this competing behaviour of NOx and particle emissions is related to the injection rate. A slow injection rate generates a longer combustion, which favours better fuel-air mixing and more complete combustion, although it also maintains the temperature in the cylinder at high values for a longer time, thus increasing NOx emissions.

Injection timing is also an example of this trade-off. Early injection generates higher combustion temperature and pressure, which provoke both higher engine efficiency and higher NOx emissions. Delaying injection allows reduces in-cylinder temperatures and generates higher emissions of particles and other products of incomplete combustion due to lower combustion temperatures and reduced time for mixing and reactions [14, 17].

The content of different compounds in the fuel, including both additives and impurities, also has an impact on particle emissions. Figure 5.7 presents the different origins of particles formed in an engine and funnel. The carbonaceous fraction, i.e., soot, is primarily formed during thermal decomposition of the fuel and lubrication oil. The organic fraction and the sulphate fraction are related to both the fuel and lubrication oil, whereas the wear of engine parts and the presence of

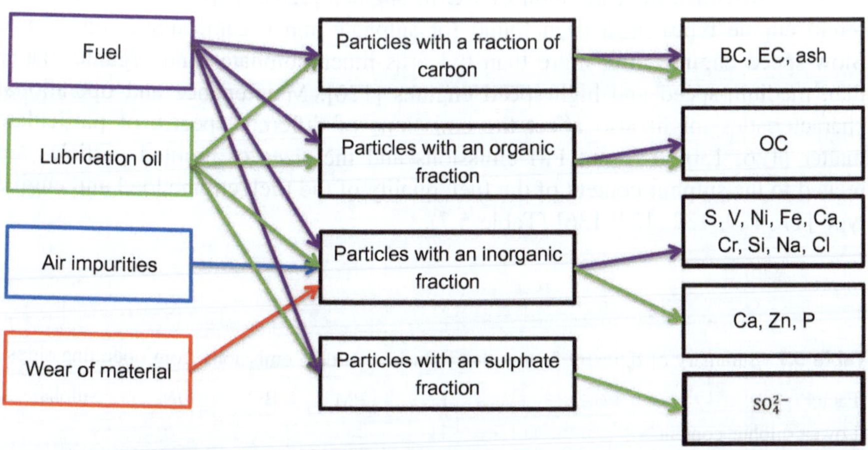

Fig. 5.7 Illustration of the origin of different particles found in particle emissions from the combustion of marine fuels. The different *arrow* colours show the relationships between the various compounds, i.e., fuel (*purple*), lubrication oil (*green*), and wear of material (*red*)

inorganic compounds in the fuel and lubrication oil contribute to the inorganic fraction of particle emissions [14]. The emitted particles from a ship consist of black carbon (BC), sulphate (SO_4), organic matter (OM)/organic carbon (OC), elemental carbon (EC), ash, metals and inorganic compounds [126, 127]. Among the inorganic compounds are vanadium (V) and nickel (Ni), which are associated with HFOs; other inorganic compounds from the fuel can generally be related to diesel fuels. Figure 5.7 also shows that certain specific inorganic compounds are related to the lubrication oil [127, 128].

Measurements conducted on operating ships and in test-bed engines emphasise that fuel type, engine load and engine type all impact different types of particles found in the emissions from an operating ship and the number and mass of the emitted particles. The impact of the different factors on particle number (PN), particle mass (PM) and the different particle types are summarised in Table 5.7. The emissions of particle-related sulphates are linearly correlated with the sulphur content of the fuel [126]. Approximately 1.3–1.4 % of the sulphur in the fuel is converted to sulphates, and the remainder of the sulphur is emitted as SO_2 (Sect. 5.3) [127, 129]. The formation of organic matter depends on the quality of the fuel and on the lubricating oil consumed in the engine, in which the consumption rate/need depends on the sulphur content in the fuel and the quality of the fuel. BC emissions are more closely related to the engine type and load. A medium-speed engine produces more than twice the emissions of a high-speed or slow-speed engine [126, 130]. Measurements emphasise the dependence on load, i.e., BC emissions are higher for lower loads than for higher engine loads [131–133]. BC emissions appear to be inversely correlated with the sulphur content in the fuel and are more closely related to the fuel quality. HFOs contain a higher content of ash and large hydrocarbon and aromatic molecules, which might enhance the formation of BC. Reducing these compounds in the fuel, which is the case for MGO, will reduce the formation of BC in engines [134]. Variations between different engine types have been found for sulphate and organic matter emissions. Slow-speed engines emit more than twice as much sulphates and organic matter than medium-speed and high speed engines [126]. Maintenance and operational characteristics might also affect the emissions of different species of particulate matter [126, 130]. PN and PM emissions and the sizes of emitted particles are related to the sulphur content of the fuel, quality of the fuel, engine load and engine type [11, 129, 132, 134–136] (Table 5.7).

Table 5.7 Summary of different factors that influence particle emissions from operating ships

Factor	PN	PM	BC	OM	Sulphate
Lower sulphur content	↑↓	↓			↓
Improved fuel quality	↑↓	↓	↓	↓	
Increased engine load	↑↓		↓		
Reduced consumption of lubrication oil				↓	

5.5.1.2 In the Plume and Atmosphere

Particles in the atmosphere can be divided into different modes according to their sizes, i.e., diameters. The two main modes are coarse particles, with diameters (Dp) > 2.5 μm, and fine particles, with Dp < 2.5 μm. Particles in the coarse mode generally originate from natural sources and are produced by different mechanical processes. Fine particles include most of the total number of particles in the atmosphere, a large portion of the particle mass, and can be further divided into three size modes: ultrafine particles (UFP) (Dp < 10 nm), Aitken nuclei particles (Dp of 10–80 nm) and accumulation mode particles (Dp of 80 nm to 1–2 μm) [66]. Particles with Dp < 50 nm are classified as nanoparticles (Fig. 5.8) [137].

The size distributions based on the number of particles from the combustion of marine diesel fuels exhibit a bimodal character, i.e., two distinct peaks can be found within the size range of the nucleation mode and accumulation mode (Fig. 5.8) [135]. Generally, the bimodal characteristic of the number-based size distribution from the combustion of diesel fuels can be expected [138]. Compared with land-based vehicles, the emitted particles from marine diesel engines are smaller and have a mean diameter of 20–40 nm. In land-based vehicles, the emitted particles are primarily produced in the accumulation mode (80 nm to 1–2 μm) [129]. This difference in particle sizes, which dominates the particle emissions from ship

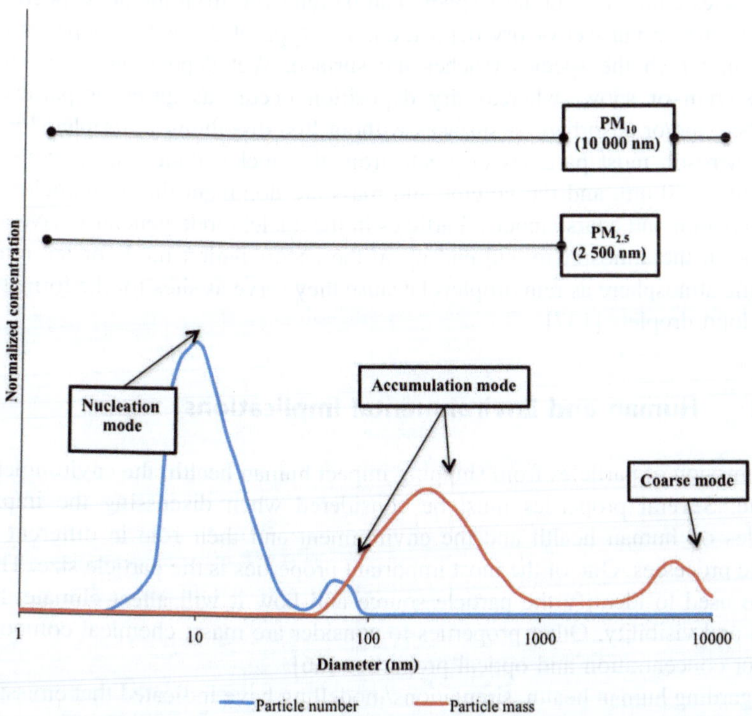

Fig. 5.8 Illustration of the particle size distributions based on number (*blue*) and mass (*red*). Three different size modes are shown; the limits for PM10 and PM2.5 are also depicted

operations and land-based vehicles, is primarily related to the sulphur content in the fuels that are used. Marine fuels have a considerably higher sulphur content (0.1–3.5 mass% S) than diesel used in land-based vehicles (max. 10 ppm S), and the sulphur in fuels enhances the formation of particles (i.e., it can be viewed as a precursor to particle formation), and consequently, a higher sulphur content leads to the enhanced formation of small particles in marine engines [126, 129]. Even the mass-based size distributions are bimodal, although the two distinct peaks are found in accumulation mode and coarse mode in this situation (Fig. 5.8) [127, 139].

Particles and gases formed in an engine and funnel are released to the air in the form of a plume, diluted with the ambient air and cooled [5]. The dilution and cooling cause a change in the chemical composition and size distribution of the particles. In the cooled exhaust, the amounts of sulphate and organic carbon increase due to the condensation of sulphuric acid (H_2SO_4) and hydrocarbons (HC). Sulphuric acid forms rapidly from SO_3 and water as the gas cools [127]. In the aging, cooling and expanding plume, the sizes/diameters of the particles increase due to the condensation of hydrocarbons and sulphates on existing particles and coagulation between particles in the nucleation, Aitken, and accumulation modes [6, 140, 141]. Small sulphuric acid particles form rapidly after being emitted from the funnel and grow via coagulation with larger particles of organic matter and black carbon [126].

Particles emitted to the atmosphere can be removed from the atmosphere to the Earth's surface via wet or dry deposition. The type of deposition depends on the phase in which the species reaches the surface. Wet deposition occurs in fog, clouds, rain or snow, whereas dry deposition occurs as gases or particles are absorbed and/or adsorbed on surfaces without first dissolving in droplets [66]. For urban aerosol, most particles originate from the nuclei mode, i.e., particles with diameters <80 nm, and the volume and mass are dominant due to particles in the accumulation and coarse modes. Particles in the nuclei mode generally have a short lifetime in the atmosphere and end up in the accumulation mode or are removed from the atmosphere as rain droplets because they serve as sites for the formation of new cloud droplets [137].

5.5.2 Human and Environmental Implications

The emission of particles from shipping impact human health, the environment and climate. Several properties must be considered when discussing the impact of particles on human health and the environment and their role in different atmospheric processes. One of the most important properties is the particle size. The size can be used to identify the particle source and how it will affect climate, human health and visibility. Other properties to consider are mass, chemical composition, number concentration and optical properties [66].

Regarding human health, simulations/modelling have indicated that emissions of particles from maritime transportation caused approximately 60,000 cardiopulmonary and lung cancer deaths per year globally in 2002, and this number will continue to increase by 40 % by 2012 due to the expected development of the maritime sector and current legislation. These models indicate that most deaths occur in coastal areas in Europe and in eastern and southern Asia [3]. An increase in emissions of especially fine and ultrafine particles could cause an increase in the number of people seeking care due to asthma and bronchitis and an increase in deaths [3, 142]. The size of the particles is one of the most important properties in discussions of particle impacts on human health. Particles in the air reach the airways and lungs via inhalation. Larger particle become caught in the mucus layer in the upper regions of the airways and are moved up and swallowed. Smaller particles (i.e., diameter less than 2.5 μm) can reach the alveolus in the lungs, which lacks a protective mucus layer [66]. Ultrafine particles (UFPs), i.e., diameters less than 10 nm, can penetrate from the lungs to the blood and be transported to other parts of the body [143]. The impact on the cardiovascular and respiratory systems due to ultrafine particle exposure has been observed in studies on both humans and animals [144]. Thus, exposure to UFPs from mobile sources, e.g., shipping, is difficult to assess and requires further monitoring system developments [145]. These ultrafine particles work as carriers for toxic compounds, e.g., transition metals and organic substances that are condensed or adsorbed to the surface of the particles [145]. Transition metals, such as vanadium and iron, have a negative impact on human health by catalysing the formation of reactive oxygen species, activating different biochemical processes [146] and causing respiratory diseases [146, 147].

The impact of particle emissions on climate is significant, and the effects can be both direct and indirect. The effects are related to an influence on radiative forcing (RF); through that process, they affect the climate. The direct effect, which has an impact on the radiative balance, is related to the absorption and scattering of light. This effect depends on the particles' optical properties and results in heating or cooling [123, 148]. The absorbing fraction of the particle emissions is BC/soot, whereas particles containing OC and sulphate will scatter light and enhance cooling. These three compounds are identified in particles in shipping emissions [123, 126, 134, 148]. The indirect effect is related to particles impact on clouds. The emitted particles affect the number of clouds, their lifetimes, and both precipitation and radiative properties [123, 148]. Emitted sulphate particles from shipping can act as cloud condensation nuclei (CCN) and impact clouds [134]. Unlike other sectors, shipping is recognised as having a net negative RF, i.e., a cooling effect on the climate. This effect is related to the high sulphur content in marine fuels, which enhances the formation of sulphate particles. The cooling effect is expected to decrease with future regulations on sulphur contents in marine fuels [149]. BC emissions from the burning of fossil fuels and other sources also impact the melting of snow and ice areas in the Arctic. BC deposited on snow and ice darkens the surface and reduces the albedo. This process enhances melting in these areas [123]. BC is also viewed as a short-lived climate forcer (SLFC); one of the main SLFCs is emitted from diesel engines. SLFCs impact climate; although they have shorter lifetimes in the atmosphere than CO_2 [150].

Particles can impact the environment primarily through acid rain and affect visibility. Visibility is reduced due to haze, which is formed by suspended particles in the air that scatter light. Acid rain is formed via the oxidation of SO_2 and NO_x to H_2SO_4 and HNO_3, which is deposited to the surface via either wet or dry deposition. Oxidation of SO_2 occurs through several processes, including the formation of secondary particles that contain H_2SO_4 [66].

5.5.3 Regulation

Particle emissions, i.e., PM, from road traffic have been regulated in Europe since 1992 through the Euro I standard. The latest version of the Euro standards, i.e., Euro VI, began being enforced in 2014 and regulates both PM and PN emissions [151]. Environmental Quality Standards (EQSs) include $PM_{2.5}$ and PM_{10} values. EQSs are established at both the European and Swedish levels (through the implementation of the EU requirements) [152, 153]. Although PM emissions from road traffic have been regulated since 1992 and the impacts on human health and the environment from particles are known, there are still no regulations that are directed against particle emissions from ocean-going ships. These emissions are viewed as indirectly regulated by Regulation 14 in MARPOL 73/78 Annex VI, which regulates the sulphur content in marine fuels because oxidised sulphur from the fuel can form new particles. Recently, the increased awareness of the impact of BC particles on climate and the melting of the snow and ice masses in the Arctic has forced IMO to further investigate BC emissions from ship traffic in the Arctic and surrounding areas for both current and future shipping activities. Shipping activities in these areas are expected to increase in the future [4, 150]. This work, with an eventual limitation on BC emissions, is in progress in the sub-committee on Pollution Prevention and Response (PPR) (the former sub-committee of Bulk and Liquid Gases (BLG)) and within the MEPC at IMO. The work in the PPR sub-committee considers how to define BC particles for ship operations, which measurement methods to use and possible control measures to advise. Because it is difficult to determine and agree on the definition of BC, as discussed in the introduction to Sect. 5.5, the work within the PPR sub-committee continues, and the PPR-2 meeting in the beginning of 2015 raised the question to the MEPC 68, which meets later in 2015, to extend the target completion year to 2017 [154].

Particle emissions from ship operations in inland waterways within the EU have been regulated through EU directives. The limitation on PM depends on the engine size and has been set to 0.2–0.5 g/kWh for 2006–2013 [80–83]. In 2014, the European Commission proposed further developments to this regulation, which will include limits for the number of particles (PN) emitted that are in line with the Euro VI standard for heavy-duty vehicles [155].

5.6 Volatile Organic Compounds

It is possible for the carbon atom to form strong bonds to itself, thereby forming long chains and ring structures, and to other compounds, such as hydrogen, nitrogen, oxygen, sulphur or halogens (chlorine and fluorine), which results in a great diversity of organic compounds. This diversity makes organic compounds the largest group of chemical compounds. Several million organic compounds are known, and this number is continually increasing. The simplest organic compounds consist of only carbon and hydrogen atoms and are often referred to as hydrocarbons. Other examples of organic compounds include the following chemical groups: alcohols, aldehydes, alkanes, aromatics, and ketones and halogenated derivatives of these substances. All of these organic compounds have different chemical and physical properties. Depending on their size and composition, organic compounds can exist as solids (e.g., tallow and stearin), liquids (e.g., methanol and gasoline) or gases (e.g., methane and propane) under atmospheric conditions. The size (weight) of the molecule is not the only factor that determines the vapour pressure and boiling point of a compound because the molecular structure and other constituents can also have a large effect. However, more complicated organic chemistry is beyond the scope of this book. The vapour pressure plays a significant role in a compound's state under atmospheric conditions and how it will behave. For comparison, the gas in a cigarette lighter consists of molecules of 3 to 4 carbon atoms, and diesel fuel consists of 10–15 carbon atoms. Organic compounds with low vapour pressures (>0.01 kPa, 20 °C) are often referred to as volatile organic compounds (VOCs) that easily vaporise and exist primarily in the gas phase under atmospheric conditions.

Figure 5.9 illustrates the relation between the vapour pressure and the number of carbon atoms in a molecule. In describing the environmental impacts of VOC emissions, the smallest hydrocarbon (i.e., methane) is often excluded, and the term

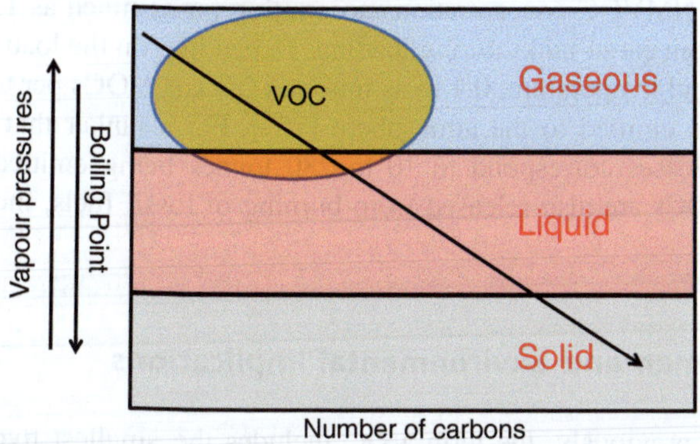

Fig. 5.9 The dependence on vapour pressure and phase on the carbon chain length for a simple hydrocarbon under atmospheric conditions. Longer carbon chains produce heavier molecules that are less likely to be found in the gas phase

non-methane VOC (NMVOC) is commonly used instead. The reason for this terminology is that methane, which consists of only one carbon, originates primarily from other sources and has different environmental implications. The environmental impact of methane emissions is primarily as a climate gas and is further described in Sect. 5.2.

The largest contribution of NMVOCs from shipping originates from cargo handling of crude oil and its products and is often described using emission factors that can vary. VOC emissions in the context of shipping arise primarily from the production, transport and storage of crude oil. As much as 2.4 million tons of VOCs, which represents hundreds of millions of euros in value, are lost each year due to the transport of crude oil and its products [20]. In addition, up to 2.4 kg of NMVOCs and 0.7 kg of methane are emitted per ton of fuel consumed for propulsion [20].

5.6.1 Sources

In addition to use as raw material in many industrial processes, NMVOCs are used as solvents and thinners in paint. These compounds are released when surfaces and equipment are cleaned and when paint is applied and the solvent evaporates to the atmosphere. However, the largest source of NVOCs in shipping is related to the handling of crude oil in tankers. Although these emissions represent a loss of considerable monetary value, the harmful consequences to the environment are estimated to be of greater importance. Crude oil contains a vast amount of different NMVOCs [156]. According to the volatile nature of these compounds during the loading, storage and transportation of crude oil on ships, the light hydrocarbons vaporise to the space between the cargo surface and the top of the tank. The amount of volatile vapours depends on the type of cargo, tank pressure, pump speed and cargo properties. These vapours are highly explosive. While a tank is filled, the vapours are often vented to the atmosphere. Experimental studies of air vented from a tanker show NMVOC concentrations accounting for as much as 13 % of the gas that escapes from cargo tanks during loading. Depending on the loading conditions, e.g., weather and temperature, 0.1 to as much as 2.8 kgs VOCs per ton of crude oil handled can be emitted to the atmosphere [157]. For a tanker that holds 100000 tonnes, these losses correspond to 10 to 280 tonnes being emitted to the atmosphere. NMVOCs are also released from burning of fossil fuels, including marine fuels.

5.6.2 Human and Environmental Implications

As mentioned previously, the term VOC includes the smallest hydrocarbon, i.e., methane; the main environmental impact of methane is its power as a greenhouse gas, which is described in Sect. 5.2. The residence time for NMVOCs in the atmosphere depends strongly on the chemical reactivity of the compound and varies from only a few hours to several years [66]. Many of the volatile compounds that evaporate from crude oil are directly harmful to human health and are considered carcinogenic [158]. In addition, VOCs are reactive and have a large influence on atmospheric chemistry. Organic molecules are oxidised by oxidative species in the atmosphere, which leads to altered chemical properties of the molecules themselves and also to the formation of secondary pollutants [66]. Examples of secondary pollutants are ground-level ozone and secondary particles that contribute to the formation of photochemical smog, which is described in Sect. 2.7.9.

5.6.3 Regulations

In addition to other emissions to the air, VOCs are regulated via MARPOL 73/78 Annex VI Regulation 15 [88]. However, this regulation only applies to tankers. Regulation 15 states that the control of VOCs emitted to the atmosphere should be achieved via the use of a vapour emission control system (VECS). Regulation 15 also states that all tankers that carry crude oil must have an approved and effectively implemented ship-specific VOC management plan [88].

5.7 Ozone-Depleting Substances (ODS)—Refrigerants

Ozone-depleting substances (ODS), such as chloroflurocarbons (CFCs) and hydroflurochlorocarbons (HCFCs), have traditionally been used as refrigerants in all types of refrigeration plants. The refrigeration plants on merchant vessels play a vital role in storing provisions and providing air conditioning. In refrigerated cargo ships, so-called reefer ships, the temperature of perishable or temperature-sensitive cargo (such as food, chemical, or liquefied gas) is controlled by the refrigeration plant on the ship. The presence of large refrigerant leakages in the refrigeration systems together with the responsibility of several of these refrigerants for the destruction of the ozone layer and their contribution to the greenhouse effect is well established and has resulted in a series of international treaties that demand a gradual phasing out of these compounds.

Approximately 60000 merchant ships have refrigerated systems for crew/passenger food supply and air conditioning, and only approximately 1200 ships are equipped with cargo-related on-board refrigeration systems (reefers) [159]. However, the transport of goods in refrigerated containers on ships has been increasing rapidly in recent years. The vast majority of marine refrigeration equipment use halocarbons (e.g., CFCs, HCFCs, and HFCs) as refrigerants. A refrigeration system can consist of either a direct or indirect system, in which indirect systems include an intermediate liquid for the transportation of cooled fluid between the refrigeration equipment and the "product" to be cooled, resulting in a smaller refrigerant charge. Approximately two-thirds of the systems used on ships

Table 5.8 Approximate refrigerant banks (tonnes) in the shipping sector (2006)

Mode of transportation	CFC	HCFC	HFC	Others	Total
Sea	2500	21,900	2730	130	27,390
Road	180	5280	18,520		23,980

Table 5.9 Approximate refrigerant emissions (tonnes/year) from the shipping sector (2006)

Mode of transportation	CFC	HCFC	HFC	Others R717/R744	Total
Sea (>100 GT)	1000	6750	270	10	7850
Road	50	1050	1650		2750

are direct systems with up to 5 tonnes of refrigerant per system, and the remaining third are indirect systems with a charge of less than 1 tonne of refrigerant [159].

The efficacy of ozone depletion is generally measured using a comparative unit known as the ozone-depletion potential (ODP), which is based on the ODP of the most common CFC refrigerant (assigned a value of unity). The most common HCFC refrigerant, i.e., HCFC-22, has an ODP of 0.05. For HFC refrigerants, the ODP is zero [160].

Approximately 80 % of the merchant fleet uses HCFC-22 as a refrigerant. Most of the remaining fleet uses HFCs, although certain fishing vessels use ammonia (R717) or ammonia/carbon dioxide (R744). However, certain CFC-based systems exist. Emissions from older direct systems are estimated to be 20–40 % per year, whereas emissions of 5–10 % per year can be achieved in indirect systems. Tables 5.8 and 5.9 present the estimated refrigerant banks and emissions from the shipping sector compared with those of the road sector, respectively [159]. As shown in these tables, the emission factor from the shipping sector is more than twice that of the road sector (29 % compared with 12 %) [159].

Reductions in emissions of ozone-depleting substances (ODSs) from ships have been achieved as a result of several international agreements, including the Montreal Protocol and MARPOL 73/78 Annex VI. Reductions in these emissions have been estimated based on figures in the 1998 and 2006 UNEP reports that indicate the following:

- CFCs—735 tonne reduction (98 %);
- HCFCs—10,900 tonne reduction (78 %);
- HFCs—415 tonne increase (315 %) [20].

The reason for the increase in HFC emissions is due to their use as a substitute for CFCs and HCFCs.

5.7.1 Sources

Halocarbons are a group of compounds that are primarily man-made gases consisting of both carbon and at least one of the halogens (fluorine, chlorine, iodine and bromine). These gases were first synthesised in 1928; since then, they have been widely used for a variety of purposes, such as propellants in aerosol cans, in the manufacture of soft and hard foams, and in refrigeration. The group used as refrigerants includes chlorofluorocarbons (CFCs), hydrochlorofluorocarbons (HCFCs) and hydrofluorocarbons (HFCs) [160].

The discovery of CFCs in the late 1920s revolutionised the refrigeration industry. CFCs are non-toxic, possess excellent thermodynamic properties, and are non-flammable [161]. For more than 50 years, CFCs were widely used as refrigerants, displacing carbon dioxide, sulphur dioxide and ammonia. In the 1950s, HCFCs (mainly a refrigerant known as HCFC-22) began to supplant CFCs in certain applications,-depletion potential (ODP) have gradually replaced CFCs and, to a certain extent, HCFCs as well [162].

Refrigerants are emitted to the atmosphere through leaks that occur during operation and in conjunction with the maintenance of equipment. Refrigerant gas might also be released when the unit is scrapped. Emissions of refrigerants from international shipping are related to four main sources [163]:

- Refrigeration plants on reefer ships
- Air conditioning and refrigeration of provisions on all types of ships
- Refrigerated containers carried on ships
- Refrigeration plants on fishing vessels

Refrigerant emissions to the atmosphere are often generically referred to as *losses* without distinguishing between the causes. However, the emission types are different, and their causes must be identified before they can be controlled [164]. The losses can be divided into six different types [165]:

- Sources of fugitive emissions that cannot be precisely located
- Tightness degradation caused by temperature variations, pressure cycling, and vibrations that can lead to unexpected and significant increases in refrigerant emission rates
- Component failures that mostly originate from poor construction or faulty assembly
- Losses during refrigerant handling, which occur primarily during charging of the system and opening of the system without previously recovering the refrigerant
- Accidental losses
- Losses during equipment disposal caused by intentional venting rather than recovery of the refrigerant at the end of system life.

5.7.2 Human and Environmental Implications

Halocarbon refrigerants impact the environment in two main ways: ozone layer depletion and global warming. The contribution to global warming was described in Sect. 5.2.

The depletion of the stratospheric ozone layer was observed in the 1970s. Because the ozone layer is a protector against harmful UV-B radiation, any damage to it could cause considerable harm to the environment and to life on earth. Exposure to increased UV-B radiation can lead to incidents of eye damage and skin cancer, reduced rates of plant growth, upsetting of the balance of ecosystems, and acceleration of the risk of disease. The refrigerants traditionally used on ships, i.e., CFCs and HCFCs, were found to be responsible substances [160].

When CFCs and HCFCs are released into the air, they rise into the atmosphere. Due to the stable structure of these chemicals, they drift into the stratosphere, where they react with ultraviolet light. In the reaction, CFC/HCFC molecules break down into atoms and smaller molecule fragments that act as ozone-depleting catalysts. The first step in the destruction of the stratospheric ozone layer occurs when a photon of sufficiently high energy breaks a carbon-chlorine bond in a CFC/HCFC molecule, which thus produces a chlorine atom. The chlorine atom, a free radical, subsequently participates in what is known as a chain reaction mechanism. The chlorine atom first combines with an ozone molecule to produce a chlorine monoxide (ClO) radical and an O_2 molecule, which destroys an ozone molecule. Next, the ClO radical destroys a second ozone molecule and recreates with the original chlorine atom, which can repeat the first reaction and continue to destroy ozone see Fig. 5.10. Chlorine atoms are notably efficient ozone-depleting catalysts because they accomplish this destruction without undergoing any permanent changes; therefore, the process can repeat extensively and continue until the atom is physically transported away fro the ozone layer [166].

It has been discvered that one chlorine atom can destroy 100,000 ozone molecules. Higher chlorine contents result in longer impacts on the ozone layer. CFCs

Stratospheric ozone formation and destruction

Naturally occurring in sunlight

Ozone formation

$O_2 + h\nu \leftarrow 2O\cdot$

$O\cdot + O_2 + M \leftarrow O_3 + M$

M can be a molecule or a solid surface

$h\nu$ = radiation of wavelength < 240 nm

Ozone destruction

$O_3 + h\nu \leftarrow O_2 + O\cdot$

$O\cdot + O_3 \leftarrow 2\,O_2$

$h\nu$ = radiation of wavelength <325 nm

Catalytic degradation

$X + O_3 \leftarrow XO + O_2$

$XO + O\cdot \leftarrow X + O_2$

X = Cl, NO, OH, O

From natural or anthropogenic sources, for example chlorine from CFCs

Fig. 5.10 Ozone depletion by CFCs: chemical reactions

have higher chlorine contents than HCFCs; therefore, CFCs have higher potential for ozone depletion. It is estimated that CFCs contribute to nearly 70 % of man-made ozone-depleting chemicals in the atmosphere [167]. HFC molecules do not contain any chlorine atoms; thus, their ozone-depletion potential is zero [164].

5.7.3 Regulations

The use of CFCs and HCFCs are regulated in Regulation 12 in MARPOL 73/78 Annex VI—"Emissions from ozone-depleting substances from refrigerating plants and fire-fighting equipment". More restrictive requirements for ozone-depleting substances have been enacted at the regional level, e.g., in the European Union, America, Australia and Japan [168].

The revised MARPOL 73/78 Annex VI bans the use of all CFCs in existing refrigeration systems if a ship was constructed on or after 19 May 2005. The term "use" is defined as the charging, topping up and removal of refrigerant from the system or equipment [88]. In conclusion, CFC refrigerants are prohibited from being used in any refrigeration systems on all ships, regardless of the flag administration [168].

HCFCs, such as HCFC-22, are currently undergoing a transitional period of legislation. Regulation 12 in MARPOL 73/78 makes an exception for HCFCs and only requires them to be prohibited in new installations by Jan 01 2020.

However, for developed countries, the Montreal Protocol has introduced a stricter regulation of HCFC fluids than MARPOL 73/78 [168]. For example, Regulation (EC) No 2037/2000 (recast as Reg (EC) 1005/2009) of the European Parliament prohibits the use of HCFC refrigerants in new equipment from 2002 onward. However, HCFCs were allowed for the maintenance of existing equipment until 2015, although only with recycled HCFCs since the beginning of 2010 [163].

MARPOL 73/78 Annex VI prohibits any deliberate emissions of ozone-depleting substances. Deliberate emissions include emissions occurring in the course of maintaining, servicing, repairing or disposing of systems or equipment, while deliberate emissions do not include minimal releases associated with the recapture or recycling of an ODS. Emissions arising from leaks of an ODS, whether or not the leaks are deliberate, might be regulated by individual Parties. Therefore, it is considered that a refrigerant leak monitoring system is required only if an ODS is in use and if the flag administration of a ship requires compliance with other statutory instruments or legislation, such as Reg (EC) 1005/2009 mentioned above or Reg (EC) No. 1137/2008 for European Union countries [168].

When servicing or decommissioning systems or equipment containing ODSs, the gases are to be duly collected in a controlled manner. If they are not to be reused onboard, they are to be landed to the appropriate reception facilities for banking or destruction. Any redundant equipment or material containing ODSs is to be landed ashore for appropriate decommissioning or disposal. The latter also applies when a ship is dismantled at the end of its service life [88, 168].

The revised Annex VI specifies that all ships must maintain a list of equipment containing ozone-depleting substances and that every ship exceeding 400 GT that has rechargeable systems must maintain an ODS record book onboard in which related supply, recharging, repair, discharge or disposal operations are recorded. This process will permit better operational control and benchmarking of emissions, increased awareness and aid for further emission reductions.

Regulation 12 does not control HFC refrigerants because these are not ODSs. However, within the EU, these substances are regulated in Regulation (EC) No 1137/2008 due to their global warming potential.

References

1. WHO. *Burden of disease from the joint effects of Household and Ambient Air Pollution for 2012*. 2014, World Health Organization: Geneva.
2. Lonati, G., Cernuschi, S. & Sidi, S., *Air quality impact assessment of at-berth ship emissions: Case-study for the project of a new freight port.* The Science of the total environment, 2010. **409**(1): p. 192-200.
3. Corbett, J. J. *et al.*, *Mortality from Ship Emissions: A Global Assessment.* Environmental Science & Technology, 2007. **41**(24): p. 8512-8518.
4. Eyring, V. *et al.*, *Transport impacts on atmosphere and climate: Shipping.* Atmospheric Environment, 2010. **44**(37): p. 4735-4771.
5. Hobbs, P. V. *et al.*, *Emissions from Ships with respect to Their Effects on Clouds.* Journal of Atmospheric Sciences, 1996. **57**: p. 2570-2590.
6. Petzold, A. *et al.*, *Experimental studies on particulate emissions from cruising ship, their characteristic properties, transformation and atmospheric lifetime in ther marine boundary layer.* Atmospheric Chemistry and Physics, 2008. **8**: p. 2387-2403.
7. Cooper, D. A. & Gustavsson, T. *SMED Methodology for calculating emissions from ships: 1. Update of emission factors*. 2004, SMHI Swedish Meterological and Hydrological Institute: Norrköping.
8. Winnes, H. *Air Pollution from Ships, Emission Measurements and Impact Assessments* Doctor of Philosophy thesis, Chalmers University of Technology, 2010.
9. Fridell, E., Jernström, M. & Segersson, D. *Översyn av sjöfartens emissioner av luftföroreningar [Overview of ship emissions to the air]*. 2008, Swedish Meteorological and Hydrological Institute: Norrköping.
10. European Parliament and the Council of the European Union, *Regulation (EC) No 595/2009 of the European Parliament and of the Council of 18 June 2009 on type-approval of motor vehicles and engines with respect to emissions from heavy duty vehicles (Euro VI) and on access to vehicle repair and maintenance information and amending Regulation (EC) No 715/2007 and Directive 2007/46/EC and repealing Directives 80/1269/EEC, 2005/55/EC and 2005/78/EC (Text with EEA relevance)* 2009, Official Journal of the European Union: Brussels.
11. Winnes, H., Moldanová, J., Anderson, M. & Fridell, E., *On-board measurements of particle emissions from marine engines using fuels with different sulphur content.* Proceedings of the Institution of Mechanical Engineers, Part M: Journal of Engineering for the Maritime Environment, 2014.
12. Anderson, M., Salo, K. & Fridell, E., *Particle- and gaseous emissions from a LNG powered ship.* Environmental Science & Technology, 2015. **49**(20): p. 12568–12575.

13. Moldanová, J. *et al.*, *Physical and chemical characterisation of PM emissions from two ships operating in European Emission Control Areas.* Atmospheric Measurement Techniques, 2013. **6**(12): p. 3577–3596.
14. Stone, R., *Introduction to Internal Combustion Engines*. 4th ed. 2012, Houndmills, Basingstoke, Hampshire and London UK: Palgrave Macmillan.
15. Woud, H. & Stapersma, D., *Design of Propulsion an Electric Power Generation Systems*. 2002, London: IMarEST, The Institute of Marine Engineering, Science and Technology.
16. Whall, C. *et al. Quantification of emissions from ships associated with ship movements between ports in the European Community, Report to the European Commission.* 2002, Entec: Norwhich, UK.
17. Heywood, J. B., *Internal combustion engine fundamentals*. 1988, USA: McGraw-Hill, Inc.
18. Miola, A. & Ciuffo, B., *Estimating air emissions from ships: Meta-analysis of modelling approaches and available data sources.* Atmospheric Environment, 2011. **45**(13): p. 2242–2251.
19. Smith, T. W. P. *et al. Third IMO GHG Study 2014*. 2014, International Maritime Organization (IMO): London, UK.
20. Buhaug, Ø. *et al. Second IMO GHG Study 2009*. 2009, International Maritime Organization (IMO): London, UK.
21. Bazari, Z. & Longva, T. *Assessment of IMO mandated energy efficiency measures for international shipping*. 2011, International Maritime Organization: London, UK.
22. Steinberger, J. K., Krausmann, F. & Eisenmenger, N., *Global patterns of materials use: A socioeconomic and geophysical analysis.* Ecological Economics, 2010. **69**(5): p. 1148-1158.
23. Krausmann, F. *et al.*, *Growth in global materials use, GDP and population during the 20th century.* Ecological Economics, 2009. **68**(10): p. 2696-2705.
24. Krausmann, F., Fischer-Kowalski, M., Schandl, H. & Eisenmenger, N., *The Global Sociometabolic Transition.* Journal of Industrial Ecology, 2008. **12**(5-6): p. 637-656.
25. Meinshausen, M. *et al.*, *Greenhouse-gas emission targets for limiting global warming to 2 C.* Nature, 2009. **458**(7242): p. 1158-1162.
26. Oberthür, S. & Ott, H. E., *The Kyoto Protocol: international climate policy for the 21st century*. 1999: Springer.
27. IPCC. *Climate Change 2014: Synthesis Report. Contribution of Working Groups I, II and III to the Fifth Assessment Report of the Intergovernmental Panel on Climate Change [Core Writing Team, R.K. Pachauri and L.A. Meyer (eds.)]*. 2014, IPCC: Geneva, Switzerland.
28. Raupach, M. R. *et al.*, *Global and regional drivers of accelerating CO2 emissions.* Proceedings of the National Academy of Sciences of the United States of America, 2007. **104** (24): p. 10288-10293.
29. Peters, G. P. *et al.*, *Rapid growth in CO2 emissions after the 2008-2009 global financial crisis.* Nature Clim. Change, 2012. **2**(1): p. 2-4.
30. Solomon, S., Plattner, G. K., Knutti, R. & Friedlingstein, P., *Irreversible climate change due to carbon dioxide emissions.* Proceedings of the National Academy of Sciences of the United States of America, 2009. **106**(6): p. 1704-1709.
31. Rosenzweig, C. *et al.*, *Attributing physical and biological impacts to anthropogenic climate change.* Nature, 2008. **453**(7193): p. 353-357.
32. Stroeve, J., Holland, M. M., Meier, W., Scambos, T. & Serreze, M., *Arctic sea ice decline: Faster than forecast.* Geophysical Research Letters, 2007. **34**(9).
33. Wang, M. & Overland, J. E., *A sea ice free summer Arctic within 30 years.* Geophysical Research Letters, 2009. **36**(7): p. L07502.
34. Schellnhuber, H. J. *et al. Turn down the heat: why a 4 degree C warmer world must be avoided.* 2012, World Bank: Washington D.C., USAs.
35. Hoegh-Guldberg, O. & Bruno, J. F., *The impact of climate change on the world's marine ecosystems.* Science, 2010. **328**(5985): p. 1523-1528.
36. Doney, S. C. *et al.*, *Climate Change Impacts on Marine Ecosystems.* Annual Review of Marine Science, 2012. **4**(1): p. 11-37.

37. Doney, S. C., Fabry, V. J., Feely, R. A. & Kleypas, J. A., *Ocean acidification: the other CO2 problem.* Marine Science, 2009. **1**.
38. Honisch, B. *et al.*, *The geological record of ocean acidification.* Science, 2012. **335**(6072): p. 1058-1063.
39. Nicholls, R. J. & Cazenave, A., *Sea-level rise and its impact on coastal zones.* science, 2010. **328**(5985): p. 1517-1520.
40. Pelling, M. & Uitto, J. I., *Small island developing states: natural disaster vulnerability and global change.* Global Environmental Change Part B: Environmental Hazards, 2001. **3**(2): p. 49-62.
41. Ho, J., *The implications of Arctic sea ice decline on shipping.* Marine Policy, 2010. **34**(3): p. 713-715.
42. Strong, A. L. *Tackling Maritime Bunker Fuel Emissions: The Evolution of Global Climate Change Policy at the International Maritime Organization* Master of Arts in Law and Diplomacy thesis, Tufts University, 2011.
43. Heitmann, N. & Khalilian, S., *Accounting for carbon dioxide emissions from international shipping: Burden sharing under different UNFCCC allocation options and regime scenarios.* Marine Policy, 2011. **35**(5): p. 682-691.
44. Kågeson, P. *Applying the principle of common but differentiated responsibility to the mitigation of greenhouse gases from international shipping*. 2011, Centre for Transport Studies: Stockholm.
45. Hackmann, B., *Analysis of the governance architecture to regulate GHG emissions from international shipping.* Int Environ Agreements, 2012. **12**(1): p. 85-103.
46. European Commission. *Maritime transport: first step to reduce emissions.* 2013, European Commission: Brussels. http://europa.eu/rapid/press-release_IP-13-622_en.htm.
47. European Commission. *WHITE PAPER: Roadmap to a Single European Transport Area – Towards a competitive and resource efficient transport system.* 2011, European Commission: Luxembourg.
48. Anderson, K. & Bows, A., *Executing a Scharnow turn: reconciling shipping emissions with international commitments on climate change.* Carbon Management, 2012. **3**(6): p. 615-628.
49. Johnson, H., Johansson, M., Andersson, K. & Södahl, B., *Will the ship energy efficiency management plan reduce CO2 emissions? A comparison with ISO 50001 and the ISM code.* Maritime Policy & Management, 2013. **40**(2): p. 177-190.
50. World Wildlife Fund and Clean Shipping Coalition. *MEPC 64/4/33 - Proposed elements for enhancing implementation requirements for SEEMP and SEEMP Guidelines*. 2012, World Wildlife Fund and Clean Shipping Coalition: Marine Environment Protection Committee, 64th session.
51. Maddox Consulting. *Analysis of market barriers to cost effective GHG emission reductions in the maritime transport sector.* 2012, European Commission: Brussels.
52. Faber, J. & Hoen, M. *Historical trends in ship design efficiency*. 2015, CE Delft: Delft.
53. Faber, J., Hoen, M., Koopman, M., Nelissen, D. & Ahdour, S. *Estimated Index Values of New Ships - Analysis of EIVs of Ships That Have Entered The Fleet Since 2009*. 2015, CE Delft: Delft.
54. Vestreng, V., Myhre, G., Fagerli, H., Reis, S. & Tarrasón, L., *Twenty-five years of continuous sulphur dioxide emission reduction in Europe.* Atmospheric Chemistry and Physics, 2007. **7**: p. 3663 - 3681.
55. Smith, S. J. *et al.*, *Anthropogenic sulfur dioxide emissions: 1850–2005.* Atmos. Chem. Phys., 2011. **11**(3): p. 1101-1116.
56. Warfvinge, P. & Bertills, U. *Naturens återhämtning från försurning – aktuell kunskap och framtidsscenarier. [The Recovery of Nature from Acidfication - Current Knowledge and Future Scenarios].* 2000, Swedish Environmental Protection Agency.
57. Klimont, Z., Smith, S. J. & Cofala, J., *The last decade of global anthropogenic sulfur dioxide: 2000–2011 emissions.* Environmental Research Letters, 2013. **8**(1): p. 014003.

58. Adams, M., Fernandez, R. & Barkman, A., *Development of Greenhouse Gas and Air Pollution Emissions* in *Air pollution and climate change: two sides of the same coin?*, Pleijel, H. (ed.). 2009, Swedish Environmental Protection Agency: Stockholm.
59. Grennfelt, P., *Common Roots of Air Pollution and Climate Change* in *Air pollution and climate change: two sides of the same coin?*, Pleijel, H. (ed.). 2009, Swedish Environmental Protection Agency: Stockholm.
60. European Environment Agency. *European Union emission inventory report 1990–2010 under the UNECE Convention on Long-range Transboundary Air Pollution (LRTAP). EEA Technical report, No 8/2012*. 2012, European Environment Agency.
61. USEPA, *Average annual emissions, all criteria pollutants in MS Excel. National Emissions Inventory (NEI) Air Pollutant Emissions Trends Data, February 2014*. 2014, United States Environmental Protection Agency.
62. IMO. *MEPC 68/3/2. Sulphur monitoring for 2014. Note by the Secretariat, 9 February 2015. Marine Environment Protection Committee, 68th session, agenda item 3. London: International Maritime Organization*. 2014.
63. Corbett, J. J. & Fischbeck, P., *Emissions from Ships*. Science, 1997. **278**(5339): p. 823-824.
64. Agrawal, H., Malloy, Q. G. J., Welch, W. A., Wayne Miller, J. & Cocker Iii, D. R., *In-use gaseous and particulate matter emissions from a modern ocean going container vessel*. Atmospheric Environment, 2008. **42**(21): p. 5504-5510.
65. Pleijel, H. *(ed.), Air pollution & Climate change Two sides of the same coin?*, 2009. Swedish Environmental Protection Agency: Stockholm.
66. Finlayson-Pitts, B. J. & Pitts Jr., J. N., *Chemistry of the Upper and Lower Atmosphere - Theory, Experiments and Applications*. 2000, USA: Academic Press.
67. Lövblad, G., Tarrasón, L., Torseth, K. & Dutchak, S. *EMEP Assessment: Part I: European Perspective*. 2004, Norwegian meteorological institute: Oslo.
68. Swedish Ministry for Foreign Affairs and Swedish Ministry of Agriculture. *Air pollution across national boundaries: the impact on the environment of sulfur in air and precipitation : Sweden's case study for the United Nations Conference on the Human Environment, 1972*. Stockholm: Norstedt.
69. *1979 Convention on Long-range Transboundary Air Pollution* 16 November 1979, entered into force 16 March 1983, Treaty Series 1302 217, 1979.
70. DeSombre, E. R., *The Global Environment and World Politics*. 2007: Continuum International Pub. Group.
71. Dalsøren, S. B. *et al.*, *Update on emissions and environmental impacts from the international fleet of ships: the contribution from major ship types and ports*. Atmos. Chem. Phys., 2009. **9** (6): p. 2171-2194.
72. Swedish Environmental Protection Agency. *Fördjupning - Nedfall av svavel. [Additional Information - Sulfur deposition]* 2015: Stockholm.
73. Svensson, E. *Sulphur Regulations for Shipping - Why a Regional approach?: Scientific and Economic Arguments in IMO Documents 1988-1997* Doctor of Philosophy thesis, Chalmers University of Technology, 2014.
74. Swedish Maritime Administration, *Sjöfartsverkets författningssamling, Sjöfartsverkets föreskrifter om villkor för miljödifferentierade farledsavgifter* [Description of the Swedish Maritime Administration's requirement for environmentally differentiated fairway dues], SJÖFS 1989:13, Norrköping, Sweden.
75. IMO. *Prevention of Air Pollution from Ships*. 2015, International Maritime Organization: London, http://www.imo.org/en/OurWork/Environment/PollutionPrevention/AirPollution/Pages/Air-Pollution.aspx, last accessed 20 May 2015.
76. IMO, *Air Pollution, Energy Efficiency and Greenhouse Gas Emissions.*, Marine Environment Protection Committee International Maritime Organization. 2015: London.
77. IMO. *Special Areas under MARPOL*. 2015, International Maritime Organization: London, http://www.imo.org/en/OurWork/Environment/SpecialAreasUnderMARPOL/Pages/Default.aspx, last accessed on 30 September 2015.

78. European Commission. *Transport & Environment: Emissions from Maritime Transport*. 2015, European Commission: Brussels, http://ec.europa.eu/environment/air/transport/ships.htm, last accessed 20 May 2015.
79. European Parliament and the Council of the European Union, *Directive 2012/33/EU of the European Parliament and of the Council of 21 November 2012 amending Council Directive 1999/32/EC as regards the sulphur content of marine fuels*. 2012, Official Journal of the European Union: Brussels.
80. European Parliament and the Council of the European Union, *Directive 97/68/EC of the European Parliament and of the Council of 16 December 1997 on the approximation of the laws of the Member States relating to measures against the emission of gaseous and particulate pollutants from internal combustion engines to be installed in non-road mobile machinery*. 1997, Official Journal of the European Communities: Brussels.
81. European Parliament and the Council of the European Union, *Directive 2001/81/EC of the European Parliament and of the Council of 23 October 2001 on national emission ceilings for certain atmospheric pollutants* 2001, Official Journal of the European Communities: Brussels.
82. European Parliament and the Council of the European Union, *Directive 2004/ 26/EC of the European Parliament and of the Council of 21 April 2004 amending Directive 97/68/EC on the approximation of the laws of the Member States relating to measures against the emission of gaseous and particulate pollutants from internal combustion engines to be installed in non-road mobile machinery* 2004, Official Journal of the European Union: Brussels.
83. European Parliament and the Council of the European Union, *Directive 2008/50/EC of the European Parliament and of the Council of 21 May 2008 on ambient air quality and cleaner air for Europe* 2008, Official Journal of the European Union: Brussels.
84. European Commission, *Proposal for a Council Decision on the position to be adopted on behalf of the European Union at the International Maritime Organization during the 66th session of the Marine Environment Protection Committee on the adoption of amendments to MARPOL Annex VI concerning the delay of the Tier III NOx emission standards* 2014, European Commission: Brussel.
85. Gauss, M., Nyiri, A., Semeena, S., Benedictow, A. & Klein, H. *Transboundary air pollution by main pollutants (S, N, O3) and PM. Sweden*. 2013, Norwegian Meteorological Institute: Oslo.
86. Zeldovich, J., *The oxidation of Nitrogen in Combustions and Explosions*. Acta Physicochimica U.R.S.S., 1946. **21**(4): p. 577-628.
87. Stone, R., *Introduction to Internal Combustion Engines*. 3rd ed. Houndmills, Basingstoke, Hampshire and London UK. 1999: Palgrave Macmillan.
88. IMO, *MARPOL Annex VI and NTC 2008 with guidelines for implementation*. 2013, London: International Maritime Organization.
89. Turns, S. R., *An Introduction to Combustion - Concepts and Applications*. 2nd ed. 2000, Singapore: McGraw-Hill.
90. Lavoie, G. A., Heywood, J. B. & Keck, J. C., *Experimental and Theoretical Study of Nitric Oxide Formation in Internal Combustion Engines*. Combustion Science and Technology, 1970. **1**: p. 313-326.
91. Henningsen, S., *Chapter 14 - Air Pollution from Large Two-Stroke Diesel Engines and Technologies to Control It*, in *Handbook of Air Pollution From Internal Combustion Engines*, Sher Eran, ed. 1998, Academic Press: San Diego. p. 477-534.
92. Pleijel, H. (ed.), *Air pollution and Climate change Two sides of the same coin?*, 2009. Swedish Environmental Protection Agency: Stockholm.
93. WHO. *Air Quality Guidelines, Global Update 2005. Particulate matter, ozone, nitrogen dioxide and sulfur dioxide*. 2006, World Health Organization: Geneva.
94. Levinton, J. S., *Marine Biology - Function, Biodiversity, Ecology*. 2nd ed. 2001, New York, Oxford: Oxford University Press.

95. IMO, *MARPOL Consolidated Edition 2006 - Articles, Protocols, Annexes, Unified Interpretations of the International Convention for the Prevention of Pollution from Ships, 1973, as modified buy the Protocol of 1978 relating thereto* 2006, London: International Maritime Organisation.
96. IMO. *Resolution MEPC.251(66). Adopted on 4 April 2014. Amendments to the Annex of the protocol of 1997 to amend the International Convention for the Prevention of Pollution from Ships, 1973, as modofied thereto*. 2014, International Maritime Organization: London.
97. HELCOM. *Helsinki Commission - Baltic Marine Environment Protection Commission, Minutes of the 33rd Meeting*. 2012.
98. HELCOM. *Helsinki Commission - Baltic Marine Environment Protection Commission, Agenda Item 4 - Matters arising from the subsidiary bodies, Report of the NECA Correspondence Group on the designation of the Baltic Sea as a NOx Emission Control Area*. 2012.
99. PBL Netherlands Environmental Assessment Agency. *Assessment of the environmental impacts and health benefits of a nitrogen emission control area in the North Sea, Policy studies*. 2012: The Hague/Bilthoven, The Netherlands.
100. Danish Ministry of the Environment - Environmental Protection Agency. *Economic Impact Assessment of a NOx Emission Control Area in the North Sea, Environmental Project no. 1427, 2012*. 2012, Miljøstyrelsen: København, Denmark.
101. European Environment Agency. *The impact of international shipping on European air quality and climate forcing, EEA Technical Report No 4/2013, ISSN 1725-2237*. 2013, European Environment Agency.
102. IMO. *MEPC 62/24, Report of the Marine Environment Protection Committee on its sixty-second session*. 2011, International Maritime Organization, Marine Environment Protection Committee; London
103. IMO. *MEPC 64/INF.8, Air Pollution and Energy Efficiency, Supplementary Information to the Interim Report of the Correspondence Group on Assessment of Technological Developments to Implement the Tier III NOx Emission Standards under MARPOL Annex VI, Submitted by the United States of America*. 2012, International Maritime Organization, Marine Environment Protection Committee: London.
104. IMO. *MEPC 65/INF.10, Air Pollution and Energy Efficiency, Supplementary Information to the final report of the Correspondence Group on Assessment of Technological Developments to Implement the Tier III NOx Emission Standards under MARPOL Annex VI, Submitted by the United States*. 2013, International Maritime Organization, Marine Environment Protection Committee: London.
105. IMO. *MEPC 65/4/27, Air Pollution and Energy Efficiency, Comments on the report of the Correspondence Group on Assessment of Technological Developments to Implement the Tier III NOx Emission Standards under MARPOL Annex VI, submitted by the Russian Federation*. 2013, International Maritime Organization, Marine Environment Protection Committee: London.
106. IMO. *MEPC 65/22, Report of the Marine Environment Protection Committee on its sixty-fifth session*. 2013, International Maritime Organization, Marine Environment Protection Committee: IMO, London.
107. IMO. *MEPC 66/6/3, Consideration and Adoption of Amendments to Mandatory Instruments, Note by the Secretariat*. 2013, International Maritime Organization, Marine Environment Protection Committee: London.
108. IMO. *Circular letter No.3370, Amendments to MARPOL and the BCH Code, the IBC Code and the NOx Technical Code 2008*. 2013, International Maritime Organization: London.
109. IMO. *MEPC, 66/INF.4, Air Pollution and Energy Efficiency, Information about the application status of Tier III compliant technologies, Submitted by EUROMOT*. 2013, International Maritime Organization, Marine Environment Protection Committee: London.
110. European Commission, *Commission Directive 2010/26/EC*. 2010, European Union: Brussels.

111. DieselNet. *EU Emission Standards for Heavy-Duty Diesel Engines*. Revision: 2012.09, http://www.dieselnet.com/standards/eu/hd.php, last accessed on 20 February 2014.
112. European Commission, *Commission Regulation (EU) No 582/2011*. 2011, European Commission: Brussel.
113. NHO - Næringslivets Hovedorganisasjon. *The NOx Fund - Affiliation*. 2014: Oslo, Norway, https://www.nho.no/Prosjekter-og-programmer/NOx-fondet/The-NOx-fund/Affiliation/, last accessed on 31 March 2014.
114. NHO - Næringslivets Hovedorganisasjon. *Participant Agreement to the Environmental Agreement on NOx*. 2014: Oslo, Norway, https://www.nho.no/siteassets/nhos-filer-og-bilder/filer-og-dokumenter/nox-fondet/participand_agreement_to_the_env.-agr.-as-of-14.15.13.pdf, last accessed on 31 March 2014.
115. NHO - Næringslivets Hovedorganisasjon. *Environmental Agreement Concerning Reduction of NOx Emissions for the Period 2011-2017*. 2014: Oslo, Norway, https://www.nho.no/siteassets/nhos-filer-og-bilder/filer-og-dokumenter/nox-fondet/hovedside-nox-fondet/english-version/affiliation/miljoavtalen-2011-2017-engelsk-110211.pdf, last accessed on 31 March 2014.
116. NHO - Næringslivets Hovedorganisasjon. *The Environmental Agreement*. 2014: Oslo, Norway, https://www.nho.no/Prosjekter-og-programmer/NOx-fondet/The-NOx-fund/The-NOx-Fund-and-the-Environmental-Agreement/The-Environmental-Agreement/, last accessed on 31 March 2014.
117. Norwegian Maritime Authority. *NOx-tax*. 2014, http://www.sjofartsdir.no/en/environment/forebyggelse-mot-forurensing-fra-skip/utslipp-til-luft/nox/nox-tax/, Last accessed on 31 March 2014.
118. UNECE. *United Nations Economic Commission for Europe, Protocol to Abate Acidification, Eutrophication and Ground-level Ozone* 2013, http://live.unece.org/env/lrtap/multi_h1.html, last accessed 17 July 2013.
119. Næringslivets NOx-fond, *Application for Support to NOx-reducing Measures - A Guide, Version 4.3, 17 February 2012*. 2012: Norway.
120. IMO. *International Maritime Organization, Marine Environment Protection Committee, MEPC 58/23/Add.1, Annex 14, RESOLUTION MEPC.177(58), Adopted on 10 October 2008, Amendments to the Technical Code on Control of Emissions of Nitrogen Oxides from Marine Diesel Engines, (NOx Technical Code 2008)*. 2008.
121. Swedish Maritime Administration, *Sjöfartsverkets författningssamling, Sjöfartsverkets föreskrifter om villkor för miljödifferentierade farledsavgift;, SJÖFS 2012:9, Norrköping, Sweden [Swedish regulations on environmentally differentiated fairway dues, only available in Swedish]*. 2012.
122. Eyring, V., *Emissions from international shipping: 1. The last 50 years*. Journal of Geophysical Research, 2005. **110**(D17).
123. Isaksen, I. S. A. *et al.*, *Chapter 12 Atmospheric Composition Change: Climate-Chemistry Interactions*, in *The Future of World's Climate*, Ann. Henderson-Sellers & Kendal. McGuffie, eds. 2012.
124. Petzold, A. *et al.*, *Recommendations for reporting "black carbon" measurements*. Atmospheric Chemistry and Physics, 2013. **13**(16): p. 8365-8379.
125. Ning, Z., Cheung, C. S. & Liu, S. X., *Experimental investigation of the effect of exhaust gas cooling on diesel particulate*. Journal of Aerosol Science, 2004. **35**(3): p. 333-345.
126. Lack, D. A. *et al.*, *Particulate emissions from commercial shipping: Chemical, physical, and optical properties*. Journal of Geophysical Research, 2009. **114**.
127. Moldanová, J. *et al.*, *Characterisation of particulate matter and gaseous emissions from a large ship diesel engine*. Atmospheric Environment, 2009. **43**(16): p. 2632-2641.
128. Amann, C. A. & Siegla, D. C., *Diesel Particulates—What They Are and Why*. Aerosol Science and Technology, 1981. **1**(1): p. 73-101.

129. Kasper, A., Aufdenblatten, S., Forss, A., Mohr, M. & Burtscher, H., *Particulate Emissions from a Low-Speed Marine Diesel Engine.* Aerosol Science and Technology, 2007. **41**(1): p. 24-32.
130. Lack, D. *et al.*, *Light absorbing carbon emissions from commercial shipping.* Geophysical Research Letters, 2008. **35**(13).
131. Petzold, A. *et al.*, *Physical Properties, Chemical Composition, and Cloud Forming Potential of Particulate Emissions from a Marine Diesel Engine at Various Load Conditions.* Environmental Science & Technology, 2010. **44**(10): p. 3800-3805.
132. Petzold, A. *et al.*, *Operation of marine diesel engines on biogenic fuels: modification of emissions and resulting climate effects.* Environ Sci Technol, 2011. **45**(24): p. 10394-10400.
133. Lappi, M., Borkowski, T. & Myskow, J. *Ship emissions and abatement technology assessment.* 2012, VTT Technical Research Centre of Finland, Maritime University of Szczecin and Clean Shipping Currents.
134. Lack, D. A. *et al.*, *Impact of Fuel Quality Regulation and Speed Reductions on Shipping Emissions: Implications for Climate and Air Quality.* Environmental Science & Technology, 2011. **45**(20): p. 9052-9060.
135. Hallquist, A. M., Fridell, E., Westerlund, J. & Hallquist, M., *Onboard measurements of nanoparticles from a SCR-equipped marine diesel engine.* Environ Sci Technol, 2013. **47**(2): p. 773-780.
136. Ushakov, S., Valland, H., Nielsen, J. B. & Hennie, E. *Particulate emission characteristics from medium-speed marine diesel engines.* in *PACIFIC 2012 International Maritime Conference.* 2012. Sydney, Australia.
137. Hinds, W. C., *Aerosol technology.* 1999: John Wiley & Sons, Inc.
138. Matti Maricq, M., *Chemical characterization of particulate emissions from diesel engines: A review.* Journal of Aerosol Science, 2007. **38**(11): p. 1079-1118.
139. Lyyränen, J., Jokiniemi, J., Kauppinen, E. I. & Joutsensaari, J., *Aerosol characterisation in medium-speed diesel engines operation with heavy fuel oils.* Journal of Aerosol Science, 1999. **30**(6): p. 771-784.
140. Petzold, A., Weinzierl, B., Fiebig, M. & Lichtenstern, M. *Particle Emissions from Ship Engines: Emissions Properties and Transformation in the Marine Boundary Layer.* in *TAC-Conference.* 2006.
141. Fridell, E., Steen, E. & Peterson, K., *Primary particles in ship emissions.* Atmospheric Environment, 2008. **42**(6): p. 1160-1168.
142. Elvingson, P. & Ågren, C., *Air and the environment.* 2004, Gothenburg: The Swedish NGO Secretariat on Acid Rain.
143. Pope, C. A. & Dockery, D. W., *Health Effects of Fine Particulate Air Pollution: Lines that Connect.* Air & Waste Management Association, 2006. **56**: p. 709-742.
144. Health Effects Institute. *Understanding the Health Effects of Ambient Ultrafine Particles.* 2013, Health Effects Institute, HEI: Boston, MA.,
145. Delfino, R. J., Sioutas, C. & Malik, S., *Potential Role of Ultrafine Particles in Associations between Airborne Particle Mass and Cardiovascular Health.* Environmental Health Perspectives, 2005. **113**(8): p. 934-946.
146. Lighty, J. S., Veranth, J. M. & Sarofim, A. F., *Combustion Aerosols: FActors Governing Their Size and Composition and Implications to Human Health.* Journal of Air & Waste Management Association, 2000. **50**: p. 1565-1618.
147. Donaldson, K. *et al.*, *Combustion-derived nanoparticles: a review of their toxicology following inhalation exposure.* Particle and fibre toxicology, 2005. **2**: p. 10.
148. Forster, P. *et al. Changes in Atmospheric Constituents and in Radiative Forcing* 2007: Cambridge University Press, Cambridge, United Kingdom and New York, NY, USA.
149. Lauer, A., Eyring, V., Corbett, J., Wang, C. & Winebrake, J. J., *Assessment of Near-Future Policy Instruments for Oceangoing Shipping: Impact on Atmospheric Aerosol Burdens and the Earth's Radiation Budget.* Environ Sci Technol, 2009. **43**: p. 5592-5598.

150. Corbett, J., Winebrake, J. J. & Green, E. H., *An assessment of technologies for reducing regional short-lived climate forcers emitted by ships with implications for Artic shipping.* Carbon Management, 2010. **1**(2): p. 207-225.
151. European Commission, *Commission Regulation (EU) No 459/2012 of 29 May 2012 amending Regulation (EC) No 715/2007 of the European Parliament and of the Council and Commission Regulation (EC) No 692/2008 as regards emissions from light passenger and commercial vehicles (Euro 6) Text with EEA relevance* 2012, Official Journal of the European Union: Brussels.
152. Miljödepartementet [Swedish Ministry of Environment], *Luftkvalitetsörordning (2010:477) [Air Quality regulation].* 2010: Svensk författningssamling.
153. The European Parliament and the Council of the European Union, *Directive 2008/50/EC of the European Parliament and of the Council of 21 May 2008 on ambient air quality and cleaner air for Europe* 2008, Official Journal of the European Union: Brussels.
154. Sub-Committee on Pollution Prevention and Response (PPR). *PPR 2-21 - Report To The Marine Environment Protection Committee (Secretariat).* 2015, IMO.
155. European Commission. *COMMISSION STAFF WORKING DOCUMENT Greening the fleet: reducing pollutant emissions in inland waterway transport. Accompanying the document Communication from the Commission to the European Parliament, the Council, the European Economic and Social Committee and the Committee of the Regions Towards quality inland waterway transport* 2013, European Commission: Brussels.
156. Tamaddoni, M., Sotudeh-Gharebagh, R., Nario, S., Hajihosseinzadeh, M. & Mostoufi, N., *Experimental study of the VOC emitted from crude oil tankers.* Process Safety and Environmental Protection, 2014. **92**(6): p. 929-937.
157. Martens, O., M, Oldervik, O. & Terje, S., *Control of VOC emissions from crude oil tankers.* Marine Technology, 2001. **38**(3): p. 208-217.
158. Kampa, M. & Castanas, E., *Human health effects of air pollution.* Environmental Pollution, 2008. **151**(2): p. 362-367.
159. UNEP. *2006 Report of the refrigeration, air conditioning and heat pumps technical options committee.* 2007, United Nations Environment Programme: Nairobi.
160. Bolaji, B. O. & Huan, Z., *Ozone depletion and global warming: Case for the use of natural refrigerant – a review.* Renewable and Sustainable Energy Reviews, 2013. **18**(0): p. 49-54.
161. Venkatarathnam, G. & Srinivasa Murthy, S., *Refrigerants for vapour compression refrigeration systems.* Reson, 2012. **17**(2): p. 139-162.
162. Benhadid-Dib, S. & Benzaoui, A., *Refrigerants and their Environmental Impact Substitution of Hydro Chlorofluorocarbon HCFC and HFC Hydro Fluorocarbon. Search for an Adequate Refrigerant.* Energy Procedia, 2012. **18**(0): p. 807-816.
163. UNEP, *2010 Report of the Refrigeration, Air Condition and Heat Pumps Technical Options Committee.* 2011, Nairobi: United Nations Environment Programme.
164. *ASHRAE Handbook Refrigeration American Society of Heating* in *Refrigerating and Air-Conditioning Engineers*, ASHRAE, ed. 2010.
165. Clodic, D., *Zero Leaks*, ed. ASHRAE Book. 1997.
166. Baird, C. & Cann, M., *Environmental Chemistry.* 2008: W. H. Freeman and Company, New York.
167. UNEP. *Handbook for International treaties for protection of the ozone layers.* 2003, United Nations Environment Programme: Nairobi.
168. Lloyds Register. *Revised MARPOL Annex VI – Regulation 12 Guidance notes.* 2010, http://www.lr.org/en/_images/213-35718_Revised_Annex_VI_Regulation_12_Guidance_Notes_formatted.pdf, last acessed 30 September 2015.
169. European Comission, Commission of the European Communities (SEC (2005) 1133) - Comission Staff working paper, Annex to: The Communication on Strategy on Air Pollution and the Directive on "Ambient Air Quality and Cleaner Air for Europe", 2005. European Commission: Brussels.

170. European Commission, COM (2014) 83: Proposal for a Council Decision on the position to be adopted on behalf of the European Union at the International Maritime Organization during the 66th session of the Marine Environment Protection Committee on the adoption of amendments to MARPOL Annex VI concerning the delay of the Tier III NO emission standards, 2014. European Commission: Brussels.
171. IMO, MEPC 64/4/16, Air Pollution and Energy Efficiency, Interim Report of the Correspondence Group on Assessment of Technological Developments to Implement the Tier III NO Emission Standards under MARPOL Annex VI, Submitted by the United States of America, 2012. International Maritime Organization: London.
172. IMO, MEPC 65/4/7, Air Pollution and Energy Efficiency, Final Report of the Correspondence Group on Assessment of Technological Developments to Implement the Tier III NO Emission Standards under MARPOL Annex VI, Submitted by the United States of America, 2013. International Maritime Organization: London.
173. Swedish Maritime Administration, Sjöfartsverkets föreskrifter om farledsavgift [Description of the Swedish Maritime Administration's fairway dues], SJÖFS 1988:12, Norrköping, Sweden.
174. European Commission. Directive 2009/30/EC of the European Parliament and of the Council of 23 April 2009 amending Directive 98/70/EC as regards the specification of petrol, diesel and gas-oil and introducing a mechanism to monitor and reduce greenhouse gas emissions and amending Council Directive 1999/32/EC as regards the specification of fuel used by inland waterway vessels and repealing Directive 93/12/EEC, 2009. Official Journal of the European Union: Brussels.

6 Anthropogenic Noise

J. Fredrik Lindgren and Magda Wilewska-Bien

Abstract

Anthropogenic noise is an issue that originates both from ships and the infrastructure that supports them, i.e., ports. Noise pollution is a known issue that can affect both humans and marine organisms. Humans are affected when ports, which are often active throughout the day and night, are located near residential areas in a city. Marine organisms are affected when noise from various activities of the shipping industry is transferred into the water. Four main sources of anthropogenic noise are generally recognised: underwater explosions, seismic explorations with high-energy systems, active sonar systems and shipping. The primary concerns regarding organisms exposed to elevated levels of anthropogenic noise include permanent or temporary hearing loss, the masking of a desired signal, and behavioural changes in response to a sound. Noise generated in port areas can affect both the staff working at the port and the neighbouring areas. Several negative health effects of noise pollution have been identified, for example, hearing and cardio-vascular disturbances, increased blood pressure, annoyance and sleep disturbance.

Keywords

Noise · Underwater explosions · Seismic exploration · Active sonar · Shipping · Health effects

J.F. Lindgren (✉) · M. Wilewska-Bien
Shipping and Marine Technology, Chalmers University of Technology, Gothenburg, Sweden
e-mail: f.lindgren@chalmers.se

M. Wilewska-Bien
e-mail: magda.w.bien@chalmers.se

K. Andersson et al. (eds.), *Shipping and the Environment*,
DOI 10.1007/978-3-662-49045-7_6

6.1 Noise

Noise pollution from the shipping industry is a known issue that can affect both humans and marine organisms. Humans are affected when ports, which often are active throughout the day and night, are located near residential areas in a city. Marine organisms are affected when noise from various activities of the shipping industry is transferred into the water.

6.1.1 Underwater Noise

The sensing of sound is very important for multiple aquatic species. In the viscous, opaque aquatic environment, sensory systems, such as vision, smell, taste, and touch, are limited by their range and speed of transmission [1]. However, sound travels nearly five times faster in water compared with air and can travel great distances at low frequencies, e.g., hundreds of kilometres. The most important factors affecting sound propagation are the frequency of the sound, density differences in the water column and the properties of the water [2]. Numerous animals depend on sound to navigate, find mates, search for food and communicate [3].

6.1.1.1 Types of Sounds/Sources

The marine environment is by nature filled with various sounds of natural origin, including underwater earthquakes, volcanic eruptions, lightning strikes, breaking waves, and precipitation at the water surface. Sound is also emitted by animals in the sea, for example, fish, mammals and snapping shrimp [4, 5]. Anthropogenic noise generated from human activities often differs from naturally generated sounds in frequency, duration, and directionality [5]. Four main sources of anthropogenic noise exist: underwater explosions, seismic explorations with high-energy systems, active sonar systems and shipping [1, 5]. Other types of anthropogenic noise include acoustic harassment devices, recreational boating, drilling and oceanographic experiments [5], although these sources are uncommon in comparison with the four main sources.

Underwater explosions are used during different activities, such as the removal of underwater structures or obstacles, naval exercises, or for ship-shock trials, where the hull integrity of ships is tested. In addition to acoustics, this type of sound produces a shockwave that can have mechanical effects on organisms, e.g., organ damage and rupture of gas-filled cavities, such as lungs, swim bladder, and ears, due to the dramatic pressure drop over a short duration [5].

In contrast with passive sonar systems that listen to and receive sounds, military active sonar systems produce and emit sound. The received echoes are then examined and interpreted. Numerous countries have these sonar systems deployed on their naval vessels and/or the ability to release them from aircraft into the ocean as floating sensors [3]. These systems often produce intense sound that can range from low (<1000 Hz) to high (>10 kHz) frequencies, reaching levels up to 210 dB.

The potential area of impact on cetaceans is estimated to be up to 3.9 million km^2 [5]. Several linkages have been observed between military exercises with active sonar and the stranding of cetaceans, e.g., in the 1980s and 2002 in the Canary Islands, 1996 in Greece, 2000 in Madeira, and 2000 in Bahamas. The most plausible explanation for these strandings is believed to be gas or fat embolism, which induces a type of decompression syndrome in supersaturated tissue, causing a behavioural response to the sonar systems [6].

High-energy seismic systems (airguns) are currently the industry standard for seismic exploration. Short, pulsed, low-frequency sounds are directed downward towards the sea floor to maximise the reflection of sound back to the surface. The reflected sound is analysed by geophysicists to determine the presence of formations commonly linked with oil and gas deposits present on the sea floor. The downward directionality of the sounds somewhat lessens the effects on marine life. Levels of sound in other directions have been less extensively examined, although significantly amounts of energy are introduced [3]. Animals known to be negatively affected by airguns include sperm whales, bowhead whales, harbour porpoises, and commercial fish species, such as pink snapper, cod and haddock [3, 7]. Alternatives and improvements to the airgun technique are in development. Improvements in the signal processing techniques that are used to analyse sound waves and identify pockets of oil and gas in the bedrock have reduced the need for low-frequency sounds. A new alternative to airguns is the marine vibrator, which uses less energy and employs lower-amplitude waves that are spread out over a longer period of time [3, 8].

In commercial shipping, in general, larger ships have greater acoustic outputs. The main source of noise from ships is the propeller, particularly through the phenomenon of cavitation. When the propeller rotates, different parts of the blades are subject to various pressures, including pressures less than the vapour pressure under certain circumstances, which generate cavities (non-spherical bubbles) on the surface of the blade. When the blade moves to regions of higher pressure, the cavities quickly collapse, generating shock waves which in turn produce a hissing sound and may damage the propeller. Propeller cavitation typically generates noise in the frequency range 10 Hz–10 kHz near the collapse point of the cavity; hundreds of kHz may be recorded. At lower frequencies, the noise spectrum is dominated by 1–5 peaks as the blade spinning frequency increases, whereas at higher frequencies, the spectrum is continuous as a result of partly random collapses. The lowest frequencies, typically 10–30 Hz, are primarily experienced as hull vibrations on board the vessel, although this hull vibration constitutes a rather effective source of low-frequency noise that is transmitted into the sea.

This emitted sound increases with ageing propellers due to the added wear or the attachment of marine fouling, e.g., barnacles. Other sounds emitted from ships include the roar of engines, the rattling of bearings and the vibrations of the outer hull. In addition, most shipping globally is short-distance sea shipping, where routes commonly run through habitats containing many marine mammal species [3]. Because commercial shipping is global, noise pollution generated by shipping has likely become inescapable for marine animals [9]. Many studies have documented the response of organisms to noise from shipping [3]. Fish species experience higher stress levels and diminished hearing ability [10]. Grey whales, humpback whales and belugas have been displaced from their habitats [3]. Manatees have been observed to change their fluke rate, heading and dive rate [11]. Narwhals have been observed to freeze in place and fall silent [3]. Methods to reduce noise from ships have been studied for several decades because every decibel of created sound is wasted energy. However, a propeller design with no cavitation is usually less efficient than a propeller with moderate cavitation. New and improved designs for propellers with sweeping blades to reduce cavitation are being studied. Noise damping from the engine using isolation mounts, dampening tiles and flexible hoses is also a possibility [3].

6.1.1.2 Organisms Affected and Impacts

Four primary concerns exist regarding organisms exposed to elevated levels of anthropogenic noise. Permanent and temporary hearing losses are modifications in the amplitude and frequency that are needed to detect sound. Masking occurs when alien sounds cover a desired signal, rendering detection more difficult, which can hinder communication, mating calls, and the search for prey. The final effect is behavioural changes in response to a sound, which can include the abandonment of an important activity, e.g., feeding, nursing or the use of an area, in response to a sound [1, 12]. The most frequently reported impacts of anthropogenic noise are related to marine mammals. However, evidence exists that indicates multiple types of organisms may also be affected, e.g., pink snapper, cod, shrimp, snow crab, and giant squid [3].

6.1.1.3 Regulations

No international law currently regulates underwater noise. However, some countries have national regulations, e.g., the Marine Mammal Protection Act in the USA and the prohibition of all active sonar systems outside the Canary Islands by Spain. In 2004, the European Parliament called for "moratoriums and restrictions of the use of high-intensity active sonars" from its member states [13]. In the same year, sixteen countries bordering the Mediterranean and Black Seas called for "a common set of guidelines" to reduce noise pollution [14].

6.1.2 Noise from Port Areas

Apart from problems with underwater noise, ports also generate noise pollution. The sound generated in different areas of a port can propagate, affecting the staff working at the port and the neighbouring area, which typically comprises the natural environment and residential areas situated in close proximity. Several negative health effects caused by noise pollution, including hearing and cardio-vascular disturbances, increased blood pressure, annoyance and sleep disturbance, have been identified [15].

Several sources of noise exist in port areas, which are generally related to berthing vessels and cargo handling operations. Thus, engines, ship ventilation fans, cargo handling equipment and warning sirens are characteristic noise generators in port areas [16]. The European port sector recognised noise pollution as their top environmental priority in 2009 [17].

While at port, the main engines of berthing ships are switched off, and the auxiliary engines produce the power needed for on-board activities (if shore-side electricity is not available). As a side effect, low-frequency noise is generated by the auxiliary engines during the full duration of the ship's stay; the intensity of the noise depends on the type of silencers, engine and design [16]. Electric power while berthed can sometimes be generated by the propeller shaft generator. As a result, the propeller turns and generates a significant amount of cavitation noise. The impact of the low-frequency sound generated by the ship may increase when resonating with various structures, such as large windows situated in close proximity [16]. In addition, ship repair operations while in port may generate noise, e.g., hull blasting.

The cargo handling operations in port, including the use of cranes, forklifts, and container lashing bars, are another source of noise. The crane engines, warning sirens, auto-backing beepers, screechy chains and wires, and crashing truck ramps also contribute to the cargo handling noise pollution [16]. Furthermore, cargo arrives and departs from the port area by road or rail transport, which also adds to the unwanted sounds in a port area.

Box 6.1 Examples of noise produced in ports

The electric motor on a container crane usually generates a sound level of approximately 110–115 dB(A).

Noise levels resulting from ship auxiliary diesel generators vary from ship to ship, generally between 100 and 115 dB(A) (L_{eq}).

Prolonged exposure to sounds in excess of 85 dB(A) may cause hearing loss in humans [18].

6.1.2.1 Regulations

The EU Environmental Noise Directive (2002/49/EC) is the legislation that affects European ports regarding noise pollution [19]. This directive provides a common basis for the problem within the European Union and applies to noise that can affect humans, particularly in urbanised areas, public parks, schools, hospitals, other noise-sensitive buildings, quiet areas in open country, or other quiet areas in an agglomeration [20].

Several other guidelines on anthropogenic noise affecting humans also exist, e.g., the World Health Organization (WHO) "Guidelines for Community Noise", the International Standard "Acoustics, description, measurement and assessment of environmental noise", the British Standard "Methods for determining the level of industrial noise affecting mixed residential and industrial areas" and the "Good Practice Guide on Port Noise Mapping and Management" of the NoMEPorts Project, formed to assist ports in meeting the requirements of the EU Noise Directive [21].

References

1. Nowacek, D., Thorne, L. H., Johnston, D. W. & Tyack, P. L., *Responses of cetaceans to anthropogenic noise.* Mammal Review, 2007. **2**: p. 81-115.
2. Urick, R. J., *Principles of underwater sound.* 1983, New York, USA: McGraw-Hill Co.
3. NRDC. *Sounding the depths II: The rising toll of Sonar, Shipping and Industrial ocean noise on marine life.* 2005.
4. Cato, D. H. *A perspective on 30 years of progress in ambient noise: Source mechanisms and the characteristics of the sound field.* 2012.
5. Weilgart, L. S., *The impacts of anthropogenic ocean noise on cetaceans and implications for management.* Canadian Journal of Zoology, 2007. **85**: p. 1091-1116.
6. Cox, T. M. *et al.*, *Understanding the impacts of anthropogenic sound on beaked whales.* Journal of Cetacean Research Management, 2006. **7**(3): p. 177-187.
7. Engås, A., Løkkeborg, S., Ona, E. & Soldal, A. V., *Effects of seismic shooting on local abundance and catch rates of Cod (Gadus morhua) and Haddock (Melanogrammus aeglefinus).* Can. J. Fish. Aquat. Sci., 1996. **53**: p. 2238-2249.
8. Weilgart, L. S. *Alternative technologies to seismic arigun surveysfor oil and gas exploration and their potential for reducing impacts on marine mammals.* 2009, Okeanos - Foundation for the Sea.
9. Picciulin, M., Sebastianutto, L., Codarin, A., Farina, A. & Ferrero, E. A., *In situ behavioural responses to boat noise exposure of Gobius cruentatus (Gmelin, 1789; fam. Gobiidae) and Chromis chromis (Linnaeus, 1758; fam. Pomacentridae) living in a Marine Protected Area.* J. Exp. Mar. Biol. Ecol., 2010. **386**(1–2): p. 125-132.
10. Haxel, J. H., Dziak, R. P. & Matsamoto, H., *Observations of shallow water marine ambient sound: The low frequency soundscape of the central Oregon coast.* Journal of Acust. Soc. Am., 2013. **133**(5): p. 2586-2596.
11. Nowacek, S. M. *et al.*, *Florida manatees, Trichechus manatus latirostris, respond to approaching vessels.* Biological Conservation, 2004. **119**(4): p. 517-523.
12. Hodgson, A. J. & Marsh, H., *Response of dugongs to boat traffic: The risk of disturbance and displacement.* J. Exp. Mar. Biol. Ecol., 2007. **340**(1): p. 50-61.
13. European Parliament, *European parliament resolution on the environmental effects of high-intensity active naval sonars*, in *B6-0018/2004*, EU, Editor. 2004.
14. ACCOBAMS, *Agreement on the conservation of Cetaceans of the Black Sea, Mediterrean Sea, and Contiguous Atlantic area, assessment and impact of man-made noise*, in *Res 2.16.* 2004.
15. van Kempen, E. E. M. M. *et al.*, *The association between noise exposure and blood pressure and ischemic heart disease: a meta-analysis.* Environmental Health Perspectives, 2002. **110**(3): p. 307-317.
16. Kalli, J., Alhosalo, M., Erkkilä, A., Åkerström, J. & Sundberg, P. *Ship originated air emissions, solid waste and wastewaters - a Feasibility Study of the New Hansa Project.* 2005, Centre for maritime studies, University of Turku: Turku, Finland.
17. ESPO. *ESPO / EcoPorts Port Environmental Review 2009. European Sea Ports Organisation's Review of Environmental Benchmark Performance in collaboration with the EcoPorts Foundation (EPF).* 2010: Brussels, Belgium.

18. Khoo, I.-H. & Nguyen, T.-H. *Study of the noise pollution at container terminals and the surroundings. Final Report*. 2011, California State University, USA.
19. ESPO. *Green Guide, Annex 2 Legislation influencing European ports*. 2012: Brussels, Belgium.
20. European Parliament. *Directive 2002/49/EC relating to the Assessment and Management of Environmental Noise. Official Journal of the European Communities 2002 No.189*. 2002.
21. ESPO. *Green Guide. Towards excellence in port environmental management and sustainability*. 2012: Brussels, Belgium.

Infrastructure, Marine Spatial Planning and Shipwrecks

7

J. Fredrik Lindgren, Karin Andersson and Hanna Landquist

Abstract

In association to the shipping industry there is a broad infrastructure that is connected with multiple activities, e.g., transportation, building and recycling. This infrastructure services the industry, for example ports, and its connected transport network, fairways and canals. The infrastructure activities also entail various environmental issues, such as land use, air emissions, noise, erosion and increased water turbidity. From an environmental perspective, ships have different impacts during the various phases of their lifetimes. Building ships is a highly energy demanding process, and various environmental issues are connected to ship yards. Moreover, when ships are scrapped, the workers typically work under very crude conditions, and often no environmental concern is exercised. To facilitate the effective use of ocean near-shore areas and avoid conflicts among stakeholders, marine spatial planning (MSP) can be applied. Marine spatial planning is a process that views a system and its potential usages from both spatial and temporal perspectives and can facilitate the implementation of ecosystem-based management plans, avoiding conflicts and creating opportunities between various actors in the area. This concept is now being introduced in many countries. Shipwrecks represent a hidden problem that must be addressed. Several thousand shipwrecks litter the ocean floor, containing massive amounts of oil and other toxic chemicals.

J.F. Lindgren (✉) · K. Andersson · H. Landquist
Shipping and Marine Technology, Chalmers University of Technology, Gothenburg, Sweden
e-mail: f.lindgren@chalmers.se

K. Andersson
e-mail: karin.andersson@chalmers.se

H. Landquist
e-mail: hanna.landquist@chalmers.se

K. Andersson et al. (eds.), *Shipping and the Environment*,
DOI 10.1007/978-3-662-49045-7_7

Keywords
Environmental impact · Ports · Fairways · Canals · Dredging · Ship construction · Ship scrapping · Shipwrecks · Marine spatial planning

Several environmental issues connected to the shipping industry are not detailed within the other chapters of this book. Instead, the issues are connected to the infrastructure that supports the shipping industry: ports, fairways, canals, dredging, and ship building and scrapping. In addition, ships occasionally become shipwrecks that lie the bottom of the ocean, which can cause environmental problems. Furthermore, the concept of marine spatial planning, which can be applied to utilise the resources of the ocean through ecosystem-based management, is introduced and explained.

Broad infrastructure is connected to the activities of the shipping industry, e.g., transportation, building and recycling. This infrastructure services the industry, e.g., ports, and the connected transport network, fairways and canals. Goods are transported from ports to the hinterland by land transport or inland waterways or to other coastal regions by short-distance sea shipping. This infrastructure must be created/built, operated and maintained, which entails environmental issues.

The IMO regulations governing the environmental issues adressed in this chapter are shown in bold in Table 7.1.

7.1 Ports

Ports play a significant role in delivering goods to a country or region; up to 90 % of the global freight volume is accommodated by shipping [1]. Ports also serve as hubs for shipping infrastructure by supplying ships with goods for transport and means to perform tasks, e.g., providing fuel and reception facilities for sludge and other wastes. Incoming goods are distributed to inland regions by truck, train or inland waterways. Further transport by ships to coastal regions is conducted using smaller vessels. Ports also promote economic growth in a region. Some of the largest ports in the world are in Singapore, Shanghai and Rotterdam, each of which have an annual throughput of goods exceeding 400 million tonnes [2].

Most ports serve all ship types and origins and are therefore divided into different terminals that serve various types of shipping, e.g., containers, bulk, mineral oils, cruise ships or roll on/roll off. Some ports are highly specialised, e.g., industrial ports connected to mines or refineries, and are only equipped to serve a few specific vessel types.

Ports and their activities have various degrees of negative impacts on the environment. First, ports require large areas of land and water to be converted into infrastructure. This area can require several hundred hectares for berths, storage areas for goods, warehouses, port related activities and rail and road infrastructure. The conversion often conflicts with the conservation of wetlands or the protection of migratory or endangered species.

Table 7.1 Various IMO conventions controlling the discharge, handling and usage of different hazardous products on ships

IMO convention	Regulates	Adoption (year)	Entered into force (year)
MARPOL 73/78	Pollution from ships	1973/1978	1983
Annex I	Oil and oily mixtures	1973	1983
Annex II	Liquid chemicals in bulk	1973	1983
Annex III	Liquid chemicals in packaged form	1973	1992
Annex IV	Sewage	1973	2003
Annex V	Garbage	1973	1988
Annex VI	Air pollution and GHG emissions	1997	2005
International convention on the control of harmful anti-fouling systems on ships	Antifouling paints	2001	2008
International convention for the control and management of ships' ballast water and sediments	Invasive species	2004	Not yet in force
Hong Kong international convention for the safe and environmentally sound recycling of ships	Recycling of ships	2009	Not yet in force
Nairobi international convention on the removal of wrecks	Removal of wrecks	2007	2015

Second, air emissions, such as SOx, NOx, ground-level ozone and particles emitted from vessels operating in port, ships leaving or entering port and the transport network serving a port, affect the air quality in the nearby region. Emissions from so-called hotelling, when the auxiliary engines of ships are running while at berth, can be substantial. In a study of NOx emissions at the Ports of Los Angeles and Long Beach, the hotelling NOx emissions were 39 % of the total emissions from ships in the port (including cruising, manoeuvring and hotelling) [3]. Another study found that hotelling emissions accounted for approximately 20 % of the total particulate matter diesel emissions in the Port of Los Angeles area [4]. These emissions can cause health problems in humans and contribute to the acidification of water and the eutrophication of water and soil. Ports are often located in areas with dense human populations, creating an additional important incentive to reduce these emissions. Particles emitted by ships through the combustion of residual fuel affect human health. The finer fractions can also stay airborne longer and travel farther. Diesel pollution has been found to be a contributor to air quality problems in cities surrounding major ports in the United States. Local air quality can also be affected by the storage of bulk goods, for example ore or coal, because dust is emitted during this process. The use of closed shipments, suction filters or devices to keep surfaces wet can limit these emissions.

In addition, ship maintenance can be a source of dust, especially in the vicinity of sand-blasting and welding activities [2].

In addition, other shipping-related environmental issues occur in ports. Noise is a problem caused by several sources in ports; noise is perceived as a nuisance to people living in the vicinity of ports, as discussed in Sect. 6.2. During bunkering operations, oil spills can occur, and ballast water exchange and failing antifouling paint can lead to the introduction of invasive species. The transport and handling of hazardous cargo within ports and into ports via ships could result in environmental problems if the hazardous cargo is handled incorrectly or if an accident occurs. These materials include tar, petrochemical products, caustic soda, nitric acid, sulphuric acid, ammonia, and fireworks. Toxic spills, explosions and fires can occur in ports; actions to handle these events must be addressed in emergency plans. Annex II of the MARPOL includes procedures and guidelines for handling hazardous material on ships or in port [2].

The environmental impacts from ports can be divided into three sectors: problems caused by the port itself, ships entering or leaving the harbour and emissions originating from the transport network serving the port. Various actions can be taken in these sectors to limit their environmental effects (see Sect. 11.13). Ports cause emissions from the trucks and transports used to move cargo, reefer trucks and containers that use diesel-powered generators for keeping the cargo cooled, and tugs that assist ships entering and leaving the port. Ships entering or leaving a harbour contribute to the same environmental problems as in other locations except that they are emitted close to land and often in densely populated areas. The manoeuvring of ships is more common in port areas, which results in more pollutants being emitted. The environmental effects of the transport network serving a port is dependent on the mode of transportation, the efficiency of the transportation and the standard fuel and vehicles being used. Transport by road is generally more detrimental to the environment than transport by railway, in-land waterways or short-sea shipping, assuming that appropriate regulations are enforced on the latter.

7.2 Fairways and Canals

Fairways are used by ships to navigate into ports or into narrow straights. Canals are used as inland waterways to transport goods or people to inland cities or lakes. The use of fairways causes shipping to be safer when entering or leaving ports, decreasing the number of accidents. The ability to use canals to transport goods inland by water instead of using trucks yields a positive environmental effect. However, the usage of fairways and canals also involves certain negative environmental impacts from, for example, ship-generated waves or the extensive usage of thrusters. The effects include the erosion of riverbanks or beaches that are located close to fairways, resuspension of contaminated sediment, and increased sediment transport and turbidity in the water column. The passage of large ships can also damage or kill fish, fish larvae and invertebrates due to shear forces or the pressure changes induced by the passage of the propulsion system. This effect is restricted to the area nearest the passing ships. Collisions with larger mammals, e.g., dolphins, whales, porpoises, manatees and dugongs, are also possible. To keep the fairways and canals navigable, dredging activities are regularly needed, which also has negative environmental consequences (see Sect. 7.3). The creation of fairways can also involve dredging activities and the clearing of shallow bedrock with explosives.

The introduction of non-native species is also a major problem when constructing and opening canals in water systems that were previously not connected, allowing organisms to pass former geographical boundaries. The impacts of invasive species can be greater than all of the direct impacts from moving vessels [5]. The opening of a canal in 1825 from the Great Lakes, on the border between Canada and United States, to the Atlantic Ocean is a good example of this phenomenon. After the opening of the canal, the sea lamprey (*Petromyzon marinus*), which is a parasitic fish that attaches to larger fish and sucks their bodily fluids, had spread throughout the entire lake system by 1947. In addition, the alewife (*Alosa pseudoharengus*) invaded the lakes in the 1930s through the canal system. Its major food was zooplankton, and it began to compete for food with other native fish. Together, these invasions contributed heavily to the decline of the large lake trout populations in the Great Lakes [6].

7.3 Dredging

Dredging involves the excavation of sediment from the bottom of the sea floor, ocean, rivers, bays and harbours and its disposal at another location. Dredging is performed to create or maintain navigable waterways for shipping because sediments accumulate at the bottom of dredged channels; the process is essential for maintaining marine traffic. Large vessels require a certain water depth to access channels and ports without running aground. Dredging can be accomplished using giant scoops or high-pressure suction. Annually, one hundred million cubic meters of marine sediment are dredged worldwide [2].

Finding a location for the large volume of excavated material that remains after dredging can be challenging. The collected material is typically a mixture of mud, rock and sand. This mixture of material is often dumped offshore because this is the least expensive option. However, the materials can also be used in construction materials (e.g., clay, bricks, and foundation fill) or to prevent erosion by rebuilding islands and replenishing sand beaches [7]. Most dredged material does not contain any toxic substances. However, dredging operations can still have negative impacts on the environment. Substrate can be physically removed together with associated plants and animals. Because large amounts of sediment are disposed at another location, seagrass, algae and corals can become buried or smothered. Smaller amounts of dredging material in the water column can increase turbidity and cloudiness in the water column, which can affect species that rely on photosynthesis for the assimilation of energy. This phenomenon is especially important for corals, where the increased turbidity can lead to sub-lethal or even lethal effects because the coral polyps become smothered, inhibiting photosynthesis and increasing respiration [8]. The dredging material can become suspended during dredging operations from the transport to the surface or from overflowing barges or leaking pipes. The severity of this effect is determined by, e.g., the quantity, frequency, duration, methodology, water depth, weather conditions and distance to ecologically sensitive areas [9].

Dredged sediments can be contaminated with both organic and inorganic pollutants. Among organic pollutants, PAHs, PCBs, TBT, and dioxins are the most common. Among inorganic pollutants, heavy metals, such as lead, chromium, zinc, cadmium, mercury and copper, are considered to be the worst pollutants [10]. Heavy metals that have been deactivated in anoxic sediments can re-enter the oxic water column and be incorporated into the marine food chain. This effect must be considered when new dredging sites are introduced and when massive amounts of material must be handled. As a consequence of dredging, e.g., due to material toxicity or increased turbidity, the total abundance and species diversity of macrobiota can be reduced. However, recovery can be rapid in some cases, allowing the richness and abundance of species to return to control levels within three months. The mean recovery time in estuaries affected by dredging operations was estimated to be 1.67–5.25 years.

Dredging has been regulated by IMO since 1996 as part of the London convention. The convention includes guidelines for preventing pollution due to the dumping of dredged material. This material may be disposed of in land-based dumps or in approved areas of the seabed. Dredging is also regulated in the European Water Framework Directive and the European Water Directive by the EU, the OSPAR Convention, within the HELCOM convention and in the USA, e.g., by the Clean Water Act [11].

7.4 Ship Construction and Scrapping

A ship is a complex product that is subject to various refurbishments throughout its long lifespan. A lifetime of 30 years is common, although large variations exist among the different types of ships and within different shipping segments [12]. The variation in lifetime may be substantial within a specific segment. A recent example is a high-speed ferry (HSS) built in 1996 that was scrapped in 2013 for economic reasons. However, many passenger ferries built in the 1970s and 1980s remain in operation in 2015, having been refurbished in various ways during the years.

The operational lifetime is not determined solely by technical lifetime. Instead, several parameters may be important, all of which are related to the economic factors affecting the operation of a ship. From a shipping economics perspective, several variables may be of importance to the shipping market that are related to supply and demand [13]. On the demand side, factors such as the world economy, political events and transports costs are important; on the supply side, the size of the world fleet, fleet productivity, the rates of building and scrapping/losses and freight rates are important. These demands and regulations may be as important as the technical state in determining when a ship should be scrapped.

From an environmental perspective, ships have different impacts during various phases of their lifetimes. From a life cycle perspective, a ship, like other products, undergoes a series of phases from material production via construction and operation to scrapping. In terms of the turnover of materials and energy, the initial design and manufacturing stages are important, including the "upstream" of materials from the extraction of raw materials via different processes creating the construction materials necessary to construct the final product. The operation phase, which is commonly the longest period of a ship's lifetime, is a very important phase from an environmental impact perspective. Refurbishments and retrofitting of equipment offer challenges and opportunities. The final step is scrapping, when the materials are recycled or turned into waste. In addition, some ships become shipwrecks on the sea floor or along the shore due to accidents, collisions or weather-related occurrences; these vessels are permanently removed from the material cycling.

The environmental impacts vary between the different phases of a ship's life cycle. However, due to the long operational time of most ships, it is not surprising that the operational phase is usually the dominating part.

7.4.1 Design Phase

Several opportunities exist to minimise the environmental impacts of a vessel from a life cycle perspective. It is easier (and thus less costly) to improve a ship's performance in the early stages of the manufacturing process, i.e., the design stage. Decisions made at the design stage may be crucial to performance. Not only the use of energy, choice of energy source and energy conversion, but also maintenance are important parameters. During the lifetime of a ship, refurbishment can be expected, which sometimes includes a complete change of use (one drastic example is the rebuilding of a road ferry to a liquid natural gas (LNG) bunker ship). Retrofitting a ship with emission abatement technology also results in substantial changes. A design that allows structured refurbishment and different types of use may also be favourable from a life cycle perspective. The possibility of structured scrapping is often determined already in the design phase. The concept of "design for the environment" or "design for recycling" is used in many industrial areas and is also valid for the scrapping of ships [14].

Thus, when the demands to minimise environmental impacts increase, the ship design process will become crucial for enabling changes during a ship's lifetime (and to foresee future requirements, e.g., LNG-ready ships). These changes should also render a ship more energy efficient and safe and reduce emissions from the beginning.

The complexity of a ship and the increased performance demands lead to a challenging, system-level design process. To provide "eco-design" products, in this case, ships, different environmental systems, analysis tools and methodologies can be used. One methodology that has been discussed and tested in the literature is life cycle assessment (LCA). LCA can be used to examine potential environmental and health impacts and the resource use of a product or service during the life cycle from raw material to waste, accounting for the production and use phases (see Sect. 9.4.1) [15]. As discussed above, a general result of all environmental impact assessments of ships (similar to other long-lived products, e.g., cars or buildings) is that the dominant impact originates from the operational phase in nearly all impact categories, e.g., climate change, acidification, eutrophication and toxicity to humans. Moreover, energy use is the largest factor in the operational phase. Only resource use is larger in the construction phase, which is somewhat counteracted by a large potential for the recycling of materials, e.g., steel in the scrapping process. Because refurbishments and maintenance occur throughout a ship's lifetime, the environmental impacts in the use phase are dependent on the initial design and the subsequent changes. The use of LCA for the entire product at the design stage may have limitations due to unknown changes in the impacts during the lifetime. For learning/investigation purposes in the design process and for decision-making regarding choices between alternatives in refurbishments, LCA may be a very useful tool.

7.4.2 Manufacturing Phase and Shipyards

Several environmental issues are involved in the construction of ships. First, the construction of ships is very energy-intensive, which can cause negative environmental impacts if the electricity used for production originates from coal or oil power plants. Second, the use of steel as a raw material releases greenhouse gases and other air pollutants and generates large volumes of wastewater during the production process. Steel production is among the world's most energy-demanding industrial processes, and the shipbuilding sector uses approximately 3 % of the global steel production.

The direct environmental impacts from shipyard activities originate primarily from ship construction and include the following processes:

- Handling of raw materials, fabrication and surface treatment;
- Joining and assembling fabricated parts into blocks;
- Erection of steel structures via the fitting and welding of blocks;
- Outfitting a ship with electronic equipment; and
- Installation of various fabricated parts.

The following maintenance and repair activities also have direct environmental impacts:

- Treatment operations and surface cleaning;
- Oil transfer operations; and
- Servicing of machinery.

During these activities, discharges of toxic chemicals to the aquatic environment from shipyards are common. Studies have observed severe sediment and water contamination in areas near shipyards, for example, contamination with lead, copper,

TBT, PCBs and PAHs. Particulate matter is also emitted into the air due to metal work and surface treatment operations. During thermal cutting, e.g., welding or plasma arc cutting, fumes are emitted that contain, e.g., heavy metals, ozone, and NOx, which contribute significantly to the overall environmental impact from shipyards. The use of antifouling paints is also a source of pollutants from shipyards. Several shipyard activities, for example construction, maintenance and repair activities, also generate large amounts of noise. Given the several thousand active shipyards around the world, pollution from the shipbuilding industry can be significant [12].

7.4.3 Operational Phase

As discussed above, the main environmental impacts of ships come from the use phase, which is similar for other long-lived products, e.g., buildings and cars. This is especially true of energy use and emissions related to energy. In most cases, the conversion of energy is also related to emissions to the environment. Studies focused on the impact of the entire fuel chain may be important as well. From an operational perspective, emissions and energy use may be minimised during ship operation by producing fuels that have low emissions and are efficient in energy conversion even though those fuels have large energy demands and emissions in the upstream fuel cycle [16, 17]. Descriptions of the various environmental issues that result from a ship's operational phase are described in Chaps. 4–6.

7.4.4 Scrapping of Ships

The final stage in a ship's life cycle is shipbreaking, which is the process of taking a ship apart, followed by scrapping and recycling. The process and steps in the process can have different names (dependent on the stakeholders), e.g., breaking, recycling, dismantling or scrapping. The shipbreaking process involves several activities, such as removing machinery and other equipment for repair/reuse or for dismantling and removing oils and chemicals for reuse or final treatment. The scrapping process also involves separating different construction materials, e.g., steel, other metals, plastics and wood. Modern ships are large structures that are primarily composed of steel. Steel is a fully recyclable material and has a considerable market value; thus, it is currently recycled to a very large extent. Approximately 95 % of the parts from a ship can be recycled and reused. The primary locations for ship recycling are Bangladesh, India, Pakistan, China and Myanmar (Burma). Approximately 200–600 ships over 2000 dwt are scrapped each year, with an increasing trend in the number of ships being recycled [12].

From an economic perspective, ship recycling not only contributes to the reuse of valuable materials but is also associated with high costs for labour and environmental protections. This has led to an increase in scrapping sites in Asia, with very few remaining ship scrapping companies in Europe and North America [18].

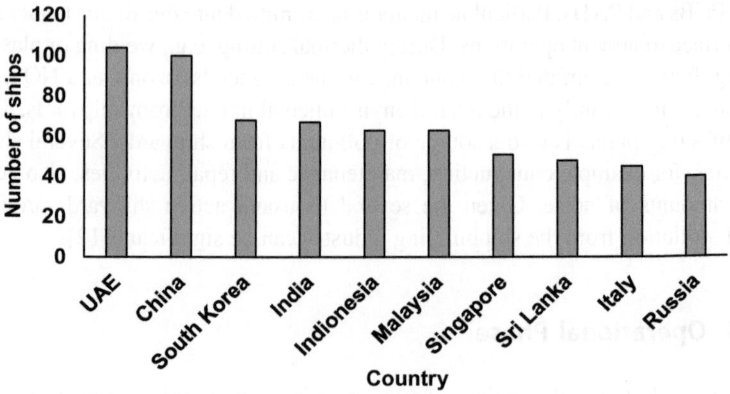

Fig. 7.1 Country of origin of the last voyage before a ship goes to scrapping. From Lloyds list [34]

Shipbreaking can be an important industry for a country without natural sources of iron ore, e.g., Bangladesh [19]. The majority of ships being scrapped do not come directly from European or American ports; instead, these vessels often come from ports in the United Arab Emirates, China, and South Korea (Fig. 7.1).

The dismantling process is not currently considered in the ship design process, where the main focus is safety and reliability. In other industries, "design for recycling" has become an important issue. In a report that focused on the transport sector, the following recommendations were made:

- "Materials and components, including hazardous substances, should be avoided or otherwise labelled and designed for disassembly, for example, to increase the quality of material fractions for recycling and energy recovery;
- PVC and copper should either be dismantled or be avoided in components that are not dismantled, for example in electrical cables, to increase the quality of material fractions for recycling and energy recovery;
- Additional design for disassembly is usually unnecessary except for a few materials and components in an optimistic dismantling scenario, i.e., large parts of monomaterials or more valuable materials;
- Plastics, including their additives and coatings, should be chosen to increase the quality of and demand for the recycled plastics; and
- Few or no guidelines for recycling are needed when assuming the best available technology [14]."

Ships contain numerous materials and chemical compounds that are potentially environmentally hazardous. The environmentally hazardous substances that may be found in ships sold for scrapping include asbestos, heavy metals, hydrocarbons and ozone-depleting substances. To identify and separate these materials from the ship and to treat them in an environmentally acceptable way involves high costs.

On many occasions, the scrapping process occurs on sandy beaches under very primitive conditions. This method of scrapping is the so-called "beaching" approach. A ship is driven aground at a high speed on a beach that is close to the users of the recycled material, e.g., steel mills. Beaching is a common fate of ocean-going ships, e.g., oil tankers, bulk carriers, general cargo carriers, container ships and passenger ships. The process is often criticised due to the many safety, health and environmental protection problems. Scrapping of ships has generated debates on occupational health and safety as well as environmental issues. Scrapping also has a social dimension because beaching is performed in developing countries. The use of beaching and the debate on health and safety issues during the process are the driving forces for implementing the Hong Kong convention.

The ship recycling business is considered one of the world's most dangerous occupations. Most accidents are caused by the falling of large heavy metal plates and by gas explosions. Furthermore, the workers typically use little or no protective equipment, directly exposing them to toxic substances and fumes. In addition, no or very few facilities exist for receiving and handling hazardous waste and toxic chemicals. Thousands of people are estimated to have died in the ship scrapping business in the last 20 years.

During the scrapping of a ship, the ship hull and structural parts are cut into large steel plates. During this process, toxic pollutants can be discharged, including heavy metals and TBT. The dismantling process also includes the removal of cargo residues, asbestos, and several tons of oil, including fuel oil lubricants, bilge oil and grease. Waste and treatment reception facilities and barriers to contain or prevent oil or other liquids from entering the marine environment are often lacking or exist in low quantities at the ship scrapping sites in Asia [12], leading to a large threat to the marine environment in the vicinity of ship scrapping sites. No exact calculations exist of the volume of discharged toxic chemicals from ship scrapping. However, the discharged oil sludge from scrapped ships is estimated at between 400,000 and 1.3 million tonnes per year, toxic antifouling paints between 6000 and 12,000 tonnes per year and the highly toxic TBT paint between 170 and 540 tonnes per year. In addition, materials that cannot be recycled, e.g., cable coatings of PVC, are often burned or dumped at the scrapping site [12].

Few studies have been performed on the impacts on the marine environment from ship scrapping activities. However, it has been shown that the amount and diversity of primary producers (phytoplankton), zooplankton, and benthic organisms near recycling areas are significantly lower compared with control areas. The number of fish species has also been found to be reduced [20]. Metals and persistent organic pollutants that have been incorporated into the sediment can bioaccumulate in the food chain.

Box 7.1 Recycling of lightweight materials

The introduction of new construction materials, such as light metals or composites, to build lighter ships that consume less energy offers new challenges in ship construction and recycling. Composite materials have been used for navy ships like corvettes and minesweepers, displacing several hundreds of tonnes. Moreover, light metals are used in overbuildings and other parts. The large leisure sector includes numerous composite boats that are reinforced by glass fibres, mainly with acrylonitrile-butadiene-styrene (ABS) plastic.

Pure metals, e.g., aluminium, and even some alloys can be easily introduced into existing recovery procedures, whereas composites may prove more difficult for reuse. Reduced energy use due to the use of lighter construction material may be offset by impairment or even prevention of material reuse at the end of the life cycle. The development and use of composites has focused on energy efficiency rather than material recovery. These materials came into use in the last 30–40 years, and numerous vessels now require scrapping. For these ships, initiatives to develop recycling have begun. Research into material recovery, e.g., of reinforcements, is under development. To recycle at least some parts of a composite, it has to be cut down and milled before being sent to a recycling facility. Reinforcement fibres can be subsequently recovered via thermal or chemical processes. This process requires a massive amount of energy and very expensive equipment. An additional problem for recycling is that composite materials may be painted or surface-treated in other ways, which introduces contaminants into the recycling steam.

7.4.5 Regulations

No international regulations exist for ship scrapping. However, the toxic and hazardous substances in ships fall under the 1989 Basel Convention on the Control of Transboundary Movements of Hazardous Wastes and their Disposal [21]. Attempts to enact international regulations for ship recycling have been made by IMO under the Hong Kong International convention for the Safe and Environmentally Sound Recycling of Ships, which was accepted by IMO in May 2009 [22]. The aim of this convention is to ensure that the recycling of ships does not pose any unnecessary risks to human health, safety or the environment. The most important requirement of the convention is that every ship must to carry an inventory of hazardous materials (IHM). The IHM must be updated throughout a ship's operational lifetime. In addition, a final survey is required prior to sending a ship for recycling.

The convention will enter into force 24 months after being ratified in which 15 states, representing 40 % of world merchant shipping by gross tonnage. A few guidelines have been developed and adopted by the MARPOL 73/78 convention to assist states in the early implementation of the technical standards of the Hong Kong Convention (Table 7.2). Furthermore, two additional guidelines have been developed and adopted to assist states in the implementation of the convention after it enters into force (Table 7.3).

Table 7.2 MARPOL 73/78 guidelines for ship recycling

2011	Guidelines for the development of the inventory of hazardous materials	MEPC.197(62)
2011	Guidelines for the development of the ship recycling plan	MEPC.196(62)
2012	Guidelines for safe and environmentally sound ship recycling	MEPC.210(63)
2012	Guidelines for the authorization of ship recycling facilities	MEPC.211(63)

Table 7.3 Guidelines to assist in the implementation of the Hong Kong convention

2012	Guidelines for the survey and certification of ships under the Hong Kong convention	MEPC.222(64)
2012	Guidelines for the inspection of ships under the Hong Kong convention	MEPC.223(64)

7.5 Shipwrecks

Another possible destiny for a ship is wreckage, which can be caused by many factors. Casualty in war, severe weather and collisions are some examples of what can damage a ship to the extent that it cannot pursue its route, ultimately ending up on the sea floor [23]. Approximately 8600 shipwrecks exist on the sea floor today, i.e., non-tank vessels of at least 400 gross tonnage (GT) (oil as fuel) and tank vessels of at least 150 GT (oil as cargo). They are estimated to contain an accumulated amount of oil exceeding 2 million tonnes [24]. However, these are only examples of shipwrecks that contain oil; the number of shipwrecks is likely even larger. A large proportion of the wrecks sank during World War II and have been deteriorating in the corrosive sea water for nearly 70 years, suggesting that corrosion of the iron structures could be well developed and ultimately cause leakage of pollutants, e.g., oil [25].

Shipwrecks can contain petroleum products, other chemicals, munitions and a wide range of various types of cargo. These substances can affect human health and the marine environment (see Sect. 4.1). Furthermore, wrecks often occur near shores in shallow waters, allowing the inevitable leakage to be more likely to cause environmental damage. Several examples exist of shipwrecks that are causing environmental problems by leaking oil into the ocean, for example *Royal Oak* in Great Britain, the *USS Arizona* in Hawaii, the *S.S. Jacob Luckenbach* in California, and the *S.S. Skytteren* in Sweden [23, 26].

Because every shipwreck is unique with respect to its potential environmental impacts (including toxic products in the wreck and the geographical area in which it might end up) and because the probability of discharge differs from wreck to wreck, objective prioritization is crucial when taking proactive measures. Moreover, the potential costs associated with future oil spills from a shipwreck are likely to exceed oil removal operations for some shipwrecks due to the lack of proactive measures. A cost-benefit analysis of oil removal operations should be conducted to determine the differences in costs between spill response and damages and pro-active oil removal operations [27]. Technology to remove oil from shipwrecks is available [28], although it is very costly and therefore it is not feasible to remediate all shipwrecks. Therefore, decision support for governments and other potential remediation actors is essential. Several risk assessment strategies intended to provide this support have been presented in the literature (Box 9.3) [23, 24, 27, 29].

7.5.1 Regulations

The International Maritime Organization adopted the Nairobi International Convention on the Removal of Wrecks in 2007 and entered into force in 2015. This convention gives a state the right to remove a wreck from their coastline if it poses a hazard by endangering, hindering or threatening to hinder navigation or may cause harmful consequences to the marine environment. Any cost in connection with the removal is the liability of the owner. However, the convention is not applicable on, for example, naval ships, excluding shipwrecks from World War II [30].

7.6 Marine Spatial Planning

Increasing demand and competition for the space and resources in near-coastal areas, e.g., offshore renewable energy, shipping, fishing and aquaculture, have led to a need for effective management and coordination of these areas. Marine spatial planning (MSP) is a process that examines a system and its potential usages from both spatial and temporal perspectives and can be used to facilitate the implementation of ecosystem-based management. This process can avoid conflict and create opportunities between various actors in the area. Marine spatial planning was developed to integrate human activities at sea and is intended for application so that human usage of the ocean is as effective and sustainable as possible. Its development has been partially catalysed by the growing usage of the ocean, particularly renewable energy development [31]. MSPs provide an opportunity not only to better understand and more effectively manage the marine environment but also to provide a long-term plan for both developers and marine managers. MSPs are not substantially different from spatial planning on land. Although the three-dimensional nature and dynamics of the ocean differ from land, many planning concepts and techniques can easily be translated to the marine environment. A country's marine spatial planning zone usually stretches from 1 nautical mile off the coast (base line) to the end of the exclusive economic zone EEZ (188 NM).

Several major points are listed below as the major benefits of implementing marine spatial planning:

- Promote investments: increased transparency and distinct regulations will stimulate development in areas of Blue Growth, e.g., renewable energy and energy grid investments and will facilitate investments in oil and gas.

- Lessen the number of conflicts and promote cooperation between actors and different sectors.
- Increased coordination: in states within the EU or counties within a country, the different authorities can use a single instrument to supervise the development of several different sea-based activities; this coordination simplifies the process and saves money.
- Improved cooperation between states in the EU, between countries in a geographical area, or between counties within a country in areas of cabling, pipelines, shipping lanes, and wind farm parks.
- Protection of the environment: by mapping the diverse usage of the sea both on and below the surface and how this affects the marine environment, problems can be discerned and measures can be taken.

To perform effective marine spatial planning there are three stages that need to be conducted in a continuous, iterative and adaptive process. *Planning and analysis* is performed to generate a plan for the protection, enhancement and sustainable use and development of the area. This step is based on research that addresses both the environmental and human processes within the area. For example, biological-economic models can be used to evaluate the trade-offs between various stakeholders within an area and assist in choosing the optimal design to minimise conflicts [31]. *Implementation* of the plan is performed by executing planned work and investments, encouragement, and through regulations, incentives and enforcements of the proposed changes and on-going activities, both on and below the sea surface. Finally, *monitoring and evaluation* are performed to assess the effectiveness of the plan, to determine ways it can be improved, and to establish review and adaption procedures. The results of the evaluation are then fed back to the first stage (planning and analysis), and the process restarts.

The final decision on what usage will be assigned to different areas is a matter of societal choice. Thus, stakeholders and the general public must be involved in the MSP process. Finally, the choice needs to be financed on a continuous basis to achieve the objectives and goals [32].

7.6.1 Regulations

For more than 35 years, marine spatial planning has been an important component of ecosystem-based management in Australia. Other states have implemented or intend to implement MSP in their marine environment. In 2014, the EU passed a legislation that introduces common guidelines with minimum standards for marine spatial planning (within the EU, this is called maritime spatial planning) in all member states; these guidelines will be implemented by 2016. By 2021, all member states shall have marine spatial plans in place. The minimum standards for designing their MSPs include adopting an ecosystem-based approach, accounting for land-sea interactions, ensuring interaction with stakeholders, cooperating between member states and using the best available data. Trans-boundary initiatives also exist within the EU marine spatial plan that encourage neighbouring countries to embed their MSPs into a broader regional context that extends beyond territorial borders, for example, OSPAR in the North Sea region and HELCOM in the Baltic Sea region [32]. Other nations have also implemented MSP guidelines, e.g., the USA, Norway, China, Canada, and Lithuania, or are planning to introduce such regulations, for example, Estonia, New Zealand, Vietnam, Indonesia, Thailand, Mexico, Bermuda and Grenada [33].

References

1. Giziakis, C. & Christodoulou, A., *Environmental awareness and practice concering maritime air emissions: the case of the Greek shipping industry.* Maritime Policy & Management, 2012. **39**(3): p. 353–368.
2. Braathen, N. A., *Environmental Impacts of International Shipping: The Role of Ports.* 2011: OECD Publishing, Paris, France.
3. Environ. *Cold Ironing - Cost effectiveness study.* 2004: Los Angeles, USA.
4. Peterson, L. P., Chavdarian, P., Islam, M. & Cayanan, C. *Tackling ship pollution from the shore.* IEEE Industry Applications Magazine, 2008. **15**(1): p. 56–60.
5. PIANC, *Considerations to reduce environmental impact of vessels: Navigation, ports, waterways.* 2008, Brussels, Belgium.
6. Mooney, H. A. & Hobbs, R. J., *Invasive species in a changing world.* 2000, Island press: Wasgington D.C., USA.
7. Limeira, J., Agullo, L. & Etxeberria, M., *Dredged marine sand in concrete: An experimental section of a harbor pavement.* Constr. Build. Mater., 2010. **24**(6): p. 863-870.
8. Erftemeijer, P. L. A., Riegl, B., Hoeksema, B. W. & Todd, P. A., *Environmental impacts of dredging and other sediment disturbances on corals: A review.* Marine Pollution Bulletin, 2012. **64**(9): p. 1737-1765.
9. Erftemeijer, P. L. A. & Lewis, R. R. R., *Environmental impacts of dredging on seagrasses: A review.* Marine Pollution Bulletin, 2006. **52**(12): p. 1553-1572.
10. Agostini, F., Skoczylas, F. & Lafhaj, Z., *About a possible valorisation in cementitious materials of polluted sediments after treatment.* Cement and Concrete Composites, 2007. **29** (4): p. 270-278.
11. OSPAR. *Assessment of the environmental impact of dredging for navigational purposes.* 2008: London, UK.
12. OECD. *Environmental and climate change issues in the shipbuilding industry.* 2010. Paris, France.
13. Stopford, M., *Maritime Economics.* 2009: Routledge, London, UK.
14. Lundqvist, U. *et al. Design for recycling in the transprot sector- future scenarions and challenges.* 2004: Centre for environmental assessment of product and material systems.
15. Baumann, H. & Tillman, A.-M. *The hitchhiker's guide to LCA : an orientation in life cycle assessment methodology and application.* 2004: Lund, Sweden.
16. Bengtsson, S., Andersson, K. & Fridell, E., *A comparative life cycle assessment of marine fuels: liquefied natural gas and three other fossil fuels.* Proceedings of the Institution of Mechanical Engineers, Part M: Journal of Engineering for the Maritime Environment, 2011. **225**(2): p. 97-110.
17. Brynolf, S., Fridell, E. & Andersson, K., *Environmental assessment of marine fuels: liquefied natural gas, liquefied biogas, methanol and bio-methanol.* Journal of cleaner production, 2014. **74**: p. 86-95.
18. Chang, Y.-C., Wang, N. & Durak, O. S., *Ship recycling and marine pollution.* Marine Pollution Bulletin, 2010. **60**(9): p. 1390-1396.

19. Khan, K., Chowdhury, H., Alam, F. & Kumar, A., *Sustainable design of ship breaking industry in developing countries.* Asian Journal of Water, Environment and Pollution, 2012. **9** (1): p. 1-11.
20. Siddiquee, N. A. *Impact of ship breaking on marine fish diversity if the Bay of Bengal.* 2004: Dhaka, Bangladesh.
21. UNEP. *Basel convention controlling transboundary movements of hazardous wastes and their disposal.* 1989: Nairobi, Kenya.
22. IMO. *Hong Kong International Convention for the Safe and Envi-ronmentally Sound Recycling of Ships.* 2009: London, UK.
23. NOAA. *Risk Assessment for Potentially Polluting Wrecks in U.S. Waters.* 2013, National Oceanic and Atmospheric Administration: Silver Spring, MD. p. 127 pp + appendices.
24. Michel, J. *et al. Potentially polluting wrecks in marine waters*, in *International oil spill conference.* 2005.
25. Johnson, D. L., *Corrosion of steel shipwreck in the marine environment: USS Arizona - Part 1.* Mater. Perform., 2005(45): p. 40-44.
26. Sayle, S., *Site assessment and risk management framework for underwater munitions.* Marine technology society journal, 2009. **43**(4): p. 41-51.
27. Schmidt Etkin, D., van Rooij, H. & McCay French, D. *Risk assessment modeling approach for the prioritization of oil removal operations from sunken wrecks*, in *Effects of oil on wildlife.* 2009. Tallin, Estonia.
28. Alcaro, L., Amato, E., Cabioch, F., Farchi, C. & Gouriou, V. *DEEPP Project, DEvelopment of European guidelines for Potentially Polluting shipwrecks.* 2007, ICRAM, Instituto Centrale per la Ricerca scientifica e tecnologica Applicata al Mare, CEDRE, CEntre de Documentation de Recherché et d'Epérimentations sur les pollutions accidentelles des eaux.
29. NOAA. *Wreck Oil Removal Program (WORP) 2009 Demonstration Project Overview.* 2009, National Oceanic and Athmospheric Administration,.
30. IMO, *Hong Kong International Convention for the Safe and Environmentally Sound Recycling of Ships.* 2007: London, UK.
31. White, C., Halpern, B. S. & Kappel, C. V., *Ecosystem service tradeoff analysis reveals the value of marine spatial planning for multiple ocean uses.* Proceedings of the National Academy of Sciences, 2012. **109**(12): p. 4696-4701.
32. Douvere, F., *The importance of marine spatial planning in advancing ecosystem-based sea use managemet.* Marine Policy, 2008. **32**: p. 762-771.
33. Ehler, C. *A guide to evaluate marine spatial plans.* 2014: Paris, France.
34. Lloyds-register (2012). *Last voyage origin of recycled ships*, 2012. Retrieved june 11th, 2013, from http://www.lloydslist.com/ll/incoming/article417465.ece/BINARY/Last+voyage+origin.pdf.

Part III
Pollution Prevention Measures

Environmental Management

8

Karin Andersson

Abstract

To ensure environmental performance at a desired level in a company, an understanding of the potential impacts and their sources and a work structure that promotes the desired outcome are necessary. Policies and strategies must be known and accepted in the company, and the organisational structure must be able to handle the challenges. Environmental management systems provide a means to manage the work process as well as communicate the company's environmental policy, goals and work processes. The majority of this book addresses impacts and their sources and technical solutions to counteract these impacts. In this chapter, the management of environmental work is discussed.

Technology that supports the minimisation of environmental impact is not sufficient for its realisation. To ensure environmental performance at a desired level in a company, an understanding of the potential impacts and their sources and a work structure that promotes the desired outcome are necessary. Policies and strategies must be known and accepted in the company, and the organisational structure must be able to handle the challenges.

K. Andersson (✉)
Shipping and Marine Technology, Chalmers University of Technology, Gothenburg, Sweden
e-mail: karin.andersson@chalmers.se

K. Andersson et al. (eds.), *Shipping and the Environment*,
DOI 10.1007/978-3-662-49045-7_8

8.1 What Is Environmental Management?

Environmental management is no different than other management activities and should not be performed by a separate system. Because environmental management is often incorporated into the work of a company after other managerial routines are implemented, it has frequently been treated as a separate system; however, in well-functioning organisations, environmental management is a part of the entire management system and must be part of the organisation. Moreover, the commitment of a company's management is essential to the outcome.

Accepting the need and benefit of an environmental management system varies with the type of company and with the demands of different stakeholders. Compliance with existing legislation and regulation is a basic demand that may be seen both as a barrier to the development of a company or as a business opportunity. The company's attitude may be described in terms of the "reactive-defensive-accommodative-proactive scale" [1]:

- Reactive companies deny responsibility and do less than required
- Defensive companies admit responsibility, although they fight it and do the least that is required
- Accommodative companies accept responsibility and do all that is required
- Proactive companies anticipate responsibility and do more than is required

The proactive strategy may seem costly at first. However, this approach also strengthens the reputation of a company among its stakeholders and prepares a company to be a "forerunner" in new technologies.

8.2 Strategies in Environmental Management

Companies are not isolated entities with respect to environmental impacts. Numerous interactions occur with the surrounding world through energy and material flows, services and communication with external stakeholders. These relationships are illustrated in Fig. 8.1. These flows and minimising their negative impacts are the primary objectives of environmental management.

To minimise negative impacts, different actions may be taken. The preferred strategy is proactive, i.e., to avoid an activity or remove the source of pollution/pollutants. When this option is not possible, "end-of-pipe" solutions, such as abatement technologies, can be applied. Additionally, when evaluating alternative actions, the result should be examined in the context of the desired function or need. Thus, an appropriate strategy for preventing environmental impacts is to investigate all viable options, as shown in Fig. 8.2.

First, in some circumstances, it is *possible to fulfil the need without environmental impact or with a greatly reduced impact*. Some qualities may not be met using the alternative approach; however, the substitute may still be an attractive solution. One example is the use of Internet meetings rather than gathering in one

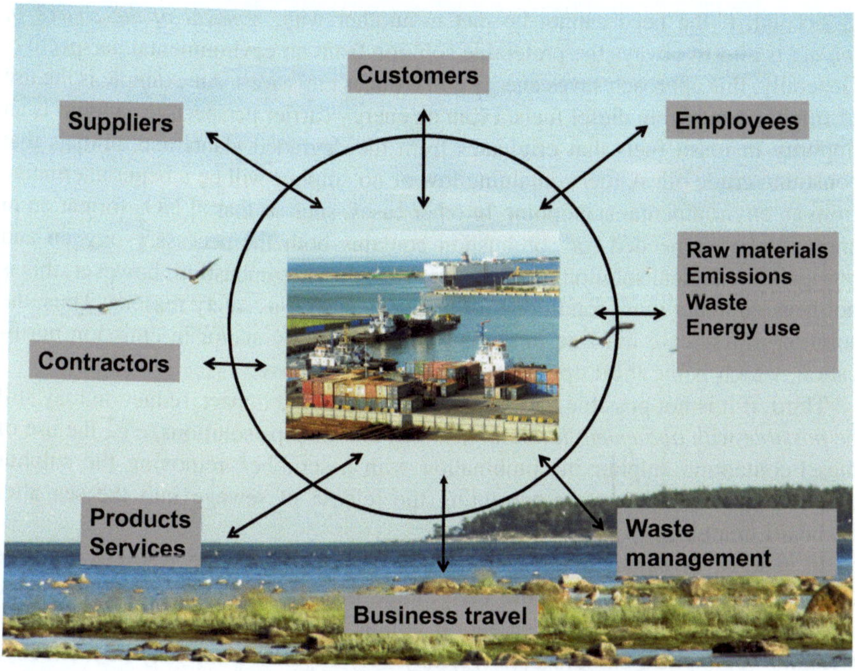

Fig. 8.1 Examples of interactions between a shipping company and the environment

Fig. 8.2 Strategies for preventing environmental impacts

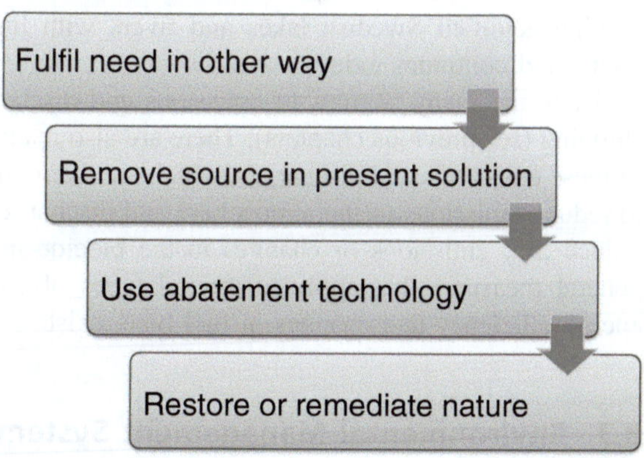

location. Some drawbacks exist to not meeting in person, e.g., difficulty in becoming acquainted with team members when beginning a project. The advantages are avoiding excess travel time and allowing more time to work. Moreover, if the need is the transportation of metal ore from Australia to China, the need can be reduced by using metal ore that is mined closer to the processing plant or by increased the transportation efficiency.

Second, if the need cannot be met in another way, *removal of the source of impact* is almost always the preferable solution from an environmental perspective. Generally, this approach saves energy and reduces emissions. An example is the use of sulphur-containing diesel fuels. From an energy carrier perspective, sulphur is an impurity in fossil fuels that originates from the degraded plants and animals that constitute crude oil. A fuel containing low or no sulphur will be a better alternative from an environmental standpoint. In other cases, such as that of NO_x formation in engines, the air needed for combustion contains both the necessary oxygen and nitrogen; the logical solution is to use pure oxygen for combustion; however, this is not feasible under most conditions for practical as well as safety reasons. Thus, the solution in this case must be in combustion technology and/or in emission purification, which is the third option in the steps listed above.

Third, if it is not possible to remove the source of the impact, reduction may still be *possible with abatement technologies*, i.e., end-of-pipe solutions, e.g., the use of diesel-containing sulphur in combination with a scrubber, removing the sulphur oxides from the exhaust, or permitting the release of sewage into the sea after on-board treatment.

In general, avoiding impacts is typically the most cost- and resource-effective approach, particularly during the early stage of product development or in the planning stage of a project. The fourth option, i.e., *restoration of nature or remediation of impacts on nature*, is typically very expensive and may be ineffective or impossible to perform. Examples of these options include the attempted remediation of oil spills in the ocean or on beaches and the attempted remediation of the negative environmental impacts of acid rain (to which sulphur oxide contributes) by treating acidified Swedish lakes and rivers with lime; this process began in the 1970s and continues today.

There are many sources of emissions and discharges to the environment from shipping (see previous chapters). There are also many ways to decrease the impacts of these emissions and discharges, which are the focus of Chap. 11. Direct methods to reduce emissions to the atmosphere and discharges to the sea, e.g., scrubbers to reduce SO_x emissions or changes in the biocide in antifouling paints, and more general measures that apply to several types of environmental impacts, such as energy efficiency and changes in fuel type, exist.

8.3 Environmental Management Systems and Standards

Formalised environmental management systems and standards have evolved from the requirements for structured environmental work and communication to stakeholders; perhaps the best known is the ISO 14000 standard series [2], which is an internationally recognised system, although other regional or national standards exist, including the European EMAS [3].

The ISO system has a structure similar to the quality management standard ISO 9000 and the recent energy management standard ISO 50000 [4, 5]. ISO 14001 provides a system for the self-declaration of an environmental management system, identification and management of potential environmental impacts, and confirmation of conformance to these standards by an external organisation, i.e., "ISO certificate". The certificate is a statement that a company has structured environmental management practices and is committed to continuous improvement; however, it does not reflect the degree of environmental impact or performance. A company that lacks a certificate may also have highly functional environmental management practices, although it is more difficult to communicate this to external stakeholders. ISO 14001 was revised in 2015, and a new edition was issued in September 2015. In this edition, further emphasis is placed on environmental management within the organisation's strategic planning, and there is more focus on leadership and the addition of proactive initiatives to prevent environmental impacts. In addition to improving environmental performance, lifecycle considerations regarding environmental aspects and a communication strategy were added [2].

The process of introducing and maintaining an environmental management system includes a general series of actions. The principle of "plan, do, check, act" is common in all processes of managing change (see also the Ship Energy Efficiency Management Plan (SEEMP) in Sect. 10.2.3). The individual steps required in formal systems may vary, although they typically involve all or most of the steps outlined in Fig. 8.3.

Fig. 8.3 Environmental management in a company

An initial review of the state of a company, including potential environmental impacts and sources of these impacts, is performed before establishing an environmental policy. This policy should be related to the specific challenges of the company and not be a general statement of "contribute to sustainable development". This policy is the high-level strategic instrument that includes environmental goals capable of being fulfilled and assessed within a predefined time scale.

After a policy is established, the corresponding organisation can be structured. People require assignments, responsibilities and positions within an organisation. Education may be necessary for the goals and routines to be understandable and accepted. The fulfilment of these goals is assessed by auditing (either internal or external). In many systems, an environmental report that is available to the public is required, whereas such a report is optional in other systems.

8.4 Environmental Reporting and the Global Reporting Initiative (GRI)

An important facet of environmental management is to facilitate internal and external communication on environmental issues. The level and quality of environmental reports varies, especially in the use of numbers as indicators of specific goals. The different cultures in companies and in different countries regarding information that should be released to the public are also factors. To enhance and standardise communication regarding all aspects of sustainability, including the economic and social aspects, a UN initiative, i.e., the Global Reporting Initiative (GRI) [6, 7], was developed to handle the communication of sustainability information from companies and organisations in a way that enhances credibility, comparability and comprehensiveness. The GRI also builds on the outcome of the Rio+20 negotiations; UNEP (United Nations Environment Programme) is among its founders, together with companies and other organisations. The GRI compiles guidelines for information on economic, environmental and societal aspects and provides a standard for reporting, although this is a voluntary process.

Regional initiatives also exist to make sustainability reporting compulsory. In December 2014, the European Union's "Directive on disclosure of non-financial and diversity information by certain large companies" [8] entered into force. It requires companies to "*disclose in their management report, information on policies, risks and outcomes as regards environmental matters, social and employee aspects, respect for human rights, anticorruption and bribery issues, and diversity in their board of directors*". This is directive is expected to become national law in two years, with the first company reports being published in 2018 for the 2017–2018 financial year. The rules will apply to companies with more than 500 employees, and approximately 6000 companies will be affected.

References

1. Kolk, A., *Economics of Environmental Management*. 2000, Harlow: Financial Times Prentice Hall.
2. ISO, *ISO 14001 Environmental Management Systems Revision*. 2015: Geneva.
3. European Commission Environment. *EMAS*. 2015, http://ec.europa.eu/environment/emas/index_en.htm, last accessed on August 31 2015.
4. ISO, *ISO 9000 Quality Management*. 2015, International Standardisation Organisation: Geneva.
5. ISO, *ISO 50000 Energy Management*. 2015, International Standardisation Organisation: Geneva.
6. UNEP. *Global Reporting Initiative (GRI)*. http://www.unep.org/resourceefficiency/Business/SustainableandResponsibleBusiness/PartnershipBuilding/GlobalReportingInitiativeGRI/tabid/78936/Default.aspx, last accessed on August 312015.
7. Visser, W., Matten, D, Pohl, M, *Global Reporting Initiative (GRI)*. 2010.
8. European Commission Banking and Finance. *Non-Financial Reporting*. 2015, http://ec.europa.eu/finance/company-reporting/non-financial_reporting/index_en.htm, last accessed on August 31 2015.

Methods and Tools for Environmental Assessment

9

Karin Andersson, Selma Brynolf, Hanna Landquist and Erik Svensson

Abstract

Selecting measures to reduce the overall environmental impact associated with shipping can be a difficult task, and a systematic approach is needed. There is risk of sub-optimisation and counteraction of different measures with one another if decisions are made based on fragmented decision support. An example of a system effect is the long lifetime of ships, which slows the introduction of new technologies. Therefore, design and retrofits must fulfil not only present but also future requirements for environmental sustainability. This chapter describes the basic details of several methods and tools that can be used in environmental assessments within the shipping industry. The methods and tools described are grouped into three categories: (1) procedural tools, (2) analytical tools and (3) aggregated tools. Examples of procedural tools are environmental impact assessment, multi-criteria decision analysis and risk management; life cycle assessment (LCA) and environmental risk assessment are examples of analytical tools. Aggregated tools include indicators, indices, and footprints.

K. Andersson (✉) · S. Brynolf · H. Landquist
Shipping and Marine Technology, Chalmers University of Technology, Gothenburg, Sweden
e-mail: karin.andersson@chalmers.se

S. Brynolf
e-mail: selma.brynolf@chalmers.se

H. Landquist
e-mail: hanna.landquist@chalmers.se

E. Svensson
Gothenburg, Sweden
e-mail: e-svensson@outlook.com

K. Andersson et al. (eds.), *Shipping and the Environment*,
DOI 10.1007/978-3-662-49045-7_9

Keywords
Systems analysis · Environmental impact assessment · Scenario analysis · Multi-criteria decision analysis · Risk management · Life cycle assessment (LCA) · Material flow analysis · Substance flow analysis · Environmental risk assessment · Cost-benefit analysis · Life cycle costing · Indicators · Indices · Footprints

The interest in finding technologies for competitive and more sustainable shipping is growing. The response to the environmental impacts from shipping in the form of international regulations, national legislation, and local rules is also of growing importance. Initiatives from individual ship owners and different consortia and organisations to find sustainable technologies and to develop a green ship or zero-emissions ship are becoming increasingly important.

The regulation of emissions and discharges from ships are covered under the same convention, MARPOL 73/78, although under different Annexes with different regulations for each pollutant, along with different dates of application. It is therefore difficult to see the complete picture. Some ship owners have expressed confusion when making decisions to fulfil upcoming or existing regulations regarding the selection of new technologies or fuels for achieving long-term sustainability. Risks associated with sub-optimisation and the use of different measures counteracting one another due to fragmented decision support are present. The long lifetime of ships also increases the importance of planning when ordering new ships. Therefore, design and retrofitting must fulfil not only present but also future sustainability requirements. A design that allows retrofitting enhances the possibility of constructing ships that are efficient over their long lifetimes.

In addition to the technical and environmental aspects, many development initiatives have been discussed under the heading of sustainable shipping without considering economic or social aspects. The concepts of sustainable development and sustainable shipping are discussed in Sect. 1.3. Many issues must be handled to obtain sustainable solutions, and the terms *sustainable* or *green* are often used without being clearly defined. Traditional sail ships are zero-emission ships, although they do not fulfil other parts of the definition of sustainability or the general demand for arrival times, work environment, or manoeuvrability in ports.

The complex nature of the decision-making process to promote sustainability in technical systems has led to the use of various tools based on a systems perspective to provide decision support within the public sector for creating policies and regulations and within private companies for investment and strategy decisions. Tools and methodologies for environmental systems analyses have been developed over the last 20 years from a need for assessment tools for environmental decision support, which is rooted in systems thinking and systems analysis.

9.1 Principles of Systems Analysis

The *systems approach* builds on a hierarchal perspective of the world as a system with connected subsystems at different levels. Interactions occur between the parts of the system, and the chain from cause to effect may be long in both time and space. A simplified model of a system is shown in Fig. 9.1. The system has components and connections and is separated from the rest of the world by so-called system boundaries; the outside areas are called the surroundings or the environment. System boundaries are established by the application or question investigated rather than by any natural law; the system is thus a social construction that is useful for a specific problem [1, 2]. Different types of systems exist, for example, machinery systems, biological systems, social systems, socio-technical systems, and nature-society-technology systems [3].

Typically, complex systems have different types of feedback loops that can enhance or decrease the effects of changes (Fig. 9.1). One very important property of a system is the existence of emergent properties, that is, the whole is greater than the sum of the individual parts. For example, the various components of a ship, including fuel and lubricants, must be used together. Seen separately, the function of a ship for transporting people and goods is difficult to conceptualise. However, if the parts are put together in the correct way, a vessel is created that can move and perform the transport service although it also emits exhaust and noise. There are subsystems within a ship, such as the energy system, where the fuel is converted to different energy types for propulsion, heating or cooling or electricity for internal systems.

Systems thinking was first described in the literature as a biological concept. The ecosystem concept described in *Fundamentals of Ecology* by Eugene Odum was first published in 1953 [4]. This publication provided a framework that was similar across disciplines and allowed multidisciplinary work. However, during World

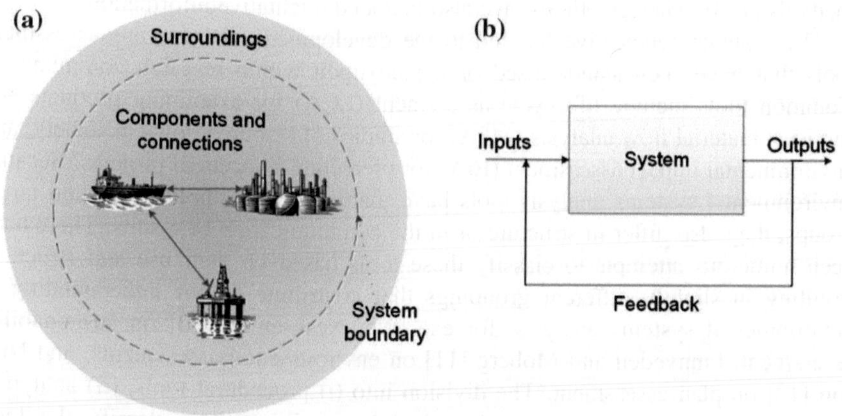

Fig. 9.1 **a** A system with components, connections and system boundaries. **b** Feedbacks affecting systems

War II, the systems concept had also been developed in logistics and resource management and in technology development [1]. The complexity of the technical systems developed during the war and the challenges related to logistics and resource management required a systematic approach. Systems analysis has since become an important tool in *systems engineering* when developing integrated and advanced products, such as cars and ships [2, 5].

Later, the concept of systems thinking was broadened for use in studies of social systems, which are termed *soft systems* in contrast to *hard systems* that can be treated in quantitative terms. The flows between the different compartments of soft systems are in the form of information rather than mass and energy, which are common in technical systems [6]. The anticipation from the establishment of soft systems analysis was to use systems analysis in soft systems in the same way as in hard systems. However, this goal has not been achieved, whereas the use of systems thinking and a systems perspective has proven useful for evaluating soft systems.

9.2 Environmental Systems Analysis

Within environmental assessments, the need for structured handling of complex systems became increasingly clear as environmental concerns related to technical systems increased. Early studies focused on energy and resource usage. However, during the 1990s, impacts on health and the environment also came into focus. Environmental systems analysis is one branch of systems analysis used for analysing, interpreting, simulating and communicating complex environmental issues from different perspectives and includes several methods and tools for the environmental assessment of anthropogenic systems using a systems perspective. Several definitions of environmental systems analysis exist. The common denominator is that they include complex decision situations and a multidisciplinary approach. Some research groups and researchers have limited the definition to include only quantitative methods [7, 8], whereas others have also included qualitative information [9].

The systems perspective has led to the development of numerous assessment tools that have been standardised or agreed upon within research communities. Common tools include life cycle assessment (LCA) for examining products and services, material flow analysis (MFA) for studies of resource flows in society, and environmental impact assessment (EIA) for predicting impacts of projects. Because environmental systems analysis tools have many different applications and target groups, they also differ in structure or in the communication of results. There have been numerous attempts to classify these tools based on their use and structure, resulting in slightly different groupings that contribute to our understanding of environmental systems analysis, for example, Ness et al. [10] on sustainability assessment, Finnveden and Moberg [11] on environmental assessments, and Nilsson [12] on plan assessment. The division into (i) procedural tools, (ii) analytical tools and (iii) aggregated tools described below is used to classify the tools described in this chapter (see Fig. 9.2). Some systems tools are procedural,

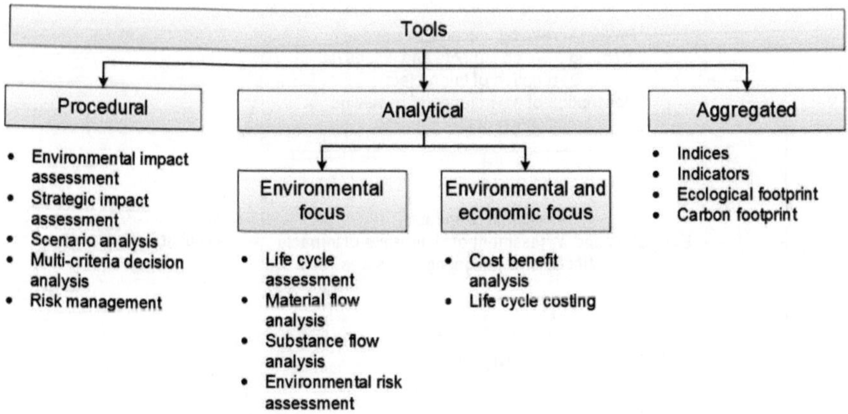

Fig. 9.2 Tools for environmental systems analysis described in this chapter

providing guidance for the assessment of a working procedure. Other tools are analytical, comprising a quantitative assessment of material and energy flows, which is often followed by aggregation and valuation. Analytical tools may be used in the decision support of procedural tools. The choice of a tool reflects the intended use of the tool and the need for communication with users. In some applications, results must be aggregated into a number that is easy to communicate, use and understand; in others, transparency is needed in the process and the quantitative information.

9.3 Procedural Tools

Procedural tools define the procedure of performing an assessment rather than being a normative description of the quantitative and/or qualitative steps in the assessment. The goal of these tools is often to include and consider many types of criteria (including environmental, economic, and social). Consensus among stakeholders regarding the recommended solution is often sought in the use of procedural tools.

9.3.1 Environmental Impact Assessment (EIA) and Strategic Environmental Assessment (SEA)

Environmental impact assessment (EIA) and strategic environmental assessment (SEA) are tools used in decision support for development projects and have the general goal of ensuring that environmental impacts are considered in decisions on projects and plans, respectively. Environmental assessment can be undertaken for individual projects, such as a dam, motorway, airport or factory. In regulatory systems, EIA was first introduced in the U.S. under the Environmental Policy Act (1969). In Europe, the basis is Directive 2011/92/EU (which is known as the EIA

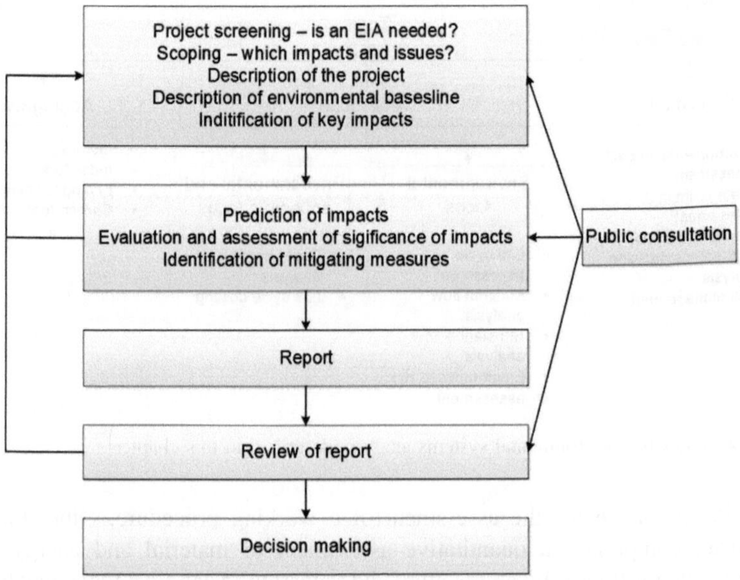

Fig. 9.3 The general environmental impact assessment (*EIA*) procedure

Directive), or for public plans or programmes, Directive 2001/42/EC (which is known as the SEA Directive).[1] The working procedure includes consultation with the public as an important step [14]. Examples of applications to shipping are EIAs in connection with permitting processes for ports and fairways where the impact from vessel traffic is considered. Currently, the use of EIA for evaluating impacts from shipping is uncommon [15].

The integration of ports and port activities into urban development is becoming increasingly more popular, including the construction of housing and recreation areas in connection with port facilities, for example, in Stockholm and Oslo. Moving container transport to ports outside cities with better road and railroad connectivity may be the result, although some sectors profit from urban ports. The port activities that are likely to remain in a central location are passenger ferries and cruise ships. Here, the requirements for port activities will increase, and shipping activities must be included in the permit process. Several impacts can be expected due to this traffic. A large flow of passengers and cars, the transportation of food to vessels, the bunkering of fuel, the discharge of waste, emissions to the atmosphere and noise from engines must be managed and their impacts minimised. In future urban port development projects, the EIA process may have a greater focus on shipping and emissions and noise from vessels.

EIAs are generally undertaken using several steps, which are summarised in Fig. 9.3. An EIA begins with a project screening, collection of background data and

[1]The EIA directive was amended in 2014 the changes entered into force on 15 May 2014 [13].

description of the environmental baseline. In addition, a follow-up procedure occurs after a decision to grant the permit is made. However, the implementation of the EIA directive varies within the EU and there are also different ways of using EIA in permitting processes throughout the world. Although the justification is to ensure that environmental aspects are considered in projects, this can be performed in different ways. The type of projects and procedures, the roles of permit applicants and authorities, and the inclusion of other criteria than environmental vary greatly. Box 9.1 describes the procedure for EIA in Sweden as an example.

Box 9.1 Environmental impact assessment (EIA) in Sweden
Under the Swedish Environmental Code, Chap. 6, a permit or licence must not be granted unless an environmental impact assessment (EIA) has been completed [16]. This mandate applies to new projects and to changes in on-going activities. The government may also demand an EIA for projects that will not need a permit according to the environmental code, e.g., for shore protection. According to the code, the purpose of an EIA is to identify and describe direct and indirect effects on humans, animals, plants, soil, water, the atmosphere, the climate, the landscape, the cultural environment, and the management of land and water resources and to identify and evaluate parameters that may affect the safety of projects involving hazardous chemicals.

An EIA must contain the following according to the environmental code:

- A description of the project with information on the location, design and extent of activities.
- A description of the measures needed to avoid, decrease or remedy adverse impacts.
- The information needed to evaluate the main impacts on people, the environment and resource management of land and water.
- A statement on alternative locations (if possible) and on alternative designs.
- A statement on the *zero alternative* that is, the consequences if the project is not realised.

In Sweden, an EIA must be prepared and financed by the permit applicant. In many other countries, it is performed by the permitting authority. The Environmental Code contains several provisions regarding the preparation of an EIA. In the very early stages of the process, the permit applicant consults with the responsible authority and the different stakeholders in the vicinity, including private parties affected by the activity. Thereafter, the authority decides whether the activity involves a significant impact on the environment. If that is the case, a more extensive procedure applies (involving additional consultations with neighbours), and a more detailed EIA must be prepared.

9.3.2 Scenario Analysis

Another procedural tool is scenario analysis, which is used to analyse possible future events by considering alternative possible outcomes or so-called scenarios. An illustration of scenario analysis and different scenario analysis methods is shown in Fig. 9.4. A scenario can be defined as "*a description of how the future may unfold based on 'if-then' propositions*" [17]. A scenario is used to evaluate the environmental impacts from different future developments of nature and society and different policy options [18]. The assessment does not predict or forecast future development, although it can provide a basis for understanding the sensitivity and potential in different future developments. Thus, scenario analysis is a useful tool for scientists to understand the future development of ecosystems and society and to guide policy makers with respect to the possible impacts of a certain policy or the possible differences between various policy alternatives [18]. Examples of scenario analyses related to ship operations include studies of the potential emissions in the Artic, where large-scale ship operations have not yet been conducted [19, 20]. For more information regarding the development in the Arctic, see Box 1.1 in Chap. 1 and Sect. 12.2.4.

Scenario analysis may also be conducted using *back-casting,* in which the possible pathways to a future final stage are modelled by beginning with the desired future state. For example, the requirements necessary to limit the atmospheric carbon dioxide (CO_2) concentration to 400 ppm by 2050 can be analysed via

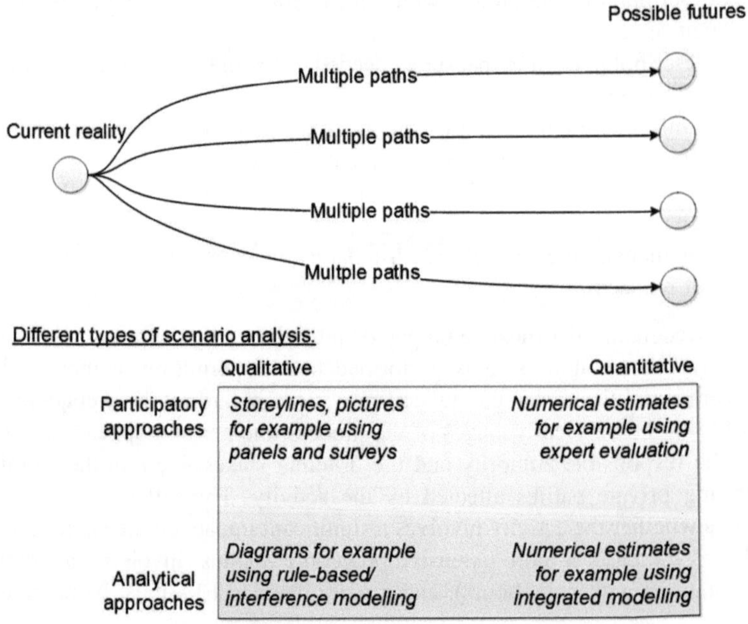

Fig. 9.4 Scenario thinking and different types of scenario analysis

back-casting. Back-casting in the context of shipping was used in a study to analyse the possibility of achieving sustainable long-distance transport in Europe by 2050, with a focus on greenhouse gas emissions and energy use [21]. The main conclusions of the study were that the goal will not be reached with *business as usual* practices, although considerable changes in technology, fuel selection and transport modes can help achieve this goal. Several possible combinations of measures can lead to an 80 % reduction in greenhouse gas emissions, indicating that the transport system is a complex and adaptive system that can be steered towards sustainability.

A scenario analysis of global marine development was applied in a study by a consortium within the shipping sector [22]. Different possible shipping industry changes through 2030 are discussed in the report. The results are used here to exemplify the five key elements of scenario analysis. The first is to create a *representation of the initial situation*, which involves describing the current situation and different prior trends that shaped the current situation, typically beginning with a base year for which sufficient data are available. This study begins with the year 2010.

The second step is to *describe drivers of change*, which are the main factors influencing the future development of nature and society. This study identifies population demographics drivers, including population growth, declining population in some regions and urbanisation; economic drivers, including trade expansion, trade blocs and cities' rules and resources and environmental drivers, such as resource demand and availability and climate change. The third step is to *describe the changes* by portraying how the driving forces are assumed to develop and interact over time. The number of time steps included in a scenario analysis is typically minimised. Only two individual years are included, i.e., 2010 and 2030. For example, the number of offshore wind turbines was 887 in 2010 and is estimated to increase to more than 65,000 in 2030 [22].

The fourth step is a *description of an image of the future*, which is usually a narrative description of the result of the end state as a result of the stepwise changes. In the study, only one time step is included; therefore, the first time step is the same as the end state. The fifth step is the description of alternative pathways to the future. A scenario analysis typically consists of several alternative pathways. In the study, three development alternatives are included: "Status Quo", "Competing Nations", and "Global Commons". To develop more than one pathway, the third and fourth steps must be repeated for each pathway.

9.3.3 Multi-Criteria Decision Analysis (MCDA)

Multi-criteria decision analysis (MCDA), which is sometimes also called multi-criteria analysis, is a structured method for decision support in complex multi-criteria situations [23]. It is used in various scientific contexts, and its use in the environmental field has grown significantly during the last decade. Examples related to short sea shipping can be found in the evaluation of multi-modal transport chains [24]. Section 10.3.1 discusses the use of fuels as a measure to reduce emissions and uses a multi-criteria perspective to evaluate different possible marine fuels.

The general principle of multi-criteria decision analysis is to establish a matrix with the different objectives and criteria versus the project options. Next, a score is given to each option to weight them in a performance matrix [25]. Multi-criteria decision analysis consists of several well-defined steps (Fig. 9.5). The first is problem identification. If a shipping company uses this method to decide which fuel to choose for a new building project planned for operation along the North American east coast in 2016, the identified problem would be fuel selection. The second step is defining alternatives and criteria. Alternatives are ways to solve the problem, which include finding combinations of fuels, engines and abatement technologies that will comply with the environmental regulations along the North American east coast in 2016 and with any other criteria specified by the company. Criteria are a set of properties that describe the performance of the alternatives. Later steps in a typical multi-criteria decision analysis are model building and model application. In model building, the alternatives are scored against the criteria. Before reaching a decision, the criteria are weighted against one another [26].

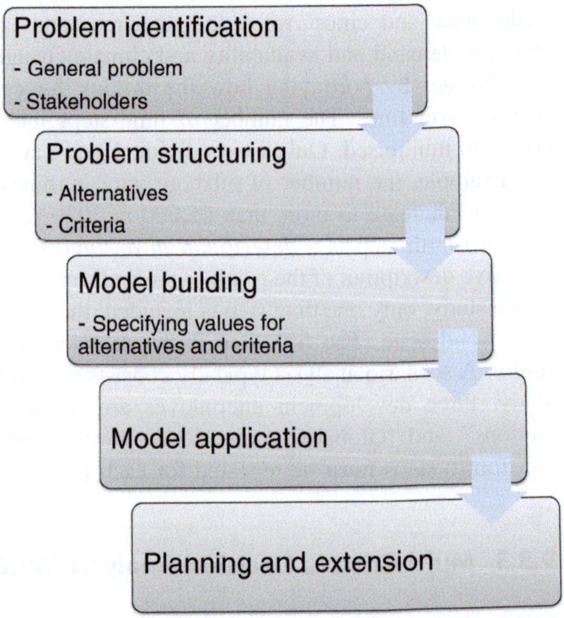

Fig. 9.5 Steps in multi-criteria decision analysis based on Belton and Stewart [27] and Linkov and Moberg [26]

9.3.4 Risk Management

Risk is the chance, within a time frame, of an adverse event with specific consequences [28].

Coping with risk, risk management, can enable an organisation to establish a basis for decision-making and assist decision-makers with taking informed actions and prioritising between options [29]. Several frameworks are used to describe the risk management process. In general, these frameworks share several basic steps (Fig. 9.6), including risk assessment, which consists of risk analysis and risk evaluation, and risk reduction and control.

An expanded framework is provided by ISO [29] (Fig. 9.7), for example, which includes a broader representation of the risk management process. In the *establishment of the context* step, an organisation sets the scope and risk criteria for the process. Within the *risk identification* step, the aim is to produce a comprehensive list of factors that might affect the possibility of meeting the objectives. *Risk analysis*, which is the process to understand the nature and the level of risk, should provide input to the risk evaluation step and act as decision support for risk treatment. Moreover, the *risk evaluation* step involves a comparison of the risk analysis results with the risk criteria to evaluate possible actions and to indicate whether the identified risks are acceptable. During the final step, which is *risk treatment*, one or more actions are chosen and implemented to mitigate the risk.

To ensure that relevant groups and people receive the proper information, communication and consultation is important during all phases of risk management. Monitoring and review should also be a part of a risk management process, including providing input to improve the process, detecting changes that should be reflected in earlier stages, and identifying emerging risks [29].

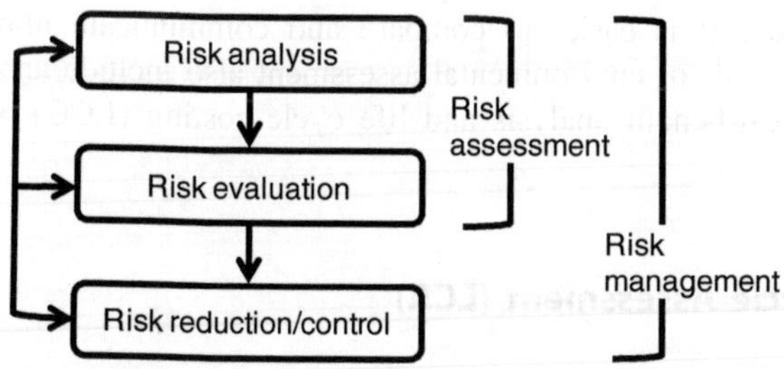

Fig. 9.6 A simplified schematic of the risk management process (adapted from IEC [30])

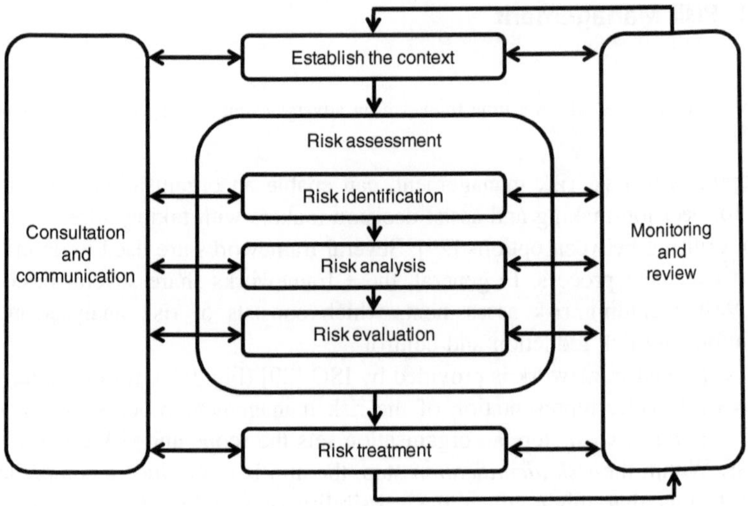

Fig. 9.7 The risk management process (adapted from ISO 31000:2009(E) [29])

9.4 Analytical Tools

The main characteristic of analytical tools is the focus on quantifying material, energy and economic flows. There are many of these calculation tools, which are often standardised, e.g., LCA is standardised in ISO 14040 and 14044 [31, 32]. Many of these tools have their origin in process engineering and energy systems modelling. The extent of aggregation of results and the use of valuation steps vary, although the principle of flow modelling is common. The limitation of analytical tools is that they can only be used for quantifiable impacts. However, this is also an advantage because it is easier to compare and communicate numbers. Several analytical tools used for environmental assessment also include an economic perspective, e.g., cost-benefit analysis and life cycle costing (LCC), which are discussed below.

9.4.1 Life Cycle Assessment (LCA)

LCA is a widely used tool that focuses on assessing the environmental impact of a product or service from a *cradle-to-grave* perspective, namely from the mining of raw materials to waste management (see Fig. 9.8).

The concept of LCA was developed from energy assessments in the early 1970s. Currently, there are attempts to broaden LCA to include life cycle cost and social aspects. LCAs that also include economic and social impacts are referred to as life cycle sustainability assessments [33]. LCA includes three main steps: *goal and*

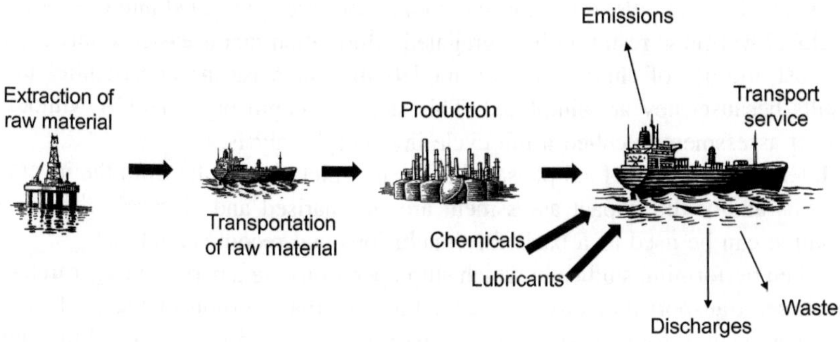

Fig. 9.8 The life cycle of marine fuels

scope definition, *inventory analysis* and *impact assessment*. A fourth step, *interpretation,* which is related to the other steps, is sometimes included.

The goal and scope definition describes the studied system and the purpose of the study. The goal should include the intended application and the reasons for the study because these aspects affect the modelling choice. Therefore, defining the goal and scope is an important component of LCA. An important modelling specification that should be stated in the goal and scope is the choice of a *functional unit*, which is a quantitative unit that represents the function of the studied system and enables comparisons between different products fulfilling the same function. A classic example is the comparison of different ways to package beverages, e.g., mineral water. Here, the function is to provide a certain amount of water to the consumer, which can be achieved using aluminium cans, disposable glass bottles, recyclable glass bottles, recyclable PET bottles and many other containers. The packages can also be of different sizes, providing different quantities of water. A comparison of the environmental impacts of a 0.33 l aluminium can, a 0.75 l glass bottle and a 1.5 l PET bottle is not very useful when selecting package. Instead, the functional unit could be 1 l of mineral water. For marine transportation, the functional unit could be one tonne of cargo transported one km with a roll-on-roll-off (ro-ro) vessel or one year of a RoPax ferry service between two ports.

The inventory analysis step consists of three parts: (i) construction of a flow model according to the system boundaries, (ii) data collection, and (iii) calculation of the resource use and emissions of the system in relation to the functional unit. Three types of system boundaries are addressed in LCA: between the technical system and the environment, between significant and insignificant processes, and between the technical system under study and other such systems. The resource flows and emissions related to each process in the system are often called elemental flows in LCA; this term is used throughout this book.

The elemental flows quantified in the inventory analysis step are classified in the impact assessment step into different impact categories and characterised, e.g., calculating the relative contributions to emissions and the consumption of resources. For example, greenhouse gas emissions are aggregated into one indicator of global warming, resulting in aggregated information that is easier to interpret. In contrast, the use of characterisation models may increase the uncertainties in the results because they are simplified. This step is compulsory; an LCA without an impact assessment is called a life cycle inventory analysis.

Interpretation is the final phase of LCA, in which the results from the inventory analysis and/or the impact assessment are summarised and discussed. This interpretation can be used as a basis for conclusions and recommendations.

When performing studies in which ship operations are a part of a larger transport chain, sea transport typically only constitutes a small portion of the total impact. LCA can be used for assessing different parts of sea transport in terms of the impact per tonne km or person km. An example related to emissions abatement is provided in Box 9.2. Marine fuels are evaluated using LCA; and the results from such an evaluation are shown in Sect. 10.4.3.

Box 9.2 Will SCR catalysts yield environmental benefits?

To evaluate whether the use of an SCR catalyst decreases the environmental impacts in a broader systems perspective, i.e., via decreased nitrogen oxides (NO_x) emissions, the environmental impacts of vessels with and without SCR have been evaluated as a case study using measured emissions and calculated average emissions for two passenger vessels during journeys from port to port. The data were evaluated using LCA. In this study, the system is broadened to also include impacts from urea production and transport (Fig. 9.9) [34].

The benefits of decreased NO_x emissions from greater than 12 g/kWh to approximately 2 g/kWh to the local environment and human health are large and support the use of SCR. Whether the impact is moved to other locations due to emissions from urea production and transport can also be addressed via this study (Figs. 9.10 and 9.11). As shown in Fig. 9.9, the decrease in NO_x

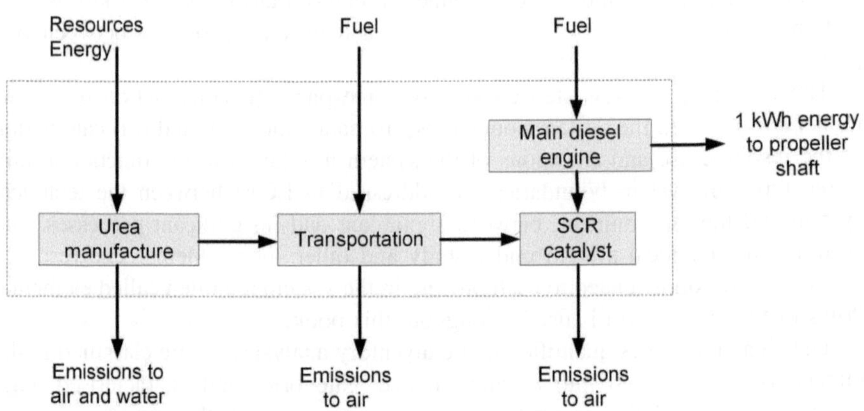

Fig. 9.9 Broadened SCR system [34]

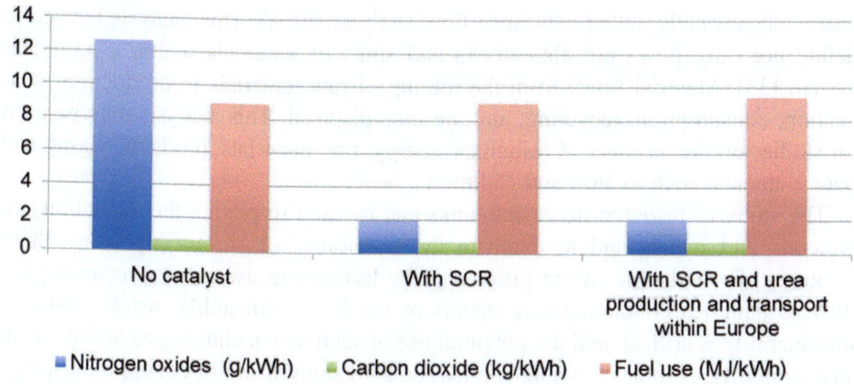

Fig. 9.10 Nitrogen oxides and carbon dioxide emissions and fuel use of ships in port-to-port journey with the effect of SCR use and added emissions from the production and transport of urea

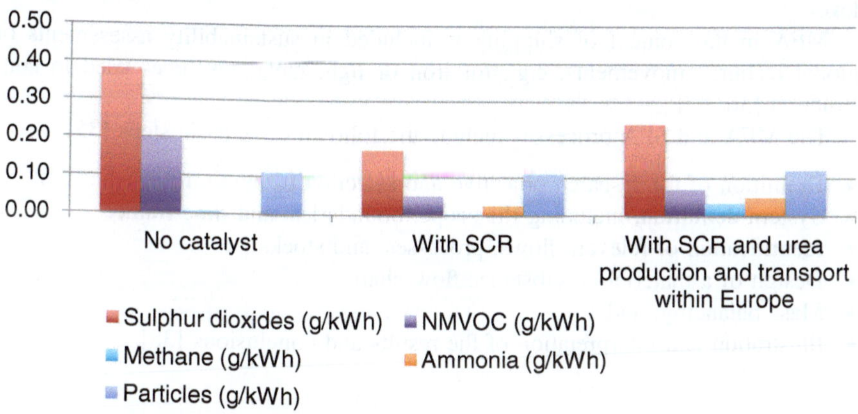

Fig. 9.11 Sulphur dioxide, non-methane volatile organic compounds (*NMVOC*), methane and ammonia emissions from ships in port-to-port journey

due to the use of a catalyst is not substantially counteracted by urea production and transport, although a possible increase in total particle emissions and an increased total energy use could result (Fig. 9.11).

9.4.2 Material Flow Analysis (MFA) and Substance Flow Analysis (SFA)

MFA is a systematic assessment of material and substance flows in society that originates from the *society's metabolism* concept. When addressing substances, the tool is occasionally called substance flow analysis (SFA). This analysis is used to define not only flows but also stocks and sinks of materials within a particular system [35]. Material flows from the mining of raw materials to production, conversion, consumption, recycling, and the final disposal. This analysis may be used for studies on the turnover of resources, energy, raw materials, products, waste, and pure elements, such as iron and cadmium.

The study of resource flows and sinks can be used to predict the availability of minerals and metals and to evaluate the possibility of sustainably using these resources. For example, when proposing the large-scale use of new technologies, such as batteries containing rare metals or catalysts with noble metals, resource sustainability is critical, and the potential use of such new technologies is dependent on availability. Another possible application is to study flows of contaminants, e.g., heavy metals, which may include lead or cadmium, or stable chemical substances, to identify accumulations and map their occurrences. The use of MFA can also be the basis for the collection of statistics on material flows at the national or regional level.

MFA in the context of shipping is included in sustainability assessments of global resource movements, e.g., for iron or light metal resources used in construction [36, 37].

The MFA and SFA processes include the following six basic steps [35]:

- Definition of the research objective and selection of monitoring indicators;
- System definition, including the scope, boundaries, and time frame;
- Identification of relevant flows, processes, and stocks;
- Design of a material or substance flow chart;
- Mass balancing; and
- Illustration and interpretation of the results and conclusions [38].

9.4.3 Environmental Risk Assessment (ERA)

Environmental risk assessment (ERA) is a tool for evaluating the environmental impacts of chemicals that are released into the environment and can be used in risk management. The potential pathways of released chemicals are modelled, and the resulting doses to ecosystems or individual species are assessed. Examples of ERA can be found in studies on the impacts of oil or chemicals released from sunken ships on single marine species or ecosystems [39, 40]. An example of an ERA in the context of shipwrecks is presented in Box 9.3.

Box 9.3 Environmental risk assessment of shipwrecks

Approximately 8000 shipwrecks lie on the sea floor, threatening to leak oil into the marine environment [41]. Oil has several effects on marine ecosystems, which were discussed in Chap. 4, and large uncertainties remain regarding shipwrecks and the probability and timing of a potential oil release.

Remediation of all shipwrecks is not feasible, e.g., due to the high costs involved [42]. Therefore, decision support is needed to prioritise wrecks that pose the greatest potential environmental risk. The cost is also more likely to be lower if proactive rather than reactive measures are taken [42].

One method for providing this type of decision support is ERA. Risk assessment methods that are specific to shipwrecks exist, although few provide a holistic risk assessment approach, e.g., few consider uncertainties that are critical for obtaining relevant results. However, based on a review of the current methods, Landquist et al. [43] suggest a holistic method for the risk assessment of shipwrecks, namely VRAKA.

VRAKA is a quantitative method that consists of several components; the first of which is a risk estimation tool. The risk estimate is based on a quantitative estimate of the probability of release. This estimate is combined with the assumed amount of oil contained within the shipwreck of interest to obtain a Tier 1 risk estimate. The risk is expressed as the expected amount of oil that could be released.

Development of the risk estimation tool VRAKA is necessary to fully include economic, social and environmental consequences. However, the results obtained from a Tier 1 risk estimation can be used to prioritise further studies of specific wrecks.

9.4.4 Cost-Benefit Analysis (CBA)

Cost-benefit analysis (CBA) is used in various ways in different disciplines. Here, the focus is on environmental cost-benefit analysis, which typically includes environmental and social costs. Cost-benefit analysis is a method to weight the costs and the benefits of a proposal, project, or policy. The social benefit is defined as an increase in human well-being, whereas the cost is defined as a reduction in human well-being [44]. Thus, a project must create greater social benefits than costs [44]. CBA is commonly used in *environmental economics* to facilitate the allocation of a society's resources [25]. CBA studies are primarily performed on a national basis, although they can be extended to include larger regions, e.g., the European Union. For example, a CBA was completed to support the impact assessment accompanying the revision of the European Directive 1999/32/EC on the sulphur content of certain liquid fuels, which also included marine fuels [45].

The impacts are quantified, and a monetary valuation is performed that considers both private (or internal) costs and external (social) costs [44]. This evaluation is performed for the entire society, that is, across societal groups, and is based on the willingness to pay (WTP) for benefits or the willingness to accept (WTA) compensation for losses. Special considerations may also be given to disadvantaged or low-income groups in the aggregation of costs and benefits due to higher marginal utilities of income. The aggregation of costs also typically includes a discounting of future costs and benefits to present values [44].

When assessing environmental assets, the cost of lost assets is also included [44]. Ecosystem services, which are the services that people obtain from ecosystems, such as fish and crops (see Sect. 1.3.3 for more information regarding ecosystem services), are one example of environmental assets. The valuation of ecosystem services is particularly difficult due to the inherent complex relationships and lack of knowledge regarding the processes involved when an ecosystem is lost or degraded. The *value of a statistical life* is considered one parameter for human health, and *years of lost lifetime* is also used in valuation.

There are many different types of sources of external costs, including air pollution, climate change, noise, accidents and infrastructure. The valuation step is essential for internalising the external costs and is sometimes debated. It is, for example, difficult to determine a precise value for human health and mortality, impacts on soil and vegetation, and traffic congestion. Several transport- and energy-related studies have examined the valuation of environmental impacts, e.g., ExternE for the European energy sector [46], ASEK for the Swedish transport sector [47] and the European CAFE project [48] for emissions within the EU.

For decision-making support based on cost-benefit analysis, other tools, such as multi-criteria decision analysis, can be used. LCA, environmental risk assessment and EIA can provide input for a cost-benefit analysis study, although the results may require conversion into economic units [49].

9.4.5 Life Cycle Costing (LCC)

LCC can also be called life cycle cost analysis and refers to a variety of methods used in different contexts. LCC can be divided into three different categories: conventional, environmental and social LCC [50]. Conventional LCC is a purely economic evaluation that considers various stages in the life cycle of a product or service. LCC does not always include internalised external costs[2] or the complete life cycle because end-of-life costs are often omitted. To address the costs at different points in time, conventional LCC typically involves discounted costs, with the discount rate selected by the decision-maker [50]. Further environmental LCC is developed to integrate LCC with LCA and uses system boundaries and functional units equivalent to those of LCA. Because the activities and flows considered in the two tools are different, modifications are required to use LCC in this context. There is a risk of double externalities and environmental impacts when combining LCC and LCA. Guidelines for performing environmental LCC and avoiding double counting can be found in Hunkeler et al. [50]. Finally, societal LCC is the

[2]External effects of transport include noise and air pollution. The costs associated with these effects are typically not borne by the transport user; therefore, they are considered to be external. External costs can be internalised directly through the regulation or indirectly by providing better incentives to transport users, including taxes, charges and emissions trading.

assessment of all costs associated with the life cycle of a product that are covered by anyone in the society (regardless of time), including environmental LCC and additional external costs that are commonly described in monetary terms [50].

9.5 Aggregated Tools

To monitor or communicate complex systems, the aggregation of information into a format that is clear and easily communicated may be necessary. As expressed by Donella Meadows in her report on sustainability indicators, people "cannot work towards sustainable development if they have no clear, timely, accurate, visible indicators of sustainable development" (p. 5) [51]. Indicators and indices help to simplify, understand and communicate complex scientific or statistical data. The information pyramid in Fig. 9.12 demonstrates that measured or collected primary data can be analysed as a first step and subsequently communicated in the form of indicators, which in turn can be aggregated into indices [52].

9.5.1 Indicators

Indicators are used in everyday contexts, often expressed as a number with or without a unit. Examples include scorecards, grades or body temperature. For use in environmental and sustainability assessments, indicators have been defined as "simple measures, most often quantitative that represent a state of economic, social and/or environmental development in a defined region" (p. 499) [10]. Measuring the complete state of a system is often not possible; only the perceived state can be determined from an indicator or several indicators. For example, the exact population of a specific species of fish in the ocean cannot be measured, whereas landed catch can be measured. Thus, the population can be estimated from the known quantity. Indicators are thus "abstractions from systems" [51].

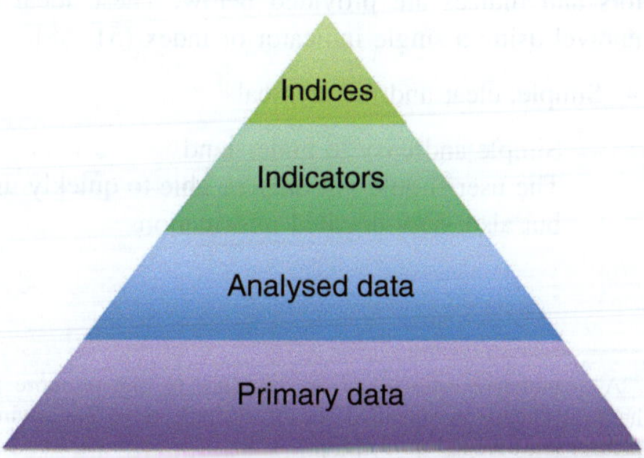

Fig. 9.12 Information pyramid (adapted from Hammond et al. [52])

An indicator measures a specific factor, e.g., tonnes of nitrogen released per year from land-based sources into the Baltic Sea. Such an indicator may not be significant by itself, although it may provide valuable insights regarding an overall phenomenon of interest [52]. Tonnes of nitrogen released per year into the Baltic Sea may not have significance to policy-makers concerned with the environmental state of the Baltic Sea. However, if the amount is expressed in relation to *critical loads*[3] of nitrogen in the Baltic Sea, a clear indication of eutrophication and a need for action can be provided. This is an example of an *environmental indicator,* which measure environmental pressures and conditions as well as societal responses. Environmental indicators can include physical, biological and chemical indicators [54]. In addition to environmental indicators, *sustainability indicators* are used to provide information relevant for targets and continued progress in sustainable development, including both environmental and societal factors that may be connected to a specific limit, time or target [54].

9.5.2 Indices

An index is a condensed description of the state of a system that is derived by aggregating several indicators or variables and is expressed using a single quantity [10, 55]. Too many individual indicators results in confusing and often conflicting results. Indices can provide clear and appealing information for decision-makers. The level of aggregation depends on the intended use and users [52]. Although indices that are too aggregated can result in high uncertainty due to the exclusion of relevant information, too much detail can drown relevant information [56]. Many indices have been developed to measure economic development, with the most well-known being the *gross national product* (GNP) and the Dow Jones Industrial Average Index [51]. Environmental indices include instruments for policy-makers, such as the *Environmental Vulnerability Index* (EVI), which represents the environmental-development pillar of sustainable development (the other two pillars are economic and social development) [57].

Examples of characteristics for ideal sustainability (and environmental) indicators and indices are provided below. These ideal characteristics are difficult to unravel using a single indicator or index [51, 58].

- Simple, clear and hierarchical
  - Simple and easy to understand
  - The user should not only be able to quickly understand the general message but also view detailed information

[3]"A quantitative estimate of an exposure to one or more pollutants below which significant harmful effects on specified sensitive elements of the environment do not occur according to present knowledge" (p. 9) [53].

- Quantification, monitoring and analysis
  - Measurable elements
  - Included elements should be capable of being monitored
  - The frequency and time period should be sufficient for trend analysis
- Scope of measurement
  - Should cover a wide spectrum of activities
  - Not too much or too little information
- Feasible
  - Measurements, analysis and monitoring at reasonable costs
- Communicative aggregation and units
  - Not too aggregated; not too much detail
  - Physical units are preferable to monetary units.
  - Public access to results
- Leading
  - Relevant information should be provided with sufficient time to take action
- Tentative
  - Sufficiently sensitive to reflect changes
  - Should have the ability to be revised

Many environmental initiatives exist in the maritime sector that focus on measuring and communicating environmental performance and are provided by an equally great diversity of stakeholders, including international organisations, governmental institutions, classification societies, individual ports, cargo owners, NGOs, and shipping and port associations. Many initiatives are characterised as environmental labels, certificates, awards and rating schemes, although some initiatives also include the term "index". Three examples of environmental indices for shipping are presented in Box 9.4.

Most initiatives are based on a set of environmental requirements or standards, in which specific installed equipment, operational measures, management aspects or compliance with environmental legislation are rewarded. Such rewards can be in points scored or in reduced port dues. Emission-based initiatives dominate the environmental incentives and awards that are commonly used in the shipping industry [59] and constitute nearly all of the initiatives inventoried by Svensson and Andersson [60], including emissions to air, CO_2 emissions, and energy efficiency. These voluntary initiatives can be important for reducing greenhouse gases in response to governmental difficulties in reaching international agreements.

Box 9.4 Three examples of environmental indices for shipping

The *Energy Efficiency Operational Indicator* (EEOI) is a tool for monitoring energy efficiency and CO_2 emissions from ships. The EEOI was developed by IMO as a voluntary instrument to accompany the mandatory instruments for energy efficiency and CO_2 reduction described in Box 5.2 in Chap. 5. The guidelines for the voluntary use of the EEOI were adopted by IMO in 2009 [61]. The EEOI is an indicator of operational efficiency for ships and is expressed as CO_2 emissions per unit of transport. Moreover, the EEOI includes the monitoring of ship-based CO_2 emissions and calculating energy efficiency for each voyage or over a specific period. Ship owners and operators (shipping companies) can use the EEOI to evaluate the operational energy efficiency of their ships and fleet, which could be applied in environmental management systems or to market the CO_2 performance of ships and shipping companies [61, 62].

The Clean Cargo Working Group (CCWG) is part of Business for Social Responsibility (BSR), which is a global network that aims to develop sustainable business strategies. The group is described by BSR as a business-to-business initiative that is open for membership by shipping companies (ocean container carriers), shippers and logistics providers [63]. The *CCWG Performance Metrics Tool* is an Excel-based tool for assessing the environmental performance of shipping companies using quantitative data from each vessel to create an environmental performance scorecard that quantifies the performance for each shipping company for six categories: CO_2, sulphur oxides and NO_x emissions, waste, water and chemicals, environmental management systems and transparency. The scorecard can be used to benchmark shipping companies against industry standards [63].

The Clean Shipping Project began as a regional initiative along the west coast of Sweden in 2006 to increase focus on the environmental issues of shipping. The project evolved into a network of large cargo owners (the Clean Shipping Network). The project developed the *Clean Shipping Index* (CSI) as a web-based tool that ranks both ships and shipping companies according to their environmental performance. The shipping companies enter information on their ships through the web-based tool; this information is recorded in the Clean Shipping Database, which cargo owners have access to by joining the network. Cargo owners can use this tool to choose providers of ship transportation based on their environmental performance. They can compare the performance of individual ships or entire fleets. Moreover, they can also choose to compare the performance in each of the five environmental areas of the index, namely SO_x and PM emissions, NO_x emissions, CO_2 emissions, chemicals and water, and waste control [64, 65].

9.5.3 Footprints

Different types of "footprints" related to human activities have been suggested and analysed, including the ecological footprint, carbon footprint and water footprint.

The concept of an ecological footprint was developed to visualise the impact of human activities on the production and resources of the planet [66], which is an attempt to convert the resource use and waste production of a population into a needed bio-productive area. If the footprint area of a region or country exceeds the available production area, the natural capital is consumed, and the area is not sustainable.

As a consequence of an increased global consensus on climate change and increased coverage in the media, companies and organisations around the world have been searching for tools to assess the contribution of their products and organisations to the release of greenhouse gases. The carbon footprint concept is one response to this demand. A carbon footprint can be calculated for a certain product by considering emissions throughout its lifecycle. The calculated carbon footprint can be used to compare the environmental performance of a product with the environmental performance of other products with similar functions. An ISO standard for carbon footprints was created in 2013 [67] that includes specific principles, requirements and guidelines for the quantification and communication of the carbon footprint of a product based on standards for LCA, quantification, environmental labels and communication.

This chapter described the basic details of several methods and tools that can be used in the environmental assessment of the shipping industry. The methods and tools described can be grouped into three categories: (1) procedural tools, (2) analytical tools and (3) aggregated tools. The primary characteristics of these methods and tools are summarised in Table 9.1.

Table 9.1 Summary of the primary characteristics of the methods and tools described in this chapter

Tool	Goal	Scope	Objects	Sources	Safeguard subjects	Formal recognition
EIA	Minimising the environmental (and social) impact of a project	Environmental (and social aspects) of a project	Project or activity	Mainly local	Human health, ecosystems, resources, culture and social aspects	Legislation and EU directives
Scenario analysis	Evaluation of possible future states	Varies	Project or activity			
MCDA	Formal handling of conflicting criteria	Measurable criteria for consequences	Project or activity	All types	Human health, ecosystems, natural resources, and social values	
Risk management	Coping with risk	Activity	Project or activity	All types	Human health, ecosystems, natural resources	ISO standard
LCA	Environmental management of product systems	Evaluation of up- and downstream flows related to a functional unit	Products and services	All types connected to function (e.g., natural environment and resources)	Ecosystems, natural resources and human health	ISO standard
MFA/SFA	Management of materials/substances through resource efficiency	Consequences of management changes for flows and stocks	Materials or substances	Only materials/substances under analysis	Resources, human health, and ecosystems	Used by governments and others
ERA	Risk management	Environmental risk to humans and ecosystems from substances	Substances	Only toxic emissions	Human health and ecosystems	OECDa, the EU in connection with REACH, and the US EPA
CBA	Economic allocation in society	Net benefit of project or activity	Project or activity	Internal and external costs (mainly local)	Human health, ecosystems, and resources	Used by governments

(continued)

Table 9.1 (continued)

Tool	Goal	Scope	Objects	Sources	Safeguard subjects	Formal recognition
LCC	Economic information	Costs and avoided costs of a product	Products and services	Internal and external costs	Human health, ecosystems, natural resources, and social values	
Indicators	Aggregated information on state	Time, limit, and target	Activity	Any kind	Sustainable development	
Indices	Aggregated information on state	Target	Activity	Any kind	Sustainable development	
Footprints	Aggregated information on state	Target	Products and services	Any kind	Sustainable development	

References

1. Flood, R. L. & Carson, E. R., *Dealing with complexity: an introduction to the theory and application of systems science*. 2nd ed. 1993, New York: Plenum Press.
2. Blanchard, B. S. & Fabrycky, W. J., *Systems Engineering and Analysis*. 1998, Upper Saddle River, NJ: Prentice-Hall Inc.
3. Ingelstam, L., *System: att tänka över samhälle och teknik [Systems - how to think about society and technology]*. 2002, Eskilstuna: The Swedish Energy Agency.
4. Odum, E. P., *Fundamentals of ecology*. 1953, Philadelphia: W. B. Saunders Company.
5. Sutton, R., *Special issue on marine systems, Editorial*. Proceedings of the Institution of Mechanical Engineers Part M-Journal of Engineering for the Maritime Environment, 2008. **222**(M2): p. I-I.
6. Checkland, P., *Systems Thinking, Systems Practice*. 1999, New York: John Wiley & Sons Inc.
7. Björklund, A., *Environmental Systems Analysis – Some Ongoing Research and Ideas About Future Developments*, in *Systems Approaches and Their Application: Examples from Sweden*, M-O Olsson & G Sjostedt, eds. 2004, Kluwer academic publishers.
8. Wageningen UR. *Environmental Systems Analysis Group*. 2015, http://www.esa.wur.nl/UK, last accessed on 30 August 2015.
9. Harou, J. J. *et al.*, *Hydro-economic models: Concepts, design, applications, and future prospects*. Journal of Hydrology, 2009. **375**(3–4): p. 627-643.
10. Ness, B., Urbel-Piirsalu, E., Anderberg, S. & Olsson, L., *Categorising tools for sustainability assessment*. Ecological Economics, 2007. **60**(3): p. 498-508.
11. Finnveden, G. & Moberg, A., *Environmental systems analysis tools - an overview*. Journal of Cleaner Production, 2005. **13**(12): p. 1165-1173.
12. Nilsson, M. *et al.*, *Analytical framework and tool kit for SEA follow-up*. Environmental Impact Assessment Review, 2009. **29**(3): p. 186-199.
13. EU. *Review of the Environmental Impact Assessment (EIA) Directive*. 2015, European Commission: Brussels, http://ec.europa.eu/environment/eia/review.htm, 19th of July 2015.
14. EU Commission. *Environmental Assessment*. 2011, http://ec.europa.eu/environment/eia/home.htm, last accessed 11 August 2011.
15. Andersson, M. A. *Syns det inte, finns det inte, eller? [If you can't see it, it does not exist, or?]*. 2011, Swedish Maritime Administration, Shipping and Marine Technology, Chalmers University of Technology: Gothenburg.
16. *Miljöbalken [The Swedish Environmental Code]* 1998:808, Government Offices of Sweden, Ministry of Environment and Energy 1998.
17. Alcamo, J. & Henrichs, T., *Chapter Two Towards Guidelines for Environmental Scenario Analysis*, in *Developments in Integrated Environmental Assessment*, Alcamo Joseph, ed. 2008, Elsevier. p. 13-35.
18. Alcamo, J., *Environmental Futures. The Practice of Environmental Scenario Analysis*. Developments in Integrated Environmental Assessment, ed. A.J Jakeman. Vol. 2. 2008, Amsterdam: Elsevier.
19. Corbett, J. J. *et al.*, *Arctic shipping emissions inventories and future scenarios*. Atmospheric Chemistry and Physics, 2010. **10**(19): p. 9689-9704.
20. Paxian, A., Eyring, V., Beer, W., Sausen, R. & Wright, C., *Present-Day and Future Global Bottom-Up Ship Emission Inventories Including Polar Routes*. Environmental Science & Technology, 2010. **44**(4): p. 1333-1339.
21. Mattila, T. & Antikainen, R., *Backcasting sustainable freight transport systems for Europe in 2050*. Energy Policy, 2011. **39**(3): p. 1241-1248.
22. QinetiQ, Lloyd's Register Group Limited & University of Strathclyde, *Global Marine Trends 2030*. 2013.
23. Henggeler Antunes, C., Almeida, L. A., Lopes, V. & Clímaco, J. N., *A decision support system dedicated to discrete multiple criteria problems*. Decision Support Systems, 1994. **12**(4-5): p. 327-335.

24. Browne, D. & Ryan, L., *Comparative analysis of evaluation techniques for transport policies.* Environmental Impact Assessment Review, 2011. **31**(3): p. 226-233.
25. Gamper, C. D., Thoni, M. & Weck-Hannemann, H., *A conceptual approach to the use of Cost Benefit and Multi Criteria Analysis in natural hazard management.* Natural Hazards and Earth System Sciences, 2006. **6**(2): p. 293-302.
26. Linkov, I. & Moberg, E., *Multi-Criteria Decision Analysis - Environmental Applications and Case Studies.* Environmental Assessment and Management, ed. Glenn W Suter II. 2012, Boca Raton: CRC Press Taylor & Francis Group.
27. Belton, V. & Stewart, T. J., *Multiple criteria decision analysis - an integrated approach.* 2003, Boston: Kluwer Academic.
28. Burgman, M., *Risk and decisions for conservation and environmental management*, ed. M Usher, A Dobson, R Peet, & D Saunders. 2005, New York: Cambridge university press.
29. ISO, *Risk management - Principles and guidelines, ISO 31000:2009(E).* 2009, International Organization for Standardization: Geneva.
30. IEC, *Dependability Management - Part 3: Application guide - Section 9: Risk analysis of technological systems.* 1995, International Electrotechnical Commission.
31. ISO, *Environmental management – Life cycle assessment – Principles and framework, ISO 14040:2006.* 2006, International Organization for Standardization: Geneva.
32. ISO, *Environmental management - Life Cycle Assessment - Requirements and guidelines, ISO 14044:2006.* 2006, International Organization for Standardization: Geneva.
33. Guinée, J. B. *et al.*, *Life Cycle Assessment: Past, Present, and Future†.* Environmental Science & Technology, 2010. **45**(1): p. 90-96.
34. Andersson, K. & Winnes, H., *Environmental trade-offs in nitrogen oxide removal from ship engine exhausts.* Proceedings of the Institution of Mechanical Engineers Part M-Journal of Engineering for the Maritime Environment, 2011. **225**(M1): p. 33-42.
35. Brunner, P. H., Rechberger, H, *Practical Handbook of Material Flow Analysis.* 2004, Florida: CRC Press.
36. Yellishetty, M., Ranjith, P. G. & Tharumarajah, A., *Iron ore and steel production trends and material flows in the world: Is this really sustainable?* Resources Conservation and Recycling. **54**(12): p. 1084-1094.
37. Bao, Z. *et al. A review of material flow analysis.* in *International Conference on Management and Service Science* 2010.
38. Huang, C. L., Vause, J., Ma, H. W. & Yu, C. P., *Using material/substance flow analysis to support sustainable development assessment: A literature review and outlook.* Resources, Conservation and Recycling, 2012. **68**: p. 104-116.
39. Yoo, J. M., Tabeta, S., Sato, T. & Jeong, S. M. *Risk assessment for the benzene leakage from a sunken ship.* in *OCEANS'08 MTS/IEEE Kobe-Techno-Ocean'08 - Voyage toward the Future, OTO'08.* 2008.
40. Hassellöv, I., Lindgren F, Landquist H, Rosén L & I, D., *Risk assessment of oil leakage from old ship wrecks, research project in progress.* 2011, Shipping and Marine Technology, Chalmers University of Technology: Gothenburg.
41. Michel, J. *et al. Potentially polluting wrecks in marine waters.* in *International oil spill conference.* 2005.
42. Schmidt Etkin, D., van Rooij, H. & McCay French, D. *Risk assessment modeling approach for the prioritization of oil removal operations from sunken wrecks.* in *Effects of oil on wildlife.* 2009. Tallin, Estonia.
43. Landquist, H., Hassellöv, I. M., Rosén, L., Lindgren, J. F. & Dahllöf, I., *Evaluating the needs of risk assessment methods of potentially polluting shipwrecks.* Journal of Environmental Management, 2013. **119**: p. 85-92.
44. Pearce, D., Atkinson, G. & Mourato, S., *Cost-Benefit Analysis and the Environment: Recent Development.* 2006, Organisation for economic co-operation and development (OECD): OECD Publishing.

45. Bosch, P. *et al. Cost Benefit Analysis to support the impact assessment accompanying the revision of Directive 1999/32/EC on the sulphur content of certain liquid fuels*. 2009, AEA: Oxfordshire, United Kingdom.
46. Bickel, P. & Friedrich, R. *ExternE - Externalities of Energy - Methodology 2005 Update*. 2005, Office for Official Publications of the European Communities: Luxemburg.
47. SIKA. *Samhällsekonomiska principer och kalkylvärden för transportsektorn: ASEK 4 [Socio-economic principles and values for the transport sector: ASEK 4]*. 2008, Statens institut för kommunikationsanalys (SIKA) [Swedish Institute for Communication Analysis]: Östersund.
48. Holland, M. P., S, Watkiss, P, Droste-Franke, B, Bickel, P. *Damages per tonne emission of PM2.5, NH3, SO2, NOx and VOCs from each EU25 Member State (excluding Cyprus) and surrounding seas*. 2005, European Commission DG Environment.
49. Pearce, D., Atkinson, G, Mourato, S, *Cost-Benefit Analysis and the Environment*. 2006, Paris: OECD.
50. Hunkeler, D. *et al.*, *Environmental life cycle costing*. SETAC Books, ed. Joseph Gorsuch. 2008, Pensacola, Florida: CRC Press.
51. Meadows, D. *Indicators and Information Systems for Sustainable Development - A report to the Balaton Group*. 1998, The Sustainability Institute: Hartland.
52. Hammond, A., Adriaanse, A., Rodenburg, E., Bryant, D. & Woodward, R. *Environmental Indicators: A Systematic Approach to Measuring and Reporting on Environmental Policy Performance in the Context of Sustainable Development*. 1995, World Resources Institute: Washington, D.C.
53. Grennfelt, P. & Nilsson, J. *Critical loads for sulphur and nitrogen: report from a workshop held at Skokloster*. 1988, Nordic Council of Ministers: Copenhagen.
54. Niemeijer, D., *Developing indicators for environmental policy: data-driven and theory-driven approaches examined by example*. Environmental Science & Policy, 2002. **5**(2): p. 91-103.
55. Ebert, U. & Welsch, H., *Meaningful environmental indices: a social choice approach*. Journal of Environmental Economics and Management, 2004. **47**(2): p. 270-283.
56. Baumann, H. & Tillman, A.-M., *The hitchhiker's guide to LCA : an orientation in life cycle assessment methodology and application*. 2004, Lund: Studentlitteratur.
57. Environmental Vulnerability Index. *The EVI in Summary*. 2015, http://www.vulnerabilityindex.net/, last accessed on August 31 2015.
58. Harger, J. R. E. & Meyer, F. M., *Definition of indicators for environmentally sustainable development*. Chemosphere, 1996. **33**(9): p. 1749-1775.
59. EMSA. *Study on Ships producing reduced quantities of ships generated waste: present situation and future opportunities to encourage the development of cleaner ships.Final Report, EMSA/OP/05/05., prepared by HPTI Hamburg Port Training Institute GmbH*. 2007, European Marine Safety Agency (EMSA): Lisaboa.
60. Svensson, E. & Andersson, K. *Inventory and Evaluation of Environmental Performance Indices for Shipping*. 2011, Department of Shipping and Marine technology, Chalmers University of Technology: Gothenburg.
61. IMO. *MEPC.1/Circ.684. Guidelines for voluntary use of the ship energy efficiency operational indicator (EEOI). Ref. T5/1.01, 17 August 2009*. 2009, International Maritime Organization: London.
62. IMO. *Description of the Package of Technical and Operational Reduction Measures for Ships Agreed by MEPC 59*. 2011, International Maritime Organization: London. http://www.imo.org/ourwork/environment/pollutionprevention/airpollution/documents/technical%20and%20operational%20reduction%20measures.pdf.
63. BSR. *Beyond the Factory Gates: How Brands Improve Supply Chain Sustainability Through Shipping and Logistics: Clean Cargo Working Group Tools for Measuring and Reducing Environmental Impacts*. 2011, Business for Social Responsibility (BSR). http://www.bsr.org/en/our-insights/report-view/beyond-the-factory-gates.

64. CSI. *Environmental opportunities for shipping*. 2015, Clean Shipping Index (CSI): Gothenburg. http://www.cleanshippingindex.com/wp-content/uploads/2015/03/CleanShippingIndex-brochure-2015.pdf.
65. CSI. *Clean shipping index. Guidance Document. Version 5.1*. 2015, Clean Shipping Index: Gothenburg, Sweden. http://www.cleanshippingindex.com/wp-content/uploads/2015/07/Guidance-doc-CLEAN-SHIPPING-INDEX-5.1-2015-07-01.pdf.
66. Wackernagel, M. & Yount, J. D., *Footprints for sustainability: The next steps*. Environment, Development and Sustainability, 2000. **2**(1): p. 21-42.
67. ISO, *Greenhouse gases - Carbon footprint of products - Requirements and guidelines for quantification and communication, ISO/TS 14067:2013*. 2013, International Organization for Standardization: Geneva.

Energy Efficiency and Fuel Changes to Reduce Environmental Impacts

10

Selma Brynolf, Francesco Baldi and Hannes Johnson

Abstract

Many different emissions from ships are directly related to a ship's fuel consumption. This is particularly true for emissions to air, which are generated during the combustion process in the engines. Hence, improving the conversion process from fuel energy to transport work can be an effective means of reducing ship emissions. Solutions for reducing ship fuel consumption are generally divided into design and operational measures. Design measures primarily include technical solutions implemented when the ship is designed, constructed, and retrofitted, such as weight reduction, hull coatings, air lubrication, improvement of hull design, optimal propulsion systems and harvesting waste energy. Operational measures are related to how the ship or the fleet is operated and include measures such as weather routing, optimal ship scheduling, improved ship logistics, and on-board energy management. Although reducing fuel consumption always generates an environmental benefit, it should be noted that the use of different fuels results in different impacts on the environment for a given energy conversion efficiency. Another way to reduce emissions is therefore related to the type of fuel used on a ship, e.g., diesel fuels, gases, alcohols and solid fuels. However, choosing a fuel is not an easy process because it is influenced by a broad range of criteria, including technical, environmental and economic criteria.

S. Brynolf (✉) · F. Baldi · H. Johnson
Shipping and Marine Technology, Chalmers University of Technology, Gothenburg, Sweden
e-mail: selma.brynolf@chalmers.se

F. Baldi
e-mail: francesco.baldi@chalmers.se

H. Johnson
e-mail: hannes.johnson@chalmers.se

K. Andersson et al. (eds.), *Shipping and the Environment*,
DOI 10.1007/978-3-662-49045-7_10

Keywords
Energy efficiency · Marine propulsion · Energy management · SEEMP · Marine fuels

The amount and type of energy or fuel used on ships greatly impact the resulting emissions, especially emissions to the air. This chapter addresses measures to reduce the amount of energy that is used on new and retrofitted ships through changes in design (Sect. 10.2) and on existing ships through operational measures (Sect. 10.3) while maintaining the same level of transport service, i.e., measures that improve energy efficiency. An overview of the different design and operational measures are shown in Fig. 10.1.

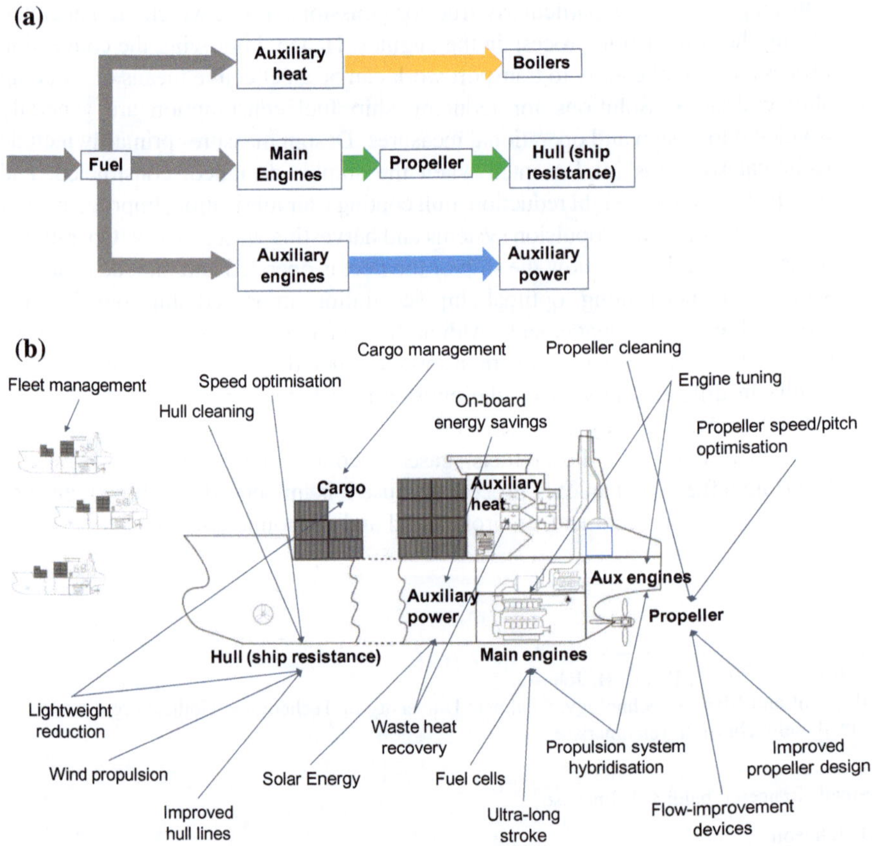

Fig. 10.1 A simplified illustration of **a** the energy flows on-board ships and **b** operational and design energy efficiency measures

These energy efficiency measures determine how much fuel is needed; the types of fuels used subsequently determine the environmental impact. The last section of this chapter addresses how changes in fuel type can reduce emissions and associated environmental impact (Sect. 10.4).

Energy efficiency and fuel type are discussed in different contexts in regard to policy making. Regulation that concerns energy efficiency is primarily devoted to strategies of reducing greenhouse gas (GHG) emissions, thereby abating the climate change impact of the shipping sector. In contrast, regulating fuels has primarily been discussed as an abatement measure to reduce sulphur oxide (SO_x), nitrogen oxide (NO_x), and particle emissions. However, changing fuels may also be necessary to yield large reductions in GHG emissions [1].

This division of matters concerning energy efficiency and fuels in a policy context can be understood as a matter of cost effectiveness. Changing fuels—for example, from heavy fuel oil (HFO) to liquefied natural gas (LNG)—to reduce GHG emissions is very costly [2]. Similarly costly is reducing SO_x emissions by improving energy efficiency: SO_x emission control areas (ECAs) resulted in a 90 % reduction in sulphur in bunker fuel. Improving energy efficiency by the same amount would be very expensive.

Improved energy efficiency, of course, also has a commercial impact for shipping companies. Indeed, as discussed in Chap. 5, because the regulation of energy efficiency in shipping remains quite lax, rising fuel costs for shipping companies can perhaps be said to be the greatest driver of improved energy efficiency in recent times. However, the impact of any actions taken today from a commercial perspective to save fuel are small compared to what will be required in a climate change context. Thus, the measures and actions being discussed from the business perspective are typically those that are cost effective to implement today and in the near future, whereas the climate change perspective requires more radical changes and support from policy instruments, particularly those that put a price or standard on GHG emissions. As discussed in Chap. 5, even substantial improvements to energy efficiency ($\sim$60 %) are not expected to yield a reduction in total emissions from shipping by 2050. The demand for shipping is simply expected to grow faster than efficiency can be improved [3].

10.1 Energy Efficiency Potential and the Energy Efficiency Gap

Energy efficiency in shipping received substantial attention from industry as oil prices rose in the late 2000s, as it had after the oil crises of the 1970s. The potential for GHG emission reductions from energy efficiency measures has been estimated to range between 10 and 50 % [4–6]. A division of these measures can be made into *technical, operational,* and *structural measures*. Many technical measures are more suitable for new ships, and some may also be suitable for retrofitting existing ships (to be discussed in Sect. 10.2). Operational measures concern actions that can be taken on ships in operation, typically without technical modification, such as finding and using the optimal trim of the vessel (see Sect. 10.3). Structural measures are those that require many stakeholders to implement, e.g., reductions in time in port to enable slower speeds at sea. In this chapter, structural measures are covered together with other operational measures (i.e., Sect. 10.3).

The seemingly slow pace of implementing these cost-effective measures has pushed political bodies to commission assessments to determine why these measures are not being implemented [7]. If they are truly cost effective, this seems to be a paradox. This phenomenon has been observed in many industrial sectors, households etc. and is often referred to as an "energy efficiency gap" (see Box 10.1) [8].

Box 10.1 The energy efficiency gap

The oil crises of the 1970s generated R&D and public policy aimed at conserving energy and reducing dependences on oil imports in many nations. Many assessments with the objective of understanding the associated costs of pursuing energy conservation strategies were completed. These assessments typically indicated that many measures that were cost effective to implement were indeed available. This gap between the assessments and reality was referred to as an "energy efficiency gap" or "energy efficiency paradox". If cost-effective measures were available, why were they not being implemented at a greater pace?

The early energy efficiency literature can (although a bit simplistically) be divided into the optimistic engineers, who asserted the existence of this large, cost-effective potential, and the pessimistic economists, who saw the non-adoption of these measures as a sign that there was no *economic* inefficiency; the markets had simply made a rational choice to allocate resources to other, more lucrative purposes [9]. The engineers developed a taxonomy of various 'barriers' to explain the slow diffusion of these measures [10]. A barrier to energy efficiency was often defined as some type of postulated mechanism that prevents a decision maker from implementing a certain measure that improves energy efficiency [11]. Economists later intervened to correct what they perceived as an inaccurate use of economic concepts: these so-called barriers could simply be "benign characteristics of well-functioning markets" where funds where simply being allocated to other more lucrative purposes [12]. Researchers thus began to adhere more closely to the economists' definitions and distinguished between market barriers and *failures*, the latter representing a subset of barriers that are cost effective for society to remove [8].

Although the discussion of the energy efficiency gap has continued between researchers, policy makers, industry and others since the oil crises in sectors other than shipping, only recently has it occurred in a maritime context. This discussion began in the aftermath of the many assessments that examined the potential to mitigate the climate impacts of shipping [4, 13]. In particular, the lack of information on the performance of vessels used by shipping organisations or markets used to monitor contractual relationships has been argued to impede the spread of energy efficiency measures [14–16]. Without access to such information, organisations may have difficulty following up on performance and assessing measures in the operations. Organisations in contractual relationships, e.g., owners, charterers, third-party managers and owners, may also have difficulties following up on contractual terms, thereby potentially leading to so-called principal-agent problems [17].

Based on similar arguments, the importance of information transparency is often raised in diplomatic discussions on energy efficiency in IMO and the EU, although challenges to actually increasing transparency remain unresolved.

10.2 Improving Energy Efficiency from a Design Perspective

This section addresses technologies aimed at reducing the input of fuels to ships. To understand the source of this potential, an understanding of how energy is used on ships is needed. Two examples of the energy flow on commercial ships are shown in Figs. 10.2 and 10.3. On the left, the input in the form of chemical energy is displayed; on the right, the output in the form of ship thrust, auxiliary power, and auxiliary heat and cooling is shown. In between, components converting energy

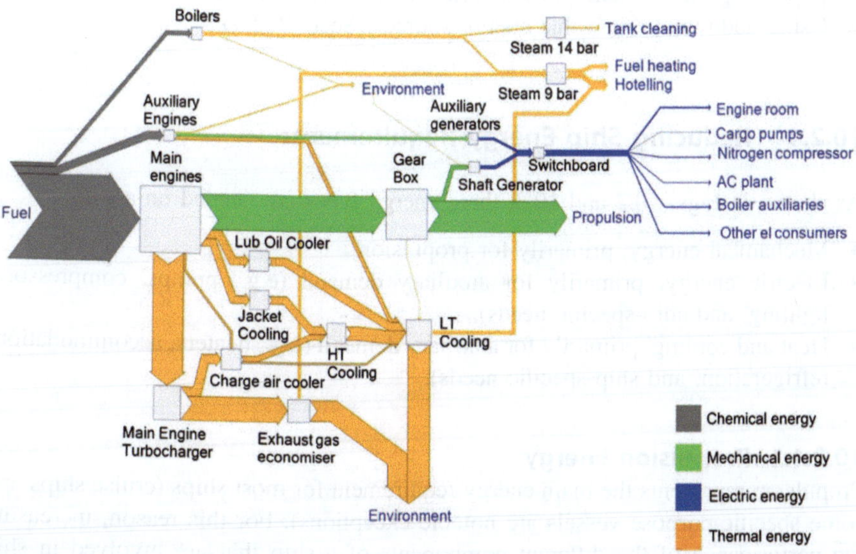

Fig. 10.2 Typical energy flows in a merchant vessel for 1 year of operations. The case of a chemical tanker

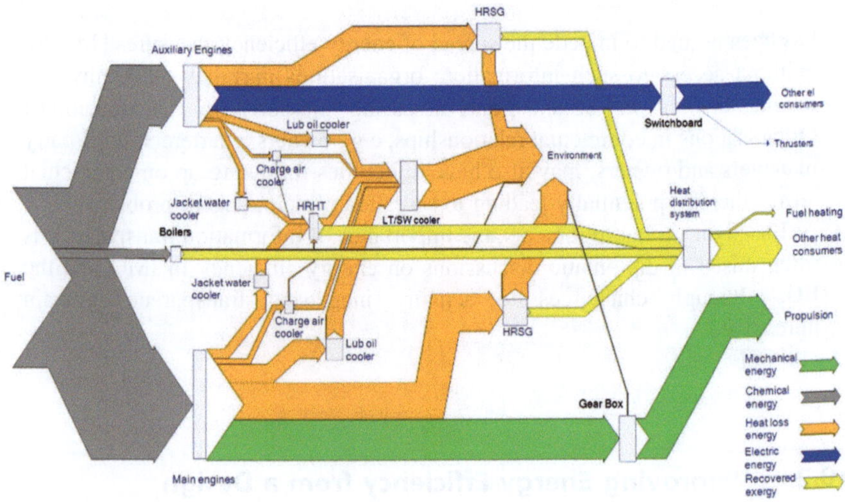

Fig. 10.3 Typical energy flows in a passenger vessel for 1 year of operation. The case of a cruise vessel in the Baltic Sea

between different forms (e.g., chemical to mechanical energy or mechanical to electrical energy) are shown. Thus, only three possibilities exist to reduce fuel consumption:

- Reducing ship energy requirements (Sect. 10.2.1)
- Improving the efficiency of converters and transmitters (Sect. 10.2.2)
- Using additional renewable energy sources (Sect. 10.2.3).

10.2.1 Reducing Ship Energy Requirements

As shown in Figs. 10.2 and 10.3, three energy types are needed on a ship:

- Mechanical energy, primarily for propulsion.
- Electric energy, primarily for auxiliary demand (e.g., pumps, compressors, lighting, and ship-specific needs).
- Heat and cooling, primarily for auxiliary demand (e.g., heaters, accommodation, refrigeration, and ship-specific needs).

10.2.1.1 Propulsion Energy

Propulsion represents the main energy requirement for most ships (cruise ships and some specific-purpose vessels are notable exceptions). For this reason, increasing the performance of the different components of a ship that are involved in ship propulsion has always been a priority. A few fuel-saving options related to the reduction of propulsion power are briefly described in this section.

Propulsion energy is needed to move a ship through water. Hence, improving hull design can lead to reduced ship resistance and a reduced need for propulsive power. Large improvements in hull design have been observed throughout the history of naval architecture, although the extensive use of computational fluid dynamics permitted by recent improvements in computational power have boosted the efforts in this field [18, 19].

- *Weight reduction*: Given ships of the same type, the propulsion power scales with displacement to the $2/3$ power. Assuming a constant cargo payload, a ship's weight can be reduced by decreasing its lightweight[1] (e.g., decreasing steel thickness, employing lighter materials, or installing less/smaller machinery) or by reducing the amount of non-profitable load (e.g., bunker fuel or spare parts). All of these measures are subject to compromises with competing interests, such as reduced safety with thinner hulls and smaller machinery, increased maintenance with fewer spare parts, and appropriate logistics with less bunkered fuel.
- *Modifications to the main dimensions*: A significant portion of ship fuel consumption can be influenced by the ship dimensions (length, breadth, and depth). These parameters are often limited by external factors, such as the water depth and berth length in ports and the width of canals, although a ship's designer still has some freedom in optimising these dimensions depending on the specific application. In particular, wave-making resistance is strongly influenced by the slenderness of the hull. Thus, fast ships (such as ferries and containerships) tend to have longer, more slender hulls. For slower ships, such as tankers and bulk carriers, for which the frictional component of resistance dominates, slender hulls provide limited benefits or possibly even reduced performance.
- *Hull coatings and air lubrication*: The frictional component of ship resistance is often dominant in determining the required propulsion power and can be decreased by reducing the roughness of the hull surface, particularly for slow ships, such as tankers and bulk carriers, for which the frictional component dominates the total ship resistance. The use of antifouling paints for preventing marine growth on a ship's hull offers substantial benefits for reducing frictional resistance, which is extensively addressed in Sect. 4.3. Different types of advanced coatings can reduce friction by up to 4 % in practical applications [20]. Air lubrication, which involves the generation of a layer of air on a portion of a ship's hull to reduce frictional resistance, is an alternative that is being tested with some success. Three air lubrication concepts have been proposed, namely, using bubbles (injection of air bubbles in different locations in the hull), air cavity (cavities in the lower part of the hull are filled with air), and air film (similar to air bubbles, although a larger film of air is used to cover the bottom of the ship). Despite lacking extensive practical applications, improvements of up to 15 % should be achievable, with the best results occurring for large, slow

[1] In marine technology, the term "lightweight" refers to the total weight of a ship without any cargo, fuel, crew, passengers, provisions, spare parts etc. on board.

ships. After the installation of such a system on selected ships, 4–8 % improvements were announced [21].

- *Improvement in hull design*: The shape and appendages of a ship's hull (e.g., rudders and bow thrusters) can be improved through an appropriate design procedure. Differences of 4–5 % in the required propulsion power for ships of the same type and size are not unusual, although these differences can reach up to 17 % [20]. Different hull designs behave differently depending on ship speed and draft. Thus, it is particularly important to design a ship for its specific operational pattern. Recently, thanks to more powerful and cheaper computers, computational fluid dynamics (CFD) simulations have been used increasingly successfully in the improvement of hull design.

Some means of converting mechanical power to thrust are needed to overcome ship resistance. The efficiency of this conversion has dramatically improved since the era of paddle wheels, and today, propellers can have efficiencies as high as 70 %. As with hull design, CFD simulations have greatly improved the efficiency of propeller design in recent years.

- *CFD-aided design*: Similar to ship hulls, propeller design can also be improved using computational methods. In particular, full-scale wake fields and hull-propeller interactions can be considered, with an estimated reduction potential of 1–4 % [20].
- *External flow-improvement devices*: Many systems have been proposed to improve propeller performance by acting on the water flow around the propeller. These devices are typically classified based on whether they are placed in front of or behind the propeller. Devices placed in front of the propeller improve propeller performance via modifications in the water inflow to the propeller by redirecting the flow and either increasing or decreasing its velocity (e.g., vortex generators and Schneekluth nozzles) or by generating a pre-swirl in the water flow (e.g., Mewis ducts and Pre-Swirl Stators). Devices placed behind the propeller recover the rotational energy of the water flow. These devices include rudder bulbs, twisted rudders, propeller boss cap fins and contra-rotating propellers.

  Improvements in fuel consumption from these devices are very difficult to estimate and can vary greatly depending on ship type and operations; a realistic estimate is in the range of 1 to 9 % [4, 20].
- *Engine-propeller interaction*: In typical design practices, a high (although difficult to estimate) potential exists for improving both propeller and engine efficiency due to the interaction of the propeller with the main engines under both design and off-design conditions.

In general, choices of the correct hull lines, propeller, and engine are not independent, and the entire propulsion system should be tailored for the specific application. This tailoring is an important aspect to consider when designing all parts of a ship, including its hydrodynamic component.

10.2.1.2 Auxiliary Energy

The term "auxiliary energy" refers to all on-board energy demands that are not strictly related to ship propulsion. Auxiliary energy is generally further divided into auxiliary electric and thermal energy demands (generally referred to as auxiliary power and heat).

A significant portion of these demands can be found on all vessels and are related to basic functions and activities of ships. Auxiliary power can refer to powering cooling pumps, engine room fans, fuel separators, lights, and navigation equipment. Auxiliary heat is generally needed for crew accommodations and the heating of fuel in the tanks prior to injection into the engines.

The remaining auxiliary energy demand includes ship-related functions and varies significantly depending on a ship's type. Some vessels, such as container-ships and bulk carriers, generally have limited ship-related auxiliary energy demands. Tankers require an extensive amount of power for cargo pumps when unloading cargo. Moreover, roll-on roll-off (RoRo) vessels demand high auxiliary power for deck ventilation, especially during cargo loading and unloading. Tankers that transport "dirty cargo", such as HFO or similar high-viscosity liquids, have high thermal energy demands for cargo heating. Cruise ships and passenger vessels have high demands for both auxiliary heat and power, primarily due to the accommodation needs of the several hundred passengers on board.

An example of the aforementioned variability can be observed in Figs. 10.2 and 10.3. Figure 10.2 presents a chemical tanker with limited auxiliary consumption that is primarily related to cargo pumps, ballast water pumps, and cleaning of the cargo holds. Figure 10.3 presents a passenger vessel, which has much higher demands for auxiliary heat and power.

Because auxiliary energy requirements largely depend on a ship's type, general recommendations are not possible. The selection of more efficient components, such as pumps, compressors, and light bulbs, should always be considered a viable solution when justified due to extensive usage. Generally, great benefits can be reached when frequency converters are applied to rotating components (e.g., pumps, compressors, and fans), allowing for the regulation of rotational speed and more efficient operation in off-design conditions. For reefers, ferries and passenger ships, better insulation can reduce heating and cooling requirements.

10.2.2 Improving the Energy Efficiency of Converters and Transmitters

As depicted in Fig. 10.2, significant energy losses occur in the components performing the conversion between different forms of energy, such as engines (chemical to mechanical), boilers (chemical to thermal), and generators (mechanical to electric). Improving the efficiency of these systems can contribute to considerable reductions in ship fuel consumption. This section provides an overview of available technologies for increasing the efficiency of energy conversion on ships, including both improvements to existing components (such as Diesel engines) and the installation of new systems (such as fuel cells and waste heat recovery systems).

10.2.2.1 Improvements in Diesel Engine Technology

As stated in Chap. 1 and as implied throughout the book, diesel engines are the most common prime movers on ships because of their high efficiency, reliability, and flexibility [22]. However, despite their high efficiency, continuous efforts are being devoted towards improving their performance. Recently, in connection with the increasing frequency of slow-steaming, development has moved towards improving not only the maximum efficiency of engines but also their performance at lower loads:

- *Increased cylinder stroke*: Increasing the cylinder stroke allows for longer expansion and increased power per cycle, resulting in decreased engine speed at a given power output, which yields higher propeller efficiency. However, these solutions are dependent on greater propeller diameters, which are often limited by other considerations related to ship draught and hull-propeller clearance.
- *Electrically controlled timing*: Optimal timing for valve opening and fuel injection vary depending on the specific engine load and operational state. Mechanically operated valves and injectors are slowly being replaced by electronically controlled systems that allow an engine to optimise valves and injection timing depending on its operational conditions. This type of technology does not sensibly affect the maximum efficiency of engines, although it does reduce the fuel penalty when operating in off-design conditions.
- *Improved turbocharging*: Engine turbocharging has been the major improvement since the introduction of the diesel engine, and all modern medium- and low-speed engines are equipped with a turbocharging system. In addition to higher-efficiency turbines and compressors, other new concepts have recently entered the turbocharger market. In multi-step turbocharging with intermediate cooling, inlet air passes through two compressors in series and through a cooler between the two, improving both the compression efficiency and final air density. Sequential turbocharging is related to the use of two turbochargers in parallel, which broadens the operational range of an engine and improves performance under off-design conditions.

10.2.2.2 Fuel Cells

Fuel cells are able to directly convert chemical energy to electricity without combustion engines and generators, which makes them very efficient for low power ranges compared to other conversion technologies. Selected fuel cell technologies can be operated on light fuels, such as methane and methanol, and can be considered for use in marine applications. Higher efficiency can be obtained if a DC electric network is used because no energy is lost in the AC/DC conversion. However, for the same power output, fuel cells are still much bulkier than all other prime movers on the market. Currently, the use of fuel cells in marine applications is limited to small vessels, such as pleasure crafts, inland ferries and submarines. Fuel cells have been tested on large commercial vessels for auxiliary power generation using a supply vessel (FellowSHIP project) and on a car carrier (METHAPU project). Both cases demonstrated the feasibility of such concepts, although installations remain rare due to the high investment costs and low operational experience.

The following two types of fuel cells were tested in the aforementioned trials on medium- to large-sized vessels; these fuel cells are expected to be the most employed in the foreseeable future:

- *Molten Carbonate Fuel Cells (MCFCs)* operate at high temperatures (greater than 600 °C) and can be fuelled by natural gas. The conversion to hydrogen occurs in the fuel cell, which does not require an external reformer. MCFCs can theoretically reach efficiencies of up to 60 %, whereas efficiencies in the range of 40–50 % have been measured in the FellowSHIP project application.
- *Solid Oxide Fuel Cells (SOFC)* also operate at very high temperatures (500–1000 °C) and can also use light hydrocarbon fuels, such as natural gas, without the need for an external reformer. Similar to MCFCs, SOFCs can theoretically achieve high efficiencies. Values of 41–43 % have been measured in the METHAPU project.

10.2.2.3 Electric Power Integration

The propulsion systems of conventional ships (see Chap. 1) primarily contain one or more engines that are mechanically connected to one or more propellers through a shaft, where auxiliary power is generated by additional engines; the propulsion and auxiliary power demands are clearly separated. However, when auxiliary demands are high (e.g., cruise ships and RoRo vessels) or when high flexibility is required (e.g., tug-boats, supply vessels, and naval applications), integration between propulsion and auxiliary power generation can become a convenient option. Several different alternatives are available depending on the degree of integration:

- *Shaft generator*: The simplest integration connects an electric generator to the main engine via a gearbox, which permits the generation of auxiliary power with high efficiency because the main engines are generally more efficient than the auxiliary engines. However, such a configuration requires the main engine(s) to operate at a constant speed, which can result in lower propulsion efficiency. The use of frequency converters or direct current distribution networks provides an effective solution to this issue.
- *Hybrid propulsion*: Most electric machines can easily be switched from generator (mechanical to electric power conversion) to motor (electric to mechanical) mode. Such a configuration is similar to the *shaft generator* mode, although this configuration permits the possibility of using the auxiliary engines to provide additional power for propulsion if they are connected to a common switchboard with the shaft generator. This solution can be useful for improving the flexibility of the system and as a redundancy measure.

- *Diesel-electric ship*: In the most advanced level of integration, no distinction exists between the main and auxiliary engines. Several engines of similar or different sizes are connected to a common switchboard, which provides energy to the electric motors for propulsion and to all auxiliary machinery systems. This arrangement provides the highest level of flexibility and redundancy and also permits the integration of additional power sources, such as waste heat recovery (WHR) systems or solar power. This system is particularly common on cruise ships, where the auxiliary demands are comparable to that of propulsion.

Therefore, the on-board integration of electric power and distribution networks provides substantial improvements to the overall ship power plant efficiency. However, when the flexibility requirements are low and auxiliary power demands are also low, the losses in the electric conversion (frequency converters, generators, and motors) become dominant. Diesel-electric propulsion is compared to traditional mechanically driven propulsion in Box 10.2.

Box 10.2 Diesel-electric versus conventional mechanically driven propulsion systems: a comparison

A numerical example can be used to compare the diesel-electric concept to a more standard propulsion system and to highlight the advantages and disadvantages of each strategy. Two alternative propulsion systems are compared: (1) a conventional propulsion system comprising a two-stroke engine that is directly connected to a fixed pitch propeller; the required auxiliary power is generated by auxiliary engines; and (2) a diesel-electric propulsion system with 4 four-stroke engines that are all coupled to the same switchboard; electric motors are connected to the propeller shafts.

These two concepts are tested using three possible ship types and operational conditions: a containership sailing at full speed, a cruise ship positioned in port, and a RoRo vessel operating at low speed. In all cases, engine efficiency is assumed to be 165 g/kWh[2] for the two-stroke engine, 175 g/kWh for the four-stroke engines, and 200 g/kWh for the auxiliary engines. The conversion efficiency for electric machines (motors and generators) is assumed to be 95 %.

Containership: A propulsion power demand of 20 MW and auxiliary power demand of 1 MW are assumed. Under these conditions, the overall efficiency of the conventional system is 50 %, whereas that of the diesel-electric system is 42 % due to the higher transmission losses and lower main engine efficiency.

Cruise ship: During the day, cruise ships are often berthed while passengers visit tourist locations. In this case, propulsion power decreases to

[2]In shipping, it is common to measure engine efficiency in g/kWh, i.e., the amount of fuel in grams required to generate one kWh of mechanical energy output. This value can therefore be considered as an efficiency assuming a constant fuel energy content.

zero, while the auxiliary power demand remains high due to the large accommodation heating/cooling requirements. Assuming 10 MW of auxiliary power demand, the efficiency of the standard system decreases to 40 %, whereas the diesel-electric system performance remains at 46 % efficiency.

RoRo ship: Because of tight schedules, ferries are normally operated at low load under standard conditions and are provided with excess power to increase speed in case of delays or to maintain speed in bad weather. In this case, a propulsion demand of 10 MW (compared with a design power of 20 MW) and an auxiliary power demand of 2 MW are assumed. The two-stroke engine, operated at less than 50 % load, now has an efficiency of 190 g/kWh, whereas the flexibility of the diesel-electric arrangement allows the engines to maintain constant efficiency. Under these conditions, the efficiencies of the two systems are comparable at 44 %.

10.2.2.4 Harvesting Waste Energy

As shown in Fig. 10.2, large quantities of energy are released to the environment from the engine as waste heat, primarily from exhaust gas, inlet air cooling, and engine cooling. This energy can be recovered in the form of heat and used on board. The concept of "large quantities" is however quite relative, especially when thermal energy is involved. Thermal energy has different qualities and can therefore be used to different extents, depending on its temperature (see Box 10.3 for additional details).

Waste heat recovery for fulfilling the on-board auxiliary demand is already common in the shipping industry. *Exhaust gas economisers* are typically installed on the exhaust line of the main engines and are used to generate steam (or, less commonly, to heat thermal oil), which is subsequently used for heating purposes (e.g., fuel or accommodation heating). The *generation of fresh water* is also often connected to the use of waste heat, especially from engine cooling, in the form of low-pressure evaporators, although other systems exist for freshwater generation that do not require the use of waste heat.

Depending on the application, a substantial amount of energy might be available for recovery even after on-board needs for auxiliary heat and freshwater generation are met. In this case, additional mechanical power can be produced if a *waste heat recovery (WHR) system*[3] is installed. There are several types of WHR systems, although Rankine cycles are most often used. According to this conversion principle, waste heat from the engine is used to generate steam, which produces

[3]There is often some confusion regarding the definition of "waste heat recovery systems". In theory, this category should include all systems employed for the recovery of waste heat, which would also include exhaust gas economisers. However, the scientific literature often refers to WHR systems only when waste heat is converted into mechanical power. This convention is used in this text.

mechanical power via expansion in a turbine. WHR systems are not commonly used in the shipping industry despite their common use in similarly sized land-based applications.

However, recent increases in fuel prices have increased the interest in WHR systems, and several manufacturers offer commercially available solutions for implementing WHR systems for increasing marine power plant efficiencies. The most common solution involves the use of both a power turbine fed directly by high-pressure exhaust gas that bypasses the turbocharger and a steam Rankine cycle that recovers heat after the turbocharger and the power turbine. Some manufacturers also offer solutions with integrated recovery of heat from cooling systems. The power turbine provides simple and inexpensive heat recovery, whereas the Rankine cycle significantly improves the efficiency.

Steam cycles are particularly effective when four-stroke engines are used as prime movers because of the higher exhaust temperature. However, in two-stroke engines low exhaust temperatures after the turbocharger (200–250 °C) complicate the utilisation of steam as a working medium from the perspective of both efficiency and material stress. Organic Rankine cycles (ORCs), which employ a working fluid that is not water/steam, have been extensively studied in the scientific literature as a method to increase the recovery efficiency for heat sources of low temperatures. However, ORCs have witnessed limited commercial applications, especially in the shipping sector.

Fuel savings declared by manufacturers range from 5 to 12 %, although these numbers have not yet been confirmed in practical applications. However, these declared fuel savings are in line with those obtained through numerical simulations in the scientific literature. WHR applications could gain increased popularity if fuels with low sulphur content, such as methane or methanol, were to see substantial development in the future. Not only do these fuels lack a heating requirement, thereby reducing auxiliary heat demands but also the absence of sulphur allows for the recovery of a greater portion of the exhaust gas energy, which is typically limited to a temperature of 160 °C to avoid the condensation of sulphuric acid and the consequent corrosion on the heat exchangers.

Other options for waste heat recovery include absorption refrigeration, which allows the use of heat for chilling purposes, and thermoelectric generation, which refers to processes based on the Seedback effect for the direct generation of electricity from a temperature difference without the need of any thermodynamic cycle. Both of these options are still in early stages of development.

For an extensive review of waste heat recovery options for ships, the interested reader is referred to Shu et al. [23].

10.2.2.5 Systems Integration

Ships are complex and isolated systems, where needs for mechanical, electric, heating, and cooling energy must be fulfilled on board without any connection to the outside world. For this reason, the system must be designed with special care to the different interactions and possible integrations.

The recent developments in ship technology related to increased fuel prices have demonstrated that new systems and components can be employed to increase efficiency, although these developments have also revealed how complex a ship is from the perspective of energy systems.

A stronger focus on systems engineering in the design of new ships, aiming at more advanced integration of the different systems and a better understanding of their interactions in different operational modes, can lead to large reductions in fuel consumption with reduced requirements for future technological development.

Box 10.3 Energy quality, heat, and exergy

Figures 10.2 and 10.3 clearly demonstrate that a large amount of energy is released to the environment in the form of thermal energy. Therefore, it seems logical that this energy could be used for other purposes, such as for conversion into mechanical or electrical energy.

However, thermal energy is consistently different than other energy forms. In contrast to mechanical and electrical energy, thermal energy is chaotic and disorganised. Converting from disorganised to organised movement, or from chaos to order, requires paying a price. As stated in the second law of thermodynamics, a given amount of thermal energy cannot be converted to an equal amount of mechanical energy. The efficiency of the conversion depends on several parameters, with the temperature of the heat source being the most relevant. Higher heat source temperatures result in higher-quality thermal energy and a higher fraction of thermal energy that can be converted into mechanical energy. Given a specific amount of thermal energy at a well-defined temperature and in a given environment, the maximum amount of mechanical energy that can be extracted by the conversion is typically referred to as the *exergy* content of the thermal energy.

These observations have a number of consequences:

- Waste heat cannot be entirely converted into work. Indeed, only a relatively small portion of the heat released by the engine to the environment can be used for auxiliary power purposes, even assuming ideal conversion machines.
- Not all sources of waste heat are of equal importance. The energy in the exhaust gas, which (depending on the engine type) is released at between 200 and 350 °C is of higher quality than that contained in the cylinder cooling water (90 °C) or in the charge air (up to 200 °C at full engine load).
- The recovery of waste heat on-board can be a particularly challenging process if the objective is to harvest it in the most efficient way. Using high-temperature exhaust gas to generate 8 bar steam corresponds to an inefficient use of the original energy flow and to a loss of energy quality. The same process occurs when 8 bar steam is used to heat fuel oil to 70 °C in the storage tanks.

– Analysing ship energy efficiency based solely on energy quantity can be misleading. A ship might recover all of its waste energy for heating purposes, which might appear efficient from an energy perspective. However, the full recovery of all waste heat does not imply that it is recovered in the most efficient way. This is the domain where exergy analysis demonstrates the greatest potential for identifying the inefficiencies of thermal systems.

10.2.3 Using Additional Renewable Energy Sources

An additional solution for reducing GHG emissions is the use of renewable energy sources to meet on-board energy demand.

10.2.3.1 Wind Propulsion

Even with the introduction of the steam engine, large sailing vessels maintained their role in global transportation for a long period of time. This era was only brought to an end by the introductions of steam turbines and diesel engines. The possibility of travelling faster and under any wind conditions ultimately determined the definitive success of propulsion systems based on fossil fuels.

Why is wind propulsion being considered again? In addition to the already mentioned interest in alternative energy sources for reducing fuel costs and environmental impacts, high fuel prices have also led to slow-steaming becoming a common practice for reducing ship fuel consumption, which makes wind propulsion a more viable solution. Moreover, improvements in materials and automation technology have enabled more efficient sails that do not require any additional crew for operation. Wind-propelled ships will still be equipped with conventional engines that will provide the required power in the event of inconvenient winds (e.g., being too strong or too weak or coming from the wrong direction) and for safety reasons.

Several different technologies that permit the integration of wind energy with traditional propulsion have been proposed to date:

- *Rigid sails*, which are also known as wing-sails, are similar to airplane wings. Depending on the arrangement, the number of sails can vary and they can be placed on different parts of a ship. Rigid sails have the highest operational experience, which is not solely due to their similarities with traditional sails. Wing-sails have been installed on merchant ships in the past, although this did not attract much attention or success due to low fuel prices. Among the different sail options, rigid sails perform the best in head winds, which is the most common situation under real operations due to the effect of the ship motion on the relative wind speed.

- *Kites* can be placed on the front of the ship and provide a significant amount of propulsion energy. A kite does not occupy space on the deck or reduce cargo space and is reasonably cheap to install. At high altitudes, where wind speeds are higher, kites can provide a very large proportion of the power needed for a ship sailing in the range of 10–15 knots. However, kites are very sensitive to wind direction and only function properly with tail winds. The practical application of kites on merchant vessels have also demonstrated limitations in terms of kite control and wear, both on the ropes and on the kite itself.
- *Flettner rotors* were invented in the early 1920s by Anton Flettner and are based on the Magnus principle. One or more large rotors (the E-ship1 vessel, which is 10,000 dwt and was launched in 2010, is equipped with four rotors with a height of 26 m and a diameter of 4 m) are installed on deck and placed in rotation; the rotation of the rotor interacts with the wind, generating a forward drag that contributes to ship propulsion. Flettner rotors work best in crosswind conditions. Therefore, these rotors are generally more operationally flexible than kites. However, Flettner rotors occupy deck space and are more expensive to manufacture.

For all types of wind-assisted technology, two main factors influence the advantage of utilising such systems. Ship speed is crucial in the estimation of the wind power contribution to the overall propulsion power because a lower ship speed means a lower total power requirement and a higher speed relative to the

Table 10.1 Average power contributions of Flettner rotor and kite technologies on the selected routes

Route	Dunkirk–Dover	London–Milford Haven	Varberg–Gillingham	Tubarao–Grimsby	Yantian–Felixstowe
Ship type	RoRo	Product tanker	General cargo	Bulk carrier	Container
Dwt	7000	8000	5500	50,000	30,000
Distance (km)	76	872	1093	9319	18,074
Ship speed (m/s)	7.4	4.9	4.5	5.8	9.5
Ship speed (knots)	14.4	9.6	8.8	11.2	18.4
Propulsion power demand (kW)	8333	1401	1014	3700	10,657
Average Flettner rotor power (kW) outgoing	338.7	193.1	206.7	200.6	246.5
Returning	372.5	205	222.1	200.7	250.1
Average kite power (kW) outgoing	182.3	260	195.3	171	126.7
Returning	300.1	405.8	460.6	236.3	160.3
Average contribution from one Flettner rotor/a kite (%)	4 %/3 %	14 %/24 %	21 %/32 %	5 %/6 %	2 %/1 %

The transient propulsion power is averaged over the course of the route and across 2011. Distance, speed, and average propulsion power demand are also shown. (Originally published by Traut et al. [24] in Applied Energy under Creative Commons License)

wind, making wind-assisted technologies more suitable for slow vessels (<15 kn) and in combination with slow-steaming. Additionally, as shown in Table 10.1, the route has a very strong impact on the amount of power that can be harvested from the wind because both wind speed and direction are critically important. For the same type of technology, the contribution from the wind can vary from 3 to 32 % on an annual basis, making wind-assisted propulsion more reliable when a vessel is operating on a fixed route where the wind speed and direction can be estimated beforehand.

Some practical issues related to the use of wind propulsion systems should also be included when considering wind propulsion as an alternative design. Conventional engines should also be installed to allow the ship to sail with non-optimal winds and to manoeuvre in poor sea conditions. If small engines are installed, only low speeds can be achieved in non-optimal winds. However, if large conventional engines are installed, they will most likely be used at low load and, therefore, at low efficiency.

10.2.3.2 Solar Energy

The cost of photovoltaic solar panels has exponentially decreased during the last 20 years, which makes them valuable for several applications. Solar panels are widely used on leisure boats as a source of auxiliary power, although they are seldom used on merchant vessels. NYK M/V Auriga Leader (car carrier) is the most notable exception. This vessel is equipped with 328 solar panels on its upper roof and has demonstrated that up to 10 % of its total electric power (propulsion and auxiliary) can be generated using solar energy. Although it is rather unlikely that solar will become a viable alternative as a stand-alone energy source for propulsion, auxiliary power can be generated for free on board. A simplified way to estimate the solar energy potential is described in Box 10.4.

Box 10.4 Estimating solar energy potential

Evaluating the potential for power production from solar panels is an intricate problem that depends on many variables, such as solar radiation, relative inclination, and panel efficiency. However, a rough estimation can be performed to determine how much power could be generated through the installation of solar panels on a ship.

For example, the chemical tanker used as a case study previously in this chapter is 186 m long and 32 m wide, leading to an available surface area of approximately 6000 m^2 (considering the surface of the ship to be a rectangle, which is a reasonable approximation for large tankers). Assuming solar irradiation of 340 W/m^2 and a conversion efficiency of 15 % for the solar panels, a total of approximately 300 kW of available power could be produced. Because the installed power on the ship is 7700 kW, this would yield less than 5 % of the total power required by the ship at full power.

10.3 Improving Energy Efficiency from an Operational Perspective

In this section, improving energy efficiency will be discussed primarily in the context of climate change as a measure to reduce GHG emissions from the shipping industry. This section will focus exclusively on energy efficiency in the operation of existing ships.

10.3.1 The Assessed Potential

Several assessments have been performed regarding the potential for improving energy efficiency in the operation of existing ships in the context of climate change, resulting the following question: how much can the shipping sector reduce its climate impact and at what costs? Typically, the assessments have demonstrated that such operational measures can decrease the climate impact at very low costs and that implementing such measures should be cost-effective and make sense from a business perspective. Table 10.2 provides details from one assessment [25].

Many of the measures listed in Table 10.2 address ship speed, and a thorough understanding of the issues requires an examination of what ship speed *is* and how it is determined.

Table 10.2 Costs and potential for increasing operational energy efficiency to reduce greenhouse gas emissions, adapted from Faber et al. [25]

Measures	Gross potential efficiency gains	Cost
Speed reduction		
Voyage optimisation, including reduced port time	0–10 %	Unknown
Bulbous bow	>10 %	Unknown
Optimisation of ballast and trim	<5 %	Unknown
Using existing larger ships	<4 %	Unknown
Increasing cargo load factor	Unknown	Unknown
Weather routing	0.1–4 %	800–1600 USD/year
Autopilot adjustment	0.5–3 %	Unknown
Increasing energy awareness	Unknown	Unknown
Regularly scheduled polishing	2–5 %	3000–5000 USD
Polishing when required	2.5–8 %	
Hull cleaning	1–10 %	35–45 USD per foot of ship

10.3.2 The Role of Ship Speed

As described in Sect. 1.4.3.1, ship speed is important when assessing emissions to air. Frictional resistance from the water in combination with wave production results in an exponential relationship between fuel usage and speed (approximately a ship's speed cubed). Thus, fuel use per distance is related through a quadratic function. Speed has been an important factor for ship economics since propulsion became a cost [26].

From a commercial perspective, choosing the appropriate speed for a fleet of ships is a complex function that considers several aspects, such as fuel costs and freight earnings. The determination of *who* chooses this optimum speed is also intricate and depends on who pays for fuel, i.e., the cargo owner or the shipping company, which in turn is dependent on the type of contract between these parties [27, 28].

The average speed of the global fleet depends primarily on the bunker price and the state of the market [29]. The economic crisis of the late 2000s provides an excellent example. This crisis caused a reduction in international trade, leading to an oversupply of vessels compared to the demand. This situation was exacerbated for several years due to the continued supply of vessels that were ordered before the crisis. Due to these orders, the size of the global fleet continued to grow for several years because the shipyards finished producing these orders; the global fleet increased by nearly 40 % in tonnage from 2008 to early 2012 [30]. The consequent reaction from most shipping companies was to slow the speed of their ships (which also occurred after the oil crises in the 1970s, having similar effects [31]). The effect on CO_2 emissions was clearly demonstrated in the Third IMO GHG report, with a 10 % reduction in emissions from 2007 to 2012 [3].

Ship speed is not a "measure" to improve energy efficiency in the usual sense. Instead, ship speed is an output of a set of commercial decisions. The risk that ships will speed up again is apparent; similarly, the risk of increased energy use and associated emissions when freight rates and inventory costs rise in times of prosperity are also evident. Indeed, as shipping companies procure more energy-efficient ships and operate them in a more energy-efficient manner, they may choose to operate them at faster speeds if it makes sense from a commercial perspective [32].

Speed could be reduced not only by adding more ships to a fleet or market, which has occurred recently but also the capacity could be increased by decreasing the time ships spend in ports, i.e., maintaining the same amount of transportation work without requiring investments for new ships. Indeed, this has been identified as a very cost-efficient measure in several reports [6, 33, 34]. However, this "cost-effective" measure is also problematic. Several costs could exist that are not acknowledged in these assessments, e.g., the costs of organizing and coordinating efforts between the shipping company, ship crew and various actors in port [35].

In summary, reduced speed is not a mitigation measure in the usual sense, although it could be seen as *a goal of* policy measures. Economic incentives for slow-steaming could be maintained by keeping energy costs at a sufficiently high level through a tax on bunker fuel or an emissions trading scheme. Reduced speed could also be legally introduced through speed limits at sea [29, 36, 37, 38].

10.3.3 Improved Energy-Management Practices

Methods for increasing the energy efficiency of organisational practices in the shipping industry have recently been explored in the literature. The list in Table 10.2 suggests that effective energy management requires the ability to prioritise measurements and implement them across organisations and in partnership with, for example, technical management and crewing companies when those functions are outsourced. When fuel costs are carried by the cargo owner and not the shipping companies, transparency should be required regarding the costs and performance in the contractual relationships between these parties. Preferably, a full Plan-Do-Check-Act is performed (see Chap. 8), requiring a means for following up on the agreement and acting on deviations from the plan.

A lack of knowledge exists regarding actual energy-management practices in shipping companies. A survey of more than 80 shipping companies indicated that only a few had achieved large (>10 %) energy savings as a result of their energy-management efforts [39]. The greatest challenges for these companies were related to internal changes in management activities, e.g., training, communication and incentive schemes, which were lacking in most companies. Moreover, most companies had only implemented simpler measures, such as slow-steaming or weather-routing plans. In particular, many companies lacked the tools and abilities with which to analyse the actual performance of vessels.

A few case studies have also been completed. For example, the energy-management programme of a shipping company was described in the literature in 1980. In particular, that company created a new role to address energy-efficient activities, with "authority commensurate with involvement in the workings of engineering, maintenance and repair, vessel operations, and purchasing; [to also enable] easy communication between the conservation manager, ship crews, and shore side personnel" (p. 73) [40].

> [The Manager of Shipboard Conservation is] engaged in frequent ship visits and maintains close liaison with vessel personnel primarily in regard to operating procedures. He is also directly involved in certain ship maintenance and repairs, necessitating frequent involvement in the business of port engineers, contractors and vendors; assessing the impact of conservation projects, specifying required work and delineating necessary follow-up action. Finally, he is dealing with the office technical staff on a daily basis, defining the need and scope of engineering design or study assignments, monitoring progress developing economic analysis, coordinating work timing, and reviewing innovative proposals. These comprise his routine responsibilities at the working level.
>
> He is also available for upper management planning and review projects. Requests for data and analyses of fleet fuel consumption trends and vessel operating trade-offs are directed to him as are technical questions related to planned vessel conversions, new construction, alternative energy sources or power cycles, and public relations aspects of fuel

consumption. His established file of energy conservation data and familiarity with current studies and operating conditions enables such demands to be handled promptly. (p. 73) [40]

In a case study regarding the implementation of a standard energy-management system in a short sea shipping company, the role of communication between the ship crew and the commercial organisation ashore, means for measuring and analysing energy performance, the roles and responsibilities for energy, and access to knowledge and competence were highlighted [41].

From a climate policy perspective, attempts to induce better energy-management practices in shipping companies have included the requirement of a Ship Energy Efficiency Management Plan (SEEMP) on ships, which was introduced in Chap. 5. In a report to IMO, a great potential for improved energy efficiency was considered possible due to the SEEMP. In total, an energy reduction of approximately 25 to 30 % could be achieved depending on a ship's trade and type. As shown in Tables 10.3 and 10.4, half of this potential was due to increased port efficiency.

In practice, the SEEMP requires very little from a shipping company compared with other typical management system standards (e.g., ISO 9001 for quality management or ISO 14001 for environmental management) to comply with the regulations. The requirements of the SEEMP are listed in Table 10.5.

Table 10.3 Technical and operational energy efficiency measures for bulk carriers and gas tankers (savings potential in %), adapted from Bazari and Longva [34]

Energy efficiency measures	Bulk carrier		Gas tanker	
	30–40 k DWT	>100 k DWT	125–155 k m^3	>175 k m^3
Engine tuning and monitoring	2.5	1.8	1.8	1.8
Hull condition	3.5	3.5	3.5	3.5
Propeller condition	1.1	0.8	1.1	0.8
Reduced auxiliary power	0.6	0.9	0.7	1
Speed reduction (increased port efficiency)	15	15	12	10
Trim/draft	0.7	0.7	1	1.4
Voyage execution	2.5	3.4	2.5	3.4
Weather routing	0.1	1	0.1	1
Advanced hull coating	3	3	3	3
Propeller upgrade and aft body flow devices	3	3	3	3
SEEMP potential considering overlaps	28.7	29.6	25.9	26.0

Table 10.4 Technical and operational energy efficiency measures for tankers, containerships and general cargo/reefers (savings potential in %), adapted from Bazari and Longva [34]

Energy efficiency measures	Tanker		Containership		General cargo/ Reefer	
	60–85 k DWT	>200 DWT	4–5 k TEU	12–14 k TEU	~3.5 k DWT	~10 k DWT
Engine tuning and monitoring	2.2	1.6	1.6	1.6	2.9	2.9
Hull condition	3.5	3.5	3.5	3.5	3.5	3.5
Propeller condition	1.1	0.8	0.8	0.8	1.1	1.1
Reduced auxiliary power	0.6	1.7	0.8	1	2.6	1.1
Speed reduction (increased port efficiency)	13	12	10	11	21	13
Trim/draft	0.7	0.7	1.7	1.7	0.7	0.7
Voyage execution	2.5	3.4	1.4	1.4	2.5	2.5
Weather routing	0.2	1	1	0.8	0.1	1
Advanced hull coating	3	3	3	3	3	3
Propeller upgrade and aft body flow devices	3	3	3	3	3	3
SEEMP potential considering overlaps	26.8	27.7	24.3	25.2	36.0	28.4

The logic of the SEEMP follows a typical Plan-Do-Check-Act cycle. In the *Planning* phase, measures should be identified in conjunction with information on who should implement them and by what means. In the *Implementation* phase, these measures should be implemented. If they are not, the reason should be recorded. A *monitoring system* should be established, including procedures for collecting data and assigning personnel, that quantitatively monitors the energy efficiency of the ship. Procedures for periodic *Self-evaluation* based on data collected through monitoring should also be developed.

The impact of the SEEMP regulation has not yet been empirically examined. However, studies on the mandatory implementation of energy-management systems in other industries have had difficulty in demonstrating any causal relationship. For example, it is difficult to determine whether an organisation would have reached the same level of savings had it continued operating as it had before [43].

The lack of strong requirements has led to suspicions regarding the extent to which this regulation can assist or force companies beyond the status quo [42]. Proposals have been submitted to IMO that have argued for further "enhancement" of the SEEMP, particularly requesting a standard for measuring and following up on measures.[4]

[4]In *Proposed elements for enhancing implementation requirements for SEEMP and SEEMP Guidelines*. Submitted by the World Wide Fund for Nature (WWF) and the Clean Shipping Coalition (CSC). MEPC 64/4/33.

Table 10.5 Requirements in the SEEMP from Johnson et al. [42]

Subject	Requirements in the SEEMP
General (3.6)	"The SEEMP should be developed as a ship-specific plan by the ship owner, operator or any other party concerned, e.g., charterer"
Measures (4.1.2)	"[…T]he specific measures for the ship to improve energy efficiency should be identified in the first place. These measures should be listed as a package of measures to be implemented, thus providing the overview of the actions to be taken for that ship"
Responsibilities (4.2.1)	"Thus, the SEEMP should describe how each measure should be implemented and who the responsible person(s) is"
Implementation (4.4.2)	"The planned measures should be carried out in accordance with the predetermined implementation system. Record-keeping for the implementation of each measure is beneficial for self-evaluation at a later stage and should be encouraged. If any identified measure cannot be implemented for any reason(s), the reason(s) should be recorded for internal use"
Monitoring (4.3.1)	"The energy efficiency of a ship should be monitored quantitatively. This should be done by an established method, preferably by an international standard"
Monitoring system (4.3.3)	"To allow for meaningful and consistent monitoring, the monitoring system, including the procedures for collecting data and the assignment of responsible personnel, should be developed"
Self-evaluation (4.4.1)	"'Self-evaluation and improvement' is the final phase of the management cycle. This phase should produce meaningful feedback for the coming first stage, i.e., planning stage of the next improvement cycle"
	"[P]rocedures for self-evaluation of ship energy management should be developed. Furthermore, self-evaluation should be implemented periodically by using data collected through monitoring"

10.4 Fuel Changes to Reduce Environmental Impacts

As discussed in Chaps. 4 and 5, the fuel used for marine propulsion can contribute to several different environmental impacts. Changing the fuels used for marine propulsion can remove several of the environmental impacts caused by shipping and address step 2 in Sect. 8.2: "removal of the source of impacts".

Fuel is a broad term that is used for a material such as coal, oil, or gas that can be converted to various usable forms of energy, such as thermal, mechanical or electric energy. In this book, a *fuel* is considered to be associated with a specific primary energy source and processing options, whereas an *energy carrier* only represents the compound or phenomenon that carries the energy. Several different fuels may therefore use the same type of energy carrier. For example, LNG and liquefied biogas (LBG) are fuels in which methane is the energy carrier. *Primary energy sources* are unrefined sources of energy that are readily found in nature, such as

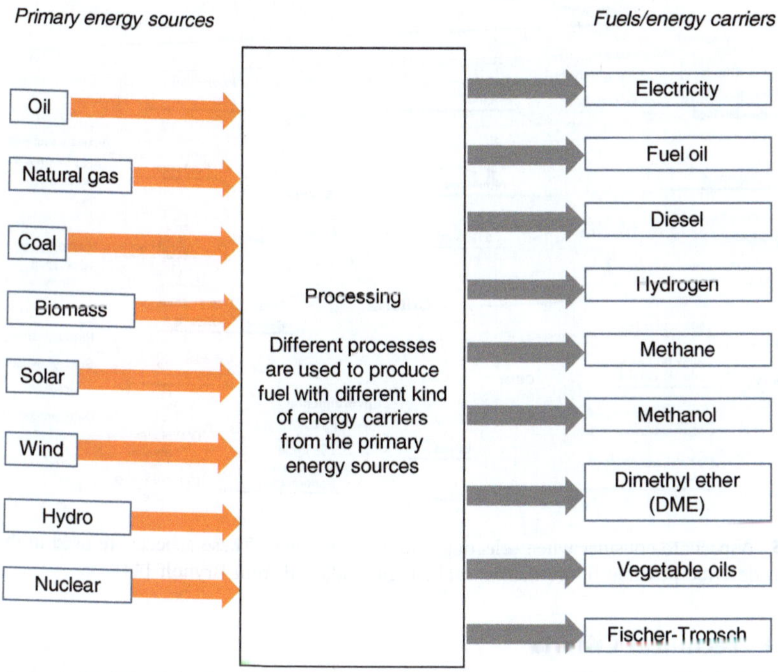

Fig. 10.4 A simplified diagram of the chain from energy resources to fuels that can be used in shipping

coal, oil and wind, and are used to produce fuels with different types of energy carriers. The type of energy carrier in the fuel will determine the prime movers that can be used to convert the chemical energy into mechanical energy in the ship's engines. Figure 10.4 show some primary energy sources and potential fuels/energy carriers.

10.4.1 Criteria for Future Marine Fuels

Several important aspects, for example, efficiency, safety, costs and environmental aspects, could be evaluated when considering future marine fuels. These factors have different weights under the many alternative perspectives, and various stakeholders have conflicting perspectives on the importance of these aspects. A set of criteria that can be useful when assessing future marine fuels is presented in Fig. 10.5, which are divided into four groups: technical, economic, environmental and other criteria. The criteria that are important will change in the future even if the criteria in Fig. 10.5 are rather generic and compiled based on a long-term perspective. The criteria that are used in this chapter describe some possible future marine fuels. Certain criteria involve minimum levels that must be satisfied, which act as boundary conditions in the selection.

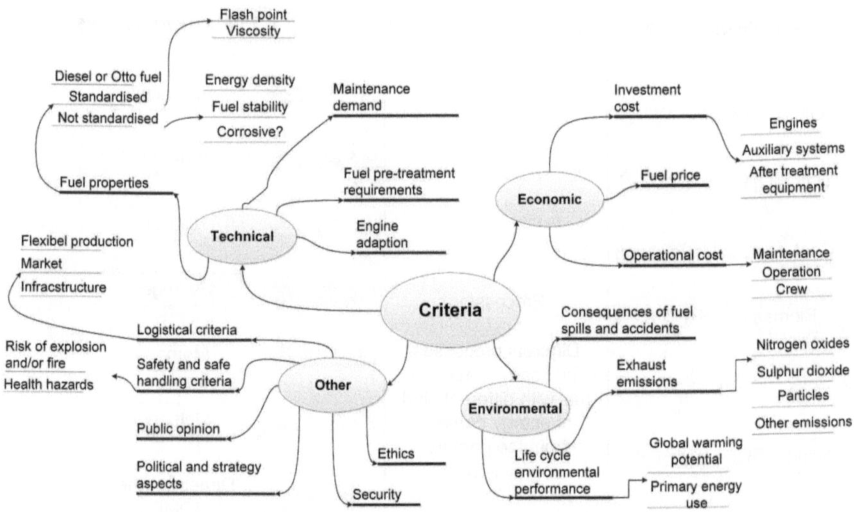

Fig. 10.5 Aspects to consider when selecting future marine fuels. These aspects are used in this chapter to describe possible future marine fuels. Figure adapted from Brynolf [44]

10.4.1.1 Technical Criteria

The technical system in the fuel chain includes the system on a ship that is associated with the fuel (e.g., engines, storage tanks, pumps, pipes and exhaust funnel), the bunkering ships and the fuel storage terminal. All of these systems must be technically feasible; thus, the boundary condition is that the construction and operation of such systems must be possible. The primary criteria under consideration are fuel properties (both standardised and non-standardised), propulsion systems and fuel pre-treatment requirements. International fuel standards have been developed to ensure the safe use of fuels in ship engines and to ensure compliance with (inter)national environmental standards. ISO 8217:2012 defines such fuel requirements as density, ash content, water content and sulphur content [45].

Fuels with good ignition qualities are advantageous because they can be used more efficiently in engines. The cetane and octane numbers represent the ignition quality for compression and spark ignition engines, respectively. The cetane number is a measure of the fuel's ignition delay between the start of injection and the start of combustion: the higher the cetane number, the shorter the delay. In simpler terms, it is the potential for smoother and more efficient operation of the engines. Higher octane numbers indicate a reduced tendency for the fuel to cause "knocking" (explosion caused by premature burning in the combustion chamber) in internal combustion engines based on the Otto cycle. The flash point is another standardised fuel property and refers to the lowest temperature at which a fuel can vaporise to form an ignitable mixture in air. The flash point is required to be at least 60 °C according to the International Convention of the Safety of Life at Sea (SOLAS). However, there are exceptions for low-flashpoint fuels, such as LNG, if special considerations are taken. The IGF code provides mandatory provisions for the arrangement, installation, control and monitoring of machinery equipment and systems using low-flash-point fuels [46].

Examples of fuel properties that are not standardised include the fuel stability, energy density and boiling point. A fuel is considered unstable when it undergoes chemical changes that produce undesirable consequences, such as deposits, acidity, or bad smells, and there are numerous test procedures for characterising fuel stability [47]. Fuels with low fuel stability can cause problems during long-term storage, which can be related to oxidative attacks, water contamination and microbial growth. The boiling point is related to the volatility of the fuel; fuels with low boiling points will readily evaporate. The energy density is a measure of the energy content per volume of fuel and provides an indication of the fuel's on-board storage demands, where a higher energy density indicate less space for fuels storage. The energy density is a limiting factor that is particularly important in ocean transportation because vessels travel long distances between bunkering.

Reciprocating internal combustion engines are currently the dominant prime mover technology. However, possible alternative propulsion systems are being considered for shipping, including fuel cells and gas and steam turbines (see Sect. 10.2).

10.4.1.2 Environmental Criteria

Emissions to air from shipping have recently received substantial attention, especially in regard to SO_x, NO_x, PM and greenhouse gas emissions. Fuel alternatives must fulfil the current environmental regulations (emissions of NO_x and SO_x). Environmental regulations will likely become stricter in the future. Thus, fuel alternatives must be able to meet tougher future environmental requirements to avoid yet another fuel change in the near future. Another important aspect is that *environmental life cycle performance* should ideally not increase when changing fuels. The environmental performance of a fuel can be assessed using life cycle assessment (see Sect. 9.4.1 for additional details about this method). Here, two aspects of the environmental life cycle perspective are considered to be most important: the life cycle energy use (measured as the total extracted energy[5]) and the life cycle climate impact [measured using the global warming potential (GWP)]. Ideally, all environmental impact categories would be important because there is no overall benefit to society if emissions are simply shifted from one stage in the life cycle to another. However, as a first set of criteria, these two categories have been considered to be the most important because the contribution from other parts of the life cycle during marine transportation is greater from greenhouse gases and energy than, for example, nitrogen oxides and sulphur dioxide. Many of the other impacts are curtailed by reducing emissions during marine transportation. Some potential marine fuels and their performance regarding these categories are presented in Fig. 10.6.

A third category involves the consequences of fuel leaks, such as those caused by accidents. Certain environmental criteria are often used to compare fuel

[5]The total extracted energy represents all fossil and renewable primary energy required to produce the product, including the energy content of the raw material extracted for the actual product.

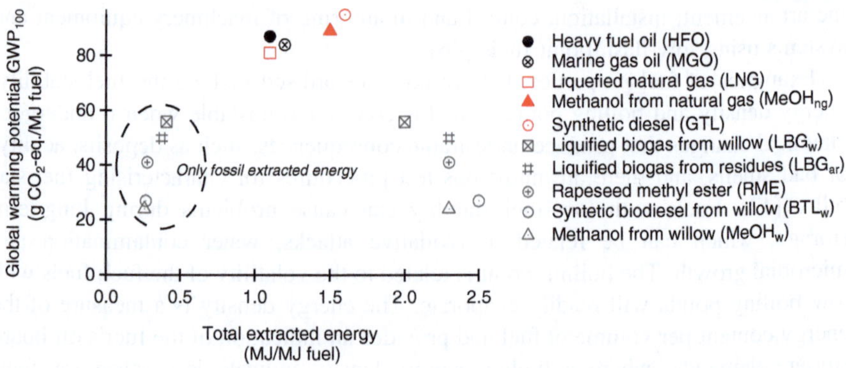

Fig. 10.6 Global warming potential versus total extracted energy of the evaluated fuels. Bio-based fuels are associated with lower emissions of greenhouse gases and higher levels of total extracted energy. However, the energy use of fossil energy is lower than that of the other evaluated fuels. The fossil energy use is, for example, based on the use of diesel and fertilisers in agriculture [44]. Figure adapted from Brynolf [44]

alternatives because they can be quantified. The consequences of fuel spills and accidents are not one of these because they depend on the characteristics of the environment that is affected. One property that affects the consequences of fuel spills is the biodegradability of the fuel.[6]

10.4.1.3 Economic Criteria

The choice of fuel in the marine industry has historically been driven by economics. However, the cost of fuel is only one factor that must be considered when evaluating fuels; other important costs include investment and operational costs.

The cost of fuel is a considerable expense in shipping. For example, "the bunker market is extremely price sensitive, with ships often basing decisions on where to bunker on the relative price of fuel available in respective ports" [49]. Depending on the type of vessel, fuel costs were estimated to represent between 30 and 54 % of the operating costs in 2006 [50]; these costs are expected to increase due to the upcoming sulphur regulations, further increasing the impact on a vessel's total operating costs. The price difference between different fuel qualities may also increase.

Operational costs can be divided into two parts: the costs directly connected to operations, which include the costs of crew, insurance and overhead, and maintenance costs, which include the service and repair of engines and the costs of spare parts. Investment costs include all costs related to the purchase of the ship and equipment installation (e.g., engine retrofitting, storage tanks and emissions abatement equipment). High investment costs act as a barrier to new investments, even if the return time is short.

[6]Blin et al. [48] compared the biodegradability of several fossil- and biofuels.

Costs are also associated with the space and weight of the propulsion system. Lowering the space and weight requirements will increase the cargo capacity of a vessel. Different types of ships are limited in their cargo capacity by different factors. For example, RoRos and containerships are primarily limited by space, whereas tankers and bulk carriers are limited by weight. These limitations are closely related to the energy density of the fuel.

10.4.1.4 Other Criteria

Other criteria include logistics, safety, security, public opinion, ethics and political and strategic aspects. Logistic criteria include requirements concerning the market, flexibility of production, economies of scale and infrastructure. The selected energy carrier must be available in a given market, and sufficient raw materials must exist to produce the desired volumes. It is an advantage if the energy carrier can be produced from a number of different raw materials, since this will increase the production flexibility.

Safety criteria include the risk of explosion and fire and the health hazards encountered by people handling the fuel. The flammability limits represent the range of concentrations in air within which a fuel is flammable. Fuels with broad limits can generate a greater potential volume of flammable gas in the atmosphere, and the time required to dilute the fuel to below the lower flammability limit may be substantial [51]. If a fuel is soluble in water, it can be diluted to a non-flammable level at ambient temperatures, removing the risk of fire or explosion. Water solubility is not of interest for gases because they disperse quickly [51]. A specific safety concern regarding gases is the potential for them to replace air in confined spaces if leaked and to act as an asphyxiant[7] if not detected. The health hazards of fuels include acute and chronic toxicity and carcinogenic properties.

Public opinion involves the attitudes expressed by the general public, such as the demand for sustainable transportation. A political and strategic aspect may involve a situation in which new jobs are created via the local production of a fuel. Ethical considerations include whether the production and use of a fuel causes hazards for certain populations. For example, the production of ethanol from corn or other food crops could result in higher food prices for citizens in poorer nations [52].

10.4.2 Present and Possible Future Marine Fuels

A list of possible future fuels for marine propulsion is shown in Fig. 10.7. Based on the type of energy carrier, the fuels are grouped into four categories: diesel-quality fuels, gases, alcohols and solid fuels.

[7] An asphyxiant gas is a gas which reduces the oxygen concentration in air.

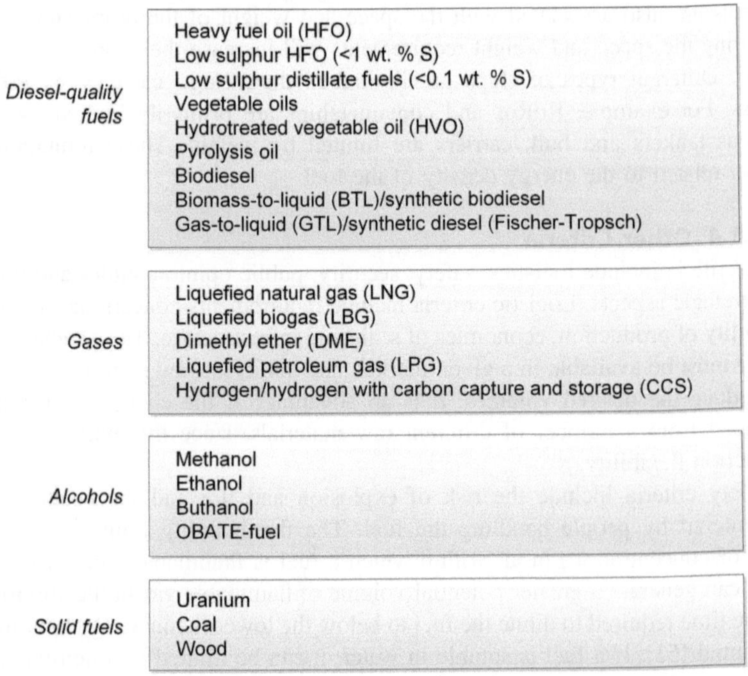

Fig. 10.7 Examples of possible future marine fuels/energy carriers. Figure adapted from Brynolf [44]

10.4.2.1 Diesel-Quality Fuels

Diesel fuels have dominated the shipping industry for the last 50 years [53]. Diesel-quality fuels can be used in compression ignition engines and are derived from both fossil and biomass resources.

HFO fulfils the technical criteria and has operated efficiently in marine diesel engines for more than 50 years. However, this fuel is also associated with various environmental issues (as discussed in Chaps. 4 and 5) and will not be in compliance with the strictest environmental regulations without exhaust abatement technologies. HFO typically has a sulphur content of at least 1 %. However, HFO can be produced from sweet (low sulphur) crude oil with a low sulphur content (0.5 wt% and lower) via technical measures performed by the refinery, such as desulphurisation [54]. The price of HFO has also significantly increased in recent years [55], and an HFO price of approximately 12 Euro/GJ has been forecasted [55].

Marine gas oil (MGO) is also commonly used. MGO is a distillate fuel with a low sulphur content, possibly less than 0.1 wt% sulphur. MGO is also a technically proven fuel with an existing infrastructure and market. Its advantage is that it will comply with the 0.1 % sulphur limits for marine fuels in the ECAs and, when combined with an SCR unit, will comply with the NO_x Tier III requirements. The environmental issues related to the use of HFO and MGO have been extensively described elsewhere in this book. For example, the problems with oil spills are discussed in Sect. 4.1, whereas those related to particle emissions are discussed in Sect. 5.5. MGO is more expensive than HFO, and future MGO prices are estimated to be in the range of 1.6–2.2 times the price of HFO [51]. The availability of both HFO and MGO will be determined by the future availability of crude oil, and these two fuels are not flexible in terms of their raw material sources.

Vegetable oils, such as jatropha oil and rapeseed oil, and animal fat can be used in diesel engines. Previous work has demonstrated that current technical methods for the storage and distribution of HFO on large ships are compatible with the use of vegetable oils [56]. Vegetable oils have very low sulphur contents, typically less than 20 wt ppm [57]. In a comparison of HFO and MGO with four vegetable oils, the emissions of NO_x were observed to be of the same order of magnitude, and the emissions of particles were significantly less than that for HFO [58]. Vegetable oils will comply with the strictest sulphur regulations, although their use must be combined with exhaust abatement technologies to comply with the NO_x Tier III requirements. These oils are biodegradable and are less harmful to the environment than are HFO and MGO in the event of a spill. Vegetable oils can replace HFO in diesel engines, although they are less compatible with diesel engines designed for MGO. *Biodiesel* may represent a better alternative for MGO-fuelled engines [59]. Biodiesel is produced by the transesterification of vegetable oils and has been tested for marine propulsion by a large container shipping company and by the United States Navy [60, 61]. The long-term storage of vegetable oils and biodiesel may be problematic due to the growth of microorganisms and oxidation [62, 63]. The increased use of vegetable oils and biodiesel has also resulted in a debate regarding the use of land and food prices (see Box 10.5).

Diesel-quality fuel can also be produced from various feedstocks using the Fischer-Tropsch process. The Fischer-Tropsch process primarily consists of three steps: synthesis gas generation, carbon synthesis and upgrading. Different feedstock can be used to generate synthesis gas consisting of hydrogen and carbon monoxide. If natural gas is used, the fuel is typically called gas-to-liquid (GTL), and if biomass is used, the fuel is called biomass-to-liquid (BTL). The downside of the Fischer-Tropsch process is that it has a relatively low efficiency. The energy efficiency of GTL production is in the range of 61–65 %, and the efficiency of BTL production is even lower [64].

Box 10.5 The food, energy, and environment trilemma
The production of biofuels has recently increased, increasing the need for arable land. Certain crops can be used as food or feed and to produce biofuels. Thus, arable land can be used to produce either food or energy crops. This phenomenon has resulted in competition for land between food and energy crops and short-term increases in food prices [65]. This competition poses a global dilemma, i.e., society's need to feed itself versus the greater monetary returns to farmers through the use of land for energy crops.

This conflict is not only a food versus energy dilemma but also a food, energy, and environment trilemma [66]. Land use is also associated with greenhouse gas emissions, and the greenhouse gas emissions related to land use conversion, forestry and current agricultural land use are at least twice the total emissions from global transport [67].

Two principles can be used to guide the production of biofuels to increase its environmental sustainability. They must be produced from feedstocks associated with much lower life cycle greenhouse gas emissions than traditional fossil fuels and with little or no competition with food production. Examples of these feedstocks are (1) perennial plants grown on degraded lands abandoned from agricultural use, (2) crop residues, such as corn stover and straw from rice and wheat, (3) sustainably harvested wood and forest residues, (4) double crops and mixed cropping systems, in which food and energy crops are grown at different seasons or simultaneously, and (5) municipal and industrial wastes [66].

10.4.2.2 Gases

Gases include all fuels that are gaseous at standard temperature and pressure, e.g., LNG, dimethyl ether (DME), liquefied petroleum gas (LPG) and hydrogen. Natural gas is already established as a fuel for domestic and commercial heating and large-scale electricity generation. Compressed natural gas (CNG) is used in short-range distribution. For long-distance distribution to be cost-effective, natural gas must be liquefied. To become a liquid, natural gas must be cooled to −162 °C, which is performed via several compression steps. LNG typically consists of methane and small proportions of ethane, propane, and nitrogen. The use of LNG as a fuel in shipping is limited to LNG carriers, which use the boil-off gas as fuel in steam turbines or as gas in dual-fuel engines, and more than 30 LNG-fuelled ships are currently operating in Norway [68]. Several examples of ships using LNG are highlighted in Box 10.6.

Box 10.6 Examples of ships using LNG as fuel

The *Glutra* entered into service in 2000 and is the first ferry to be powered by LNG. This vessel has four lean-burn gas engines. The Glutra receives LNG from the Statoil plant Tjeldbergodden and is loaded by truck.

The *Bit Viking* is the first LNG-fuelled product tanker. This vessel was converted to LNG propulsion during 2011 and has been in operation since October 2011. The two main engines have been converted to four-stroke dual-fuel engines. The main fuel for the vessel is LNG, although the dual-fuel solution enables the vessel to run on traditional fuel (MDO) as a backup. MDO is a fuel typically blended from HFO and MGO.

The *Viking Grace* began operations in January 2013 and is the first large passenger vessel in the world that uses LNG as fuel. This vessel is equipped with four main dual-fuel engines and can run on heavy fuel oil, marine gas oil, and LNG.

The LNG-propelled ships in operation in Norway are either equipped with spark-ignited (SI) lean-burn gas engines or dual-fuel (DF) engines (see Box 10.7 for more information about these engines). One of the downsides of LNG is the complicated and costly retrofits required for existing engines. LNG has gained substantial interest as a marine fuel because it can comply with the strictest environmental regulations that are currently enforced. It has a sulphur content of only a few ppm, and four-stroke SI and DF engines will comply with the NO_x Tier III regulations. However, one potential problem associated with the use of LNG is the potential leakage of methane, which is a stronger greenhouse gas than CO_2 (see Sects. 2.2.1 and 10.4.3).

Liquefied biogas (LBG), which is produced from biomass, can also be used instead of LNG. LBG consists primarily of methane and does not contain any substantial proportions of other hydrocarbons, such as ethane and propane. The most common way to produce LBG is via anaerobic digestion, although it can also be produced from the gasification of biomass. In terms of the engine, LNG and LBG are nearly identical; only the upstream portion of their fuel life cycles are different, i.e., the raw material that is used, the production process, the price and the availability. Estimates of future prices for LBG and LNG range from 11 to 50 Euro/GJ [69] and 7–12 Euro/GJ [55] and suggest that LBG will be more costly than LNG.

LNG and LBG have lower energy densities than do traditional marine fuels and must be stored in cryogenic tanks [70]. This non-traditional storage requirement may have an impact on the on-board space requirements. An LNG retrofit reduced the cargo capacity of a feeder container vessel by 4 %, whereas a retrofit of an ocean-going tanker vessel did not result in a reduced cargo capacity [70]. The actual space requirement will vary from vessel to vessel and is difficult to estimate. The space requirement must be considered for each new ship or retrofitting scenario.

DME and LPG are similar in their handling and storage requirements [71]. Both are gases at standard temperatures and pressures and are compressed to form liquids. Unlike LNG and LBG, these gases do not require cooling. DME is a colourless gas at standard temperature and atmospheric pressure and is considered to be a redundant fuel in compression-ignition diesel engines. DME is an oxygenated gas with a low auto-ignition temperature, which leads to low NO_x emissions and minimal soot production [72]. DME is produced from synthesis gas, whereas LPG is composed of hydrocarbon gases, typically a mixture of propane and butane. LPG is produced during the refining of crude oil or is extracted during natural gas production. A large engine manufacturer is promoting a two-stroke marine engine that runs on LPG and that can be adapted to DME propulsion [73]. DME has also received much attention as a future road fuel [74, 75].

Hydrogen is seen as a potential future fuel, and if the hydrogen is produced from renewable sources, no fossil CO_2 is emitted; moreover, if the hydrogen is used in fuel cells, the only local emission is water vapour. The life cycle performance of hydrogen as a fuel depends primarily on raw material extraction, fuel production and distribution; the selected pathways affect the environmental impact. Many problems that have not yet been resolved are associated with hydrogen. For example, storage and transport represent challenges. Even in its liquid form, hydrogen has a very low energy density, approximately 10,000 MJ/m^3 [76] (compared with LNG ≈ 16,000 MJ/m^3 and methanol ≈ 22,000 MJ/m^3) [77]. Hydrogen has not been extensively discussed as a future marine fuel, although the use of liquid hydrogen in internal combustion engines in a long-haul feeder container ship has been evaluated [78]. It is thought that the use of hydrogen for ship propulsion cannot be implemented within the next ten years on a large scale, although the use of hydrogen in fuel cells for auxiliary power is possible.

Box 10.7 Gaseous fuels in internal combustion engines: gas and dual-fuel engines

Several new fuels are expected to enter the shipping bunker market in the near future in the wake of increasing prices of fuels based on crude oil and stricter environmental regulations. The utilisation of gaseous fuels in conventional engines poses challenges and opportunities from the perspective of engine technology. Three main concepts for the combustion of gaseous fuels in internal combustion engines are commercially available:

DF-diesel cycle: A standard diesel principle can be applied when gaseous fuels are used if a sufficiently large pilot injection of diesel fuel is provided. This technology is most often employed in large two-stroke engines and provides the highest fuel flexibility. However, because of the larger quantity of injected pilot fuel and the higher combustion temperatures, these engines are generally associated with higher NO_x, SO_x and PM emissions compared with the other types of gas engines described below (although the CH_4 emissions are lower). In addition, the gas must be compressed at very high pressures before injection (>300 bar), yielding considerable safety issues.

DF-Otto cycle: Given the combustion properties of most gaseous fuels, the Otto cycle is often the preferred solution for dual-fuel engines, especially for medium-speed four-stroke engines. In this technology, the engine is operated according to the Otto principle, i.e., fuel is premixed with air before entering the cylinder. A pilot injection of diesel fuel is used to trigger the combustion, which is dominated by the premixed fuel. This type of engine requires less pilot fuel, allows lower combustion temperatures, and can comply with Tier III NO_x emission regulations. In addition, the gaseous fuel does not require compression and is more easily handled. However, as a consequence of the low compression ratio, these engines are less efficient when operated in diesel mode and running on HFO or MDO fuel.

Gas engines: Pure gas engines can only be operated on gaseous fuel and function according to the lean-burn Otto principle, which uses spark ignition and no pilot fuel. Despite the high mean effective pressures (which are limited to 18 bars versus the 25 bars in modern marine diesel engines), the lean-burn Otto combustion principle prevents knocking and allows low combustion temperatures, which reduce NO_x emissions. Because they are specifically designed for gas operations, pure gas engines also feature the best environmental performance among current technologies.

10.4.2.3 Alcohols

Of the possible alcohols that could be used as fuels in shipping, methanol has received the greatest attention. Ethanol is also a possibility, and it is currently used primarily as a car fuel to replace petrol [79]. Ethanol may be produced from agricultural feedstock, such as sugarcane grown in Brazil.

Methanol is the simplest alcohol and has traditionally been used as a chemical base material. Currently, methanol is primarily produced from natural gas, although it can also be produced from coal, which is primarily conducted in China. [80] Methanol displays many desirable combustion and emission characteristics, thus making it a good fuel for premixed combustion in Otto engines. Methanol can also be used in the diesel process, where a glow plug or pilot fuel is used as the ignition source. The conversion of engines to a methanol-diesel configuration is easier and less expensive than the premixed diesel fuel configuration; the former is seen as a possible intermediate-term solution [81]. Methanol can be used as a marine fuel in fuel cells (see Sect. 10.2.2.2). Methanol will comply with the strictest sulphur regulations and, depending on the selected engine technology, may comply with the NO_x Tier III requirements. There is a risk of increased emissions of formaldehyde, which requires consideration. Methanol will most likely be associated with lower impacts from fuel spills because it is not persistent in the environment and biodegrades quickly. However, methanol is toxic at high concentrations; thus, local effects before dilution could occur [82]. From 1975 to 2008, the average wholesale price of methanol fluctuated between 4 and 15 Euro/GJ. The few peaks in methanol prices during this period resulted from increased demand, production problems and high natural gas prices [83]. Methanol can be produced from several energy sources, e.g., biomass, coal and natural gas, making it a flexible fuel with respect to production. Methanol can also be produced using electricity to split water into hydrogen, which is then combined with CO_2 to form an electrofuel (see Box 10.8).

Another potential alcohol that has received attention is glycerol, which is also called glycerine. Glycerol is a simple alcohol that is commonly used in the cosmetics and pharmaceutical industries and is a by-product of biodiesel production. The supply of glycerol has grown with the growth in biodiesel production. One potential use of glycerol is as a fuel. It is non-toxic and non-volatile. A combustion cycle that allows the use of glycerol in compression ignition engines has been developed [3]. Glycerol is expected to yield low NO_x and PM emissions and no SO_2 emissions. The global supply of glycerol was 3.2 million tonnes in 2008 [84], and glycerol can also be produced by *Dunalieila* algae in saline water [85]. The GLEAMS project in UK intends to develop marine engines that can use glycerol and comply with the strictest environmental regulations that are currently enforced.

Box 10.8 Electrofuels

Electrofuel is an umbrella term for carbon-based fuels that are produced using electricity as the primary source of energy. The carbon in the fuel is from carbon dioxide and can be captured from the air, the sea or exhaust gases. Electrofuels can also assist in balancing energy production when production is intermittent, as in electricity generation from wind and solar sources. Therefore, electrofuels may be produced during periods of excess energy and used later for various purposes in the energy system [86].

Electrofuels offer feedstock flexibility and potential high-efficiency synthesis of fuels from renewable energy resources without competition for arable land and scarce water resources. However, the production of electrofuels is still in its infancy, and many challenges must be overcome before these products are brought to market on a large scale [87].

Methanol is produced from geothermal electricity, water, and carbon dioxide originating from the same geothermal source in Iceland by Carbon Recycling International. This firm operates a pilot plant and a commercial plant. The commercial plant has operated since late 2011 and has the capacity to produce 5 million litres of methanol per year [88].

10.4.2.4 Solid Fuels

Solid fuels can also be used. Coal was previously used as a ship fuel to produce steam for steam turbines [53]. Coal was also recommended as the future marine fuel in 1980 in a report by the Maritime Transportation Research Board [89].

Nuclear reactor propulsion has been used by many military vessels, e.g., submarines and aircraft carriers, and by a few civilian vessels, e.g., icebreakers, for more than 50 years [90]. However, nuclear reactors were and still are extensively employed in non-merchant applications due to their high autonomy (several years can pass without refuelling) and lack of a need for combustion air. Compared with conventional energy sources for shipping, a nuclear-powered propulsion system is much smaller in size, nearly emission-free and essentially insensitive to fuel price fluctuations.

Nuclear merchant ships have been economically difficult to operate, although higher fuel prices may change this. Before the Fukushima accident, far-reaching discussions had occurred regarding the construction of new nuclear merchant ships. However, due to security issues, the International Atomic Energy Agency (IAEA) is concerned about the use of nuclear reactors in merchant shipping and in research [91], and the introduction of new nuclear vessels may be slow due to these constraints and due to public opinion in port states.

Today, the development of cheaper and smaller reactors could lead to further development of this technology, especially where public opinion is not an issue. Nuclear reactors have relatively high rated power, making them suitable for large vessels operating over long distances and at high speeds. Similar to nuclear power plants, the very high investment cost can be offset by the low operating expenses.

10.4.3 Life Cycle Assessment of Marine Fuels

Life cycle assessment (LCA) is a widely used tool that focuses on assessing the environmental impact of a product or service from a *cradle-to-grave* perspective, namely from the mining of raw materials to waste management (see Sect. 9.4.1 and Fig. 9.8). Assessing the life cycle performance of marine fuels is important when changing fuels to avoid situations wherein more emissions are generated at other parts of the life cycle. Therefore, results from an LCA of three types of energy carriers (diesel-quality fuels, methanol and methane-based fuels) produced using three different raw materials (crude oil, natural gas and biomass) will be highlighted in this section. The fuels and their abbreviation are briefly presented in Table 10.6.

The scope of the LCA is geographically limited to ship operations in the northern part of Europe but do include the entire fuel cycle, from raw material extraction to combustion in marine engines, excluding the production of capital goods (e.g., ships, terminals, and exhaust abatement technology) (Fig. 10.8). For more information about the goal and scope and about the data choices made in this assessment, see Brynolf [44].

It has been shown that all the fuels outperform HFO in nearly all of the assessed impact categories, implying that a change to any alternative marine fuel will have an overall positive effect on the environmental impacts associated with shipping (see Fig. 10.9). The use of exhaust abatement technologies (open scrubbers and SCR units) in combination with the diesel-quality fuels can reduce the local and regional environmental impacts by reducing the NO_x and SO_2 emissions from the engines, with only small increases in energy use and climate impact in the life cycle.

The life cycle environmental performance is correlated with the types of energy carrier and primary energy source that are employed. These two factors have a significant impact on the results. Biomass-based fuels generally have a lower climate change impact compared with fuels based on crude oil and natural gas, whereas the opposite pattern is observed in terms of the total extracted energy. Fuels based on crude oil are associated with the lowest total extracted energy. Methane-based fuels have the lowest impact in terms of photochemical ozone formation, particulate matter formation, acidification and eutrophication potential. The impact of methanol-based fuels lies between that of methane-based and diesel

Table 10.6 Summary of fuels for which a life cycle performance is presented

Full name (abbreviation)	Energy carrier	Primary energy source	Sulphur content by mass.	Production method
Heavy fuel oil (HFO)	Diesel	Crude oil	1 %	Refining
Marine gas oil (MGO)	Diesel	Crude oil	0.1 %	Refining
Synthetic diesel (GTL)	Diesel	Natural gas	≈0 %	Autothermal reforming and Fischer-Tropsch synthesis
Rapeseed methyl ester (RME)	Diesel	Biomass (i.e., rapeseed)	≈0 %	Oil extraction and transesterification
Synthetic biodiesel (BTL_w)	Diesel	Biomass (i.e., willow)	≈0 %	Biomass gasification and Fischer-Tropsch synthesis
Liquefied natural gas (LNG)	Methane	Natural gas	≈0 %	Liquefaction
Liquefied biogas (LBG_{ar})	Methane	Biomass (i.e., agricultural residues, manure and municipal organic waste)	≈2.5 ppm	Anaerobic digestion and liquefaction
Liquefied biogas (LBG_w)	Methane	Biomass (i.e., willow)	≈3 ppm	Biomass gasification and liquefaction
Methanol ($MeOH_n$)	Methanol	Natural gas	≈0 %	Autothermal reforming and methanol synthesis
Methanol ($MeOH_w$)	Methanol	Biomass (i.e., willow)	≈0 %	Biomass gasification and methanol synthesis

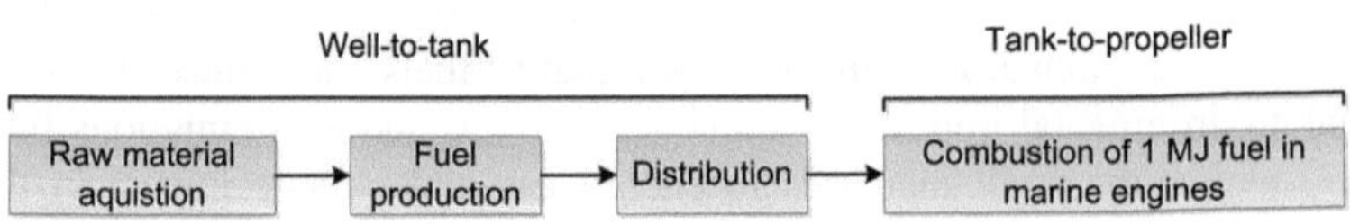

Fig. 10.8 A simplified flowchart of the life cycle of marine fuels. Figure adapted from Brynolf [44]

quality fuels. Two factors associated with the type of energy carrier are the NO_x emissions from marine engines and the leakage of methane, particularly the methane slip from gas engines using methane-based fuels. These factors are critical to the overall life cycle environmental performance. The impact on climate of methane leakage when using LNG as a fuel is further described in Box 10.9.

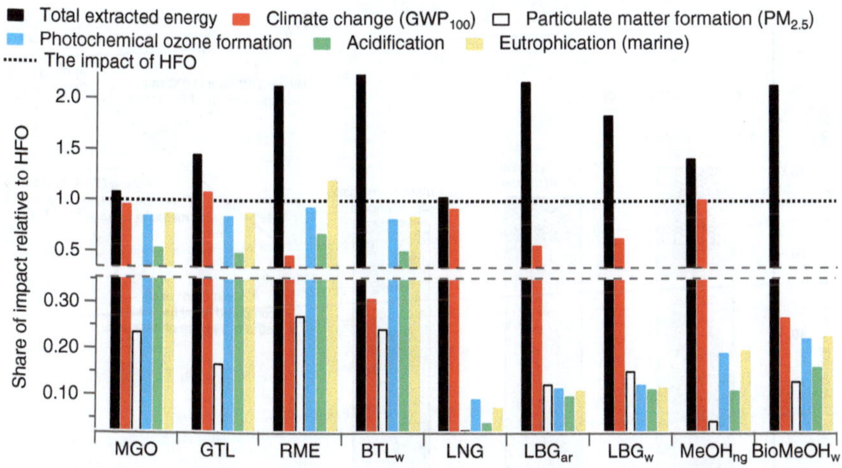

Fig. 10.9 Summary of the investigated impact categories for all fuels compared with the environmental impact of HFO as a marine fuel (represented by the *dashed line*). Abbreviations: *HFO* heavy fuel oil, *MGO* marine gas oil, *RME* rapeseed methyl ester, BTL_w synthetic diesel produced from willow, *LNG* liquefied natural gas, LBG_{ar} liquefied biogas from residues, LBG_w liquefied biogas from methanation of willow, $MeOH_{ng}$ methanol produced from natural gas, $BioMeOH_w$ methanol produced from willow, and GWP_{100} global warming potential over a time horizon of 100 years. Figure adapted from Brynolf [44]

Box 10.9 Carbon Footprint of LNG Dependent on Engine Choice and Methane Leakage

Methane leakages in the life cycles of methane-based fuels have a significant effect on the climate impact associated with these fuels. Methane is a strong greenhouse gas; even small emissions can affect the GWP of methane-based fuels, particularly when considering short time horizons because methane has a short residence time in the atmosphere [92]. Large uncertainty is also associated with the actual methane slip from gas engines during operation, and only a few measurements of these emissions have been published. Methane slip was also observed to differ depending on the type of engine that is employed, the engine load and when the engine was produced. The first LNG engines used in Norway have significant methane slip, particularly at low loads [93]. An example of how methane slip impacts the life cycle climate impact of LNG is shown in Fig. 10.10.

To summarise, this chapter shows that the potential to decarbonise the shipping industry through energy efficiency and changing fuels is very promising both the short and long term. A change of fuel can also contribute to reducing several other environmental impacts. Box 10.10 includes suggested further readings for some of the areas discussed in this chapter.

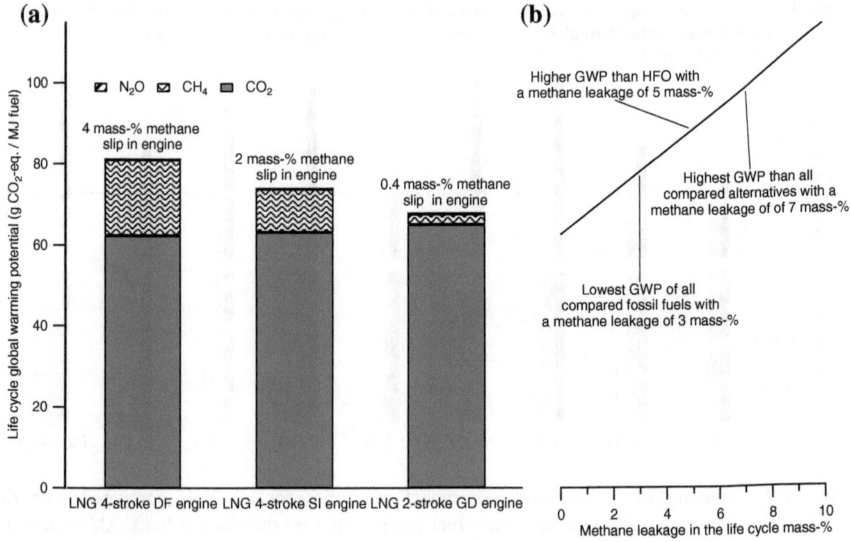

Fig. 10.10 The climate performance of LNG burned in various types of engines and the impact of methane leakage during the life cycle. The data are presented per MJ of combusted fuel; the results do not indicate that the two-stroke GD engine is more efficient than the four-stroke engines. Figure adapted from Brynolf [44]

Box 10.10 Recommended further readings

Woud, H. K. & Stapersma, D., *Design of Propulsion and Electric Power Generation Systems*. 2003: Imarest Publications.

Bertram, V., *3.7 Fuel-Saving Options*, in *Practical Ship Hydrodynamics*. 2011, Butterworth Heinemann: Oxford, UK.

Shu, G. et al., *A review of waste heat recovery on two-stroke IC engine aboard ships*. Renewable and Sustainable Energy Reviews, 2013. **19**: p. 385-401

Brynolf, S. *Environmental assessment of present and future marine fuels* Doctor of Philosophy thesis, Chalmers University of Technology, 2014.

References

1. Anderson, K. & Bows, A., *Executing a Scharnow turn: reconciling shipping emissions with international commitments on climate change*. Carbon Management, 2012. **3**(6): p. 615-628.
2. Eide, M. S., Endresen, Ø., Skjong, R., Longva, T. & Alvik, S., *Cost-effectiveness assessment of CO2 reducing measures in shipping*. Maritime Policy & Management, 2009. **36**(4): p. 367-384.

3. Smith, T. W. P. *et al. Third IMO GHG Study 2014*. 2014, International Maritime Organization (IMO): London, UK.
4. Buhaug, O. *et al. Second IMO GHG Study 2009*. 2009, International Maritime Organization (IMO): London, UK.
5. AEA Energy Environment. *Low carbon commercial shipping, Report for the UK Department of Transport. Report No ED05465*. 2007.
6. Eide, M. S., Longva, T., Hoffmann, P., Endresen, Ø. & Dalsøren, S. B., *Future cost scenarios for reduction of ship CO2 emissions*. Maritime Policy & Management, 2011. **38**(1): p. 11-37.
7. Maddox Consulting. *Analysis of market barriers to cost effective GHG emission reductions in the maritime transport sector*. 2012, European Commission: Brussels.
8. Jaffe, A. B. & Stavins, R. N., *The energy-efficiency gap - what does it mean?* Energ Policy, 1994. **22**(10): p. 804-810.
9. Sanstad, A. H. & Howarth, R. B., *'Normal' markets, market imperfections and energy efficiency*. Energ Policy, 1994. **22**(10): p. 811-818.
10. Blumstein, C., Krieg, B., Schipper, L. & York, C., *Overcoming social and institutional barriers to energy conservation*. Energy, 1980. **5**(4): p. 355-371.
11. Weber, L., *Some reflections on barriers to the efficient use of energy*. Energ Policy, 1997. **25** (10): p. 833-835.
12. Sutherland, R. J., *Market barriers to energy-efficiency investments*. The Energy Journal, 1991: p. 15-34.
13. Skjølsvik, K. O., Andersen, A. B., Corbett, J. J. & Skjelvik, J. M. *Study of Greenhouse Gas Emissions from Ships (MEPC 45/8 Report to International Maritime Organization on the outcome of the IMO Study on Greenhouse Gas Emissions from Ships)*. 2000, MARINTEK Sintef Group: Trondheim, Norway.
14. Johnson, H. & Andersson, K., *Barriers to energy efficiency in shipping*. WMU J Marit Affairs, 2014. p. 1–18.
15. Agnolucci, P., Smith, T. & Rehmatulla, N., *Energy efficiency and time charter rates: Energy efficiency savings recovered by ship owners in the Panamax market*. Transportation Research Part A: Policy and Practice, 2014. **66**: p. 173-184.
16. Jafarzadeh, S. & Utne, I. B., *A framework to bridge the energy efficiency gap in shipping*. Energy, 2014. **69**: p. 603-612.
17. Jensen, M. C. & Meckling, W. H., *Theory of the firm: Managerial behavior, agency costs and ownership structure*. Journal of Financial Economics, 1976. **3**(4): p. 305-360.
18. Larsson, L., Stern, F. & Visonneau, M., *Numerical Ship Hydrodynamics - An assessment of the Gothenburg 2010 Workshop*. 2014, Netherlands: Springer.
19. Nowacki, H., *Five decades of Computer-Aided Ship Design*. Computer-Aided Design, 2010. **42**(11): p. 956-969.
20. Bertram, V., *Resistance and propulsion*, in *Practical Ship Hydrodynamics*. 2011, Butterworth Heinemann: Oxford, UK.
21. Chew, G. *Further recognition for air lubrication system*. Sustainable Shipping, 2013.
22. Woud, H. K. & Stapersma, D., *Design of Propulsion and Electric Power Generation Systems*. 2003: Imarest Publications.
23. Shu, G. et al., *A review of waste heat recovery on two-stroke IC engine aboard ships*. Renewable and Sustainable Energy Reviews, 2013. 19: p. 385–401
24. Traut, M. *et al.*, *Propulsive power contribution of a kite and a Flettner rotor on selected shipping routes*. Applied Energy, 2014. **113**(0): p. 362-372.
25. Faber, J., Wang, H., Nelissen, D., Russell, B. & Amand, D. *Marginal Abatement Costs and Cost Effectiveness of Energy-Efficiency Measures*. 2011, International Maritime Organization: London, UK.,
26. Lowe, J., *Economic marine propulsion*. Journal of the American Society of Naval Engineers, 1891. **3**(2): p. 218-221.

27. Pirrong, S. C., *Contracting practices in bulk shipping markets: a transactions cost explanation.* Journal of Law and Economics, 1993. **36**(2): p. 937-976.
28. Psaraftis, H. N. & Kontovas, C. A., *Speed models for energy-efficient maritime transportation: A taxonomy and survey.* Transportation Research Part C: Emerging Technologies, 2013. **26**: p. 331-351.
29. Faber, J., Nelissen, D., Hon, G., Wang, H. & Tsimplis, M. *Regulated slow steaming in maritime transport: An assessment of options, costs and benefits.* 2012, CE Delft: Delft.
30. UNCTAD. *Review of Maritime Transport.* 2012, UNCTAD: Geneva, Switzerland.
31. Ronen, D., *The effect of oil price on the optimal speed of ships.* Journal of the Operational Research Society, 1982: p. 1035-1040.
32. Smith, T. W. P., *Technical energy efficiency, its interaction with optimal operating speeds and the implications for the management of shipping's carbon emissions.* Carbon Management, 2012. **3**(6): p. 589-600.
33. Faber, J. *et al. Technical support for European action to reducing Greenhouse Gas Emissions from international maritime transport.* 2009, CE Delft.
34. Bazari, Z. & Longva, T. *Assessment of IMO mandated energy efficiency measures for international shipping.* 2011, International Maritime Organization: London, UK.,
35. Johnson, H. & Styhre, L., *Increased energy efficiency in short sea shipping through decreased time in port.* Transportation Research Part A: Policy and Practice, 2015. **71**: p. 167-178.
36. Corbett, J. J., Wang, H. & Winebrake, J. J., *The effectiveness and costs of speed reductions on emissions from international shipping.* Transportation Research Part D: Transport and Environment, 2009. **14**(8): p. 593-598.
37. Cariou, P. & Cheaitou, A., *The effectiveness of a European speed limit versus an international bunker-levy to reduce CO2 emissions from container shipping.* Transportation Research Part D: Transport and Environment, 2012. **17**(2): p. 116-123.
38. Lindstad, H., Asbjørnslett, B. E. & Strømman, A. H., *Reductions in greenhouse gas emissions and cost by shipping at lower speeds.* Energ Policy, 2011. **39**(6): p. 3456-3464.
39. Kühnbaum, J. *Energy Management Study 2014.* 2014, DNV-GL: Hamburg, Germany.
40. Sweeney, J. J. *A comprehensive programme for shipboard energy conservation.* in *Shipboard energy conservation '80.* 1980. New York: Society of Naval Architects and Marine Engineers (SNAME).
41. Johnson, H., Johansson, M. & Andersson, K., *Barriers to improving energy efficiency in short sea shipping: an action research case study.* Journal of Cleaner Production, 2014. **66**(1): p. 317-327.
42. Johnson, H., Johansson, M., Andersson, K. & Södahl, B., *Will the ship energy efficiency management plan reduce CO2 emissions? A comparison with ISO 50001 and the ISM code.* Maritime Policy & Management, 2013. **40**(2): p. 177-190.
43. Stenqvist, C. & Nilsson, L. J., *Energy efficiency in energy-intensive industries—an evaluation of the Swedish voluntary agreement PFE.* Energy Efficiency, 2011: p. 1-17.
44. Brynolf, S. *Environmental assessment of present and future marine fuels* Doctor of Philosophy thesis, Chalmers University of Technology, 2014.
45. ISO, *Petroleum products - Fuels (class F) - Specifications of marine fuels, ISO 8217:2012.* 2012, International Organization for Standardization: Geneva.
46. IMO. *Gas and low-flashpoint fuels code adopted by IMO.* 2015, International Maritime Oragization: London, http://www.imo.org/en/MediaCentre/PressBriefings/Pages/26-MSC-95-ENDS.aspx, last accessed on 31 August 2015.
47. University of Idaho. *Biodiesel education. Biodiesel shortcourse - Fuel stability.* 2015, http://web.cals.uidaho.edu/biodiesel/biodiesel-shortcourse-fuel-stability/, last accessed on 11 May 2015.
48. Blin, J., Volle, G., Girard, P., Bridgwater, T. & Meier, D., *Biodegradability of biomass pyrolysis oils: Comparison to conventional petroleum fuels and alternatives fuels in current use.* Fuel, 2007. **86**(17–18): p. 2679-2686.

49. Notteboom, T. E. & Vernimmen, B., *The effect of high fuel costs on liner service configuration in container shipping.* Journal of Transport Geography, 2009. **17**(5): p. 325-337.
50. Kalli, J., Karvonen, T. & Makkonen, T. *Sulphur content in ships bunker fuel in 2015 - A study on the impacts of the new IMO regulations on transportation costs* 2009, Ministry of Transport and Communications Finland.
51. Astbury, G. R., *A review of the properties and hazards of some alternative fuels.* Process Safety and Environmental Protection, 2008. **86**(6): p. 397-414.
52. Mueller, S. A., Anderson, J. E. & Wallington, T. J., *Impact of biofuel production and other supply and demand factors on food price increases in 2008.* Biomass and Bioenergy, 2011. **35** (5): p. 1623-1632.
53. Corbett, J. J., *Marine Transportation and Energy Use*, in *Encyclopedia of Energy*, J. Cleveland Cutler, ed. 2004, Elsevier: New York. p. 745-758.
54. Avis, M. J. & Birch, C. H. *Impacts on the EU refining industry & markets of IMO specification changes & other measures to reduce the sulphur content of certain fuels.* 2009, Purvin & Gerts Inc.
55. Danish Maritime Authority. *North European LNG Infrastructure Project - A feasibility study for an LNG filling station infrastructure and test of recommendations.* 2012: Copenhagen. www.dma.dk.
56. Jiménez Espadafor, F., Torres García, M., Becerra Villanueva, J. & Moreno Gutiérrez, J., *The viability of pure vegetable oil as an alternative fuel for large ships.* Transportation Research Part D: Transport and Environment, 2009. **14**(7): p. 461-469.
57. He, B. B., Gerpen, J. H. V. & Thompson, J. C., *Sulfur Content in Selected Oils and Fats and their Corresponding Methyl Esters.* 2009. **25**(2).
58. Petzold, A. *et al.*, *Operation of Marine Diesel Engines on Biogenic Fuels: Modification of Emissions and Resulting Climate Effects.* Environmental Science & Technology, 2011. **45**(24): p. 10394-10400.
59. Opdal, O. A. & Hojem, J. F. *Biofuels in ships, A project report and feasibility study into the use of biofuels in the Norwegian domestic fleet.* 2007, Zero Emission Resource Organisation (ZERO): Oslo.
60. Gallagher, T. L. *Maersk to Run Biodiesel Test.* Journal of Commerce Online, 2010.
61. Bruckner-Menchelli, N. *US Navy targets 2016 for green fuel use.* Sustainable shipping, 2011.
62. Bengtsson, S., Fridell, E. & Andersson, K., *Environmental assessment of two pathways towards the use of biofuels in shipping.* Energ Policy, 2012. **44**(0): p. 451-463.
63. *Chapter 4 Oxidation Stability of Biodiesel*, in *Biodiesel: Production and Properties.* 2012, The Royal Society of Chemistry. p. 51-79.
64. Edwards, R., Larivé, J.-F., Rickeard, D. & Weindorf, W. *Well-to-wheels analysis of future automotive fuels and powertrains in the European context - Well-to-Tank (WTT) Report, Version 4.* 2013, JEC - Joint Research Centre-EUCAR-CONCAWE collaboration: Luxembourg: Publications Office of the European Union.
65. Rathmann, R., Szklo, A. & Schaeffer, R., *Land use competition for production of food and liquid biofuels: An analysis of the arguments in the current debate.* Renewable Energy, 2010. **35**(1): p. 14-22.
66. Tilman, D. *et al.*, *Beneficial Biofuels—The Food, Energy, and Environment Trilemma.* Science, 2009. **325**(5938): p. 270-271.
67. Baumbert, K. A., Herzog, T. & Pershing, J. *Navigating the Numbers. Greenhouse Gas Data and International Climate Policy.* 2005, World Resource Institute.
68. Blikom, L. P. *The LNG fuelled fleet will double every second year.* Energy of the Future 2012, DNV GL, http://blogs.dnvgl.com/lng/2012/12/the-lng-fuelled-fleet-will-double-every-second-year/, last accessed on 2 April 2014.
69. Florentinus, A., Hamelinck, C., Bos, A. v. d., Winkel, R. & Cuijpers, M. *Potential of biofuels for shipping.* 2012, ECOFYS Netherlands B.V.: Utrecht.

70. Gullberg, M. & Gahnström, J. *North European LNG Infrastructure Project - A feasibility study for an LNG filling station infrastructure and test of recommmendations - Baseline report*. 2011, ÅF Industry AB & SSPA Sweden AB: Gothenburg.
71. Verbeek, R. & Weide, J. v. d. *Global Assessment of Dimethyl-Ether: Comparison with Other Fuels*. in *International Spring Fuels & Lubricants Meeting*. 1997. Dearborn Michigan: SAE The Engineering Society for Advancing Land Sea Air and Space.
72. Müller, M. & Hübsch, U., *Dimethyl Ether*, in *Ullmann's Encyclopedia of Industrial Chemistry*. 2000, Wiley-VCH Verlag GmbH & Co. KGaA.
73. MAN Diesel & Turbo. *Press release - New Liquid ME-GI Engine Signals Expansion of MAN B&W Gas Portfolio*. 2011, http://www.mandieselturbo.com/files/news/filesof15559/Master_PR%20LPG%20ME-LGI%20engine.pdf, last accessed on 17 February 2012.
74. Arcoumanis, C., Bae, C., Crookes, R. & Kinoshita, E., *The potential of di-methyl ether (DME) as an alternative fuel for compression-ignition engines: A review*. Fuel, 2008. **87**(7): p. 1014-1030.
75. Salsing, H. *DME combustion in heavy duty diesel engines,* Doctor of Philosophy thesis, Chalmers University of Technology, 2011.
76. Mazloomi, K. & Gomes, C., *Hydrogen as an energy carrier: Prospects and challenges*. Renewable and Sustainable Energy Reviews, 2012. **16**(5): p. 3024-3033.
77. Bengtsson, S. *et al. Criteria for future marine fuels*. in *The IAME 2012 conference, 6-8 September*. 2012. Taipei, Taiwan.
78. Veldhuis, I. J. S., Richardson, R. N. & Stone, H. B. J., *Hydrogen fuel in a marine environment*. International Journal of Hydrogen Energy, 2007. **32**(13): p. 2553-2566.
79. Goldemberg, J., *Ethanol for a sustainable energy future*. Science, 2007. **315**(5813): p. 808-810.
80. Yang, C.-J. & Jackson, R. B., *China's growing methanol economy and its implications for energy and the environment*. Energ Policy, 2012. **41**(0): p. 878-884.
81. Fagerlund, P. & Ramne, B. *WP 9 Summary Report - EffShip Project Summary and Conclusions*. 2013: Gothenburg.
82. Clary, J. J., *The toxicology of methanol*, ed. John J Clary. 2013, Hoboken, N.J, New Jersey: John Wiley & Sons.
83. Olah, G. A., Goeppert, A. & Prakash, G. K. S., *Beyond oil and gas: the methanol economy*. 2009, Weinheim: Wiley-VCH.
84. Day, P., McNeil, J. & Sirovski, F., *(E)mission impossible?* The Chemical Engineer, 2011. **2011**(839): p. 33-35.
85. Ben-Amotz, A., Sussman, I. & Avron, M., *Glycerol production by Dunaliella*. Experientia, 1982. **38**(1): p. 49-52.
86. Nikoleris, A. & Nilsson, L. J. *Elektrobränslen - en kunskapsöversikt*. 2013, Lunds tekniska högskola: Lund.
87. Conrado, R., Haynes, C., Haendler, B. & Toone, E., *Electrofuels: A New Paradigm for Renewable Fuels*, in *Advanced Biofuels and Bioproducts*, James W. Lee, ed. 2013, Springer New York. p. 1037-1064.
88. Carbon Recycling International. *Plants*. 2014, http://www.carbonrecycling.is/, last accessed on 18 February 2014.
89. Committee on Alternative Fuel for Maritime Use. *Alternative fuels for maritime use*. 1980, Maritime Transportation Research Board, Commission on Sociotechnical Systems, National Research Council.: Washington. p. 342,
90. World Nuclear Association. *Nuclear-Powered Ships*. 2014, January 2014, http://www.world-nuclear.org/info/Non-Power-Nuclear-Applications/Transport/Nuclear-Powered-Ships/, last accessed on 25 February 2014.
91. IAEA. *Fuelling Research Reactors - Experts Meet to Discuss Research Reactor Fuel Management*. 2010, International Atomic Energy Agency, http://www.iaea.org/NewsCenter/News/2010/fuellingresreactors.html, last accessed on 19 March 2012.

92. IPCC, *Climate Change 2007: The Physical Science Basis. Contribution of Working Group I to the Fourth Assessment. Report of the Intergovernmental Panel on Climate Change.*, S. Solomon *et al.*, eds. 2007: Cambridge University Press, Cambridge, United Kingdom and New York, NY, USA. p. 996 pp.
93. Nielsen, J. B. & Stenersen, D. *Emission factors for CH_4, NO_X, particulates and black carbon for domestic shipping in Norway, revision 1.* 2010, Marintek: Klima og forurensningsdirektoratet.

Measures to Reduce Discharges and Emissions

11

Magda Wilewska-Bien, J. Fredrik Lindgren, Mathias Magnusson, Maria Zetterdahl, Kent Salo, Cecilia Gabrielii, Lena Granhag and Selma Brynolf

Abstract

Discharges and emissions from shipping can be reduced through different technical measures, many of which apply similar principles, e.g., filtration or absorption. Ballast water treatment systems can be used to limit the spread of invasive species. Selective catalytic reduction units and exhaust gas recirculation can be used to reduce nitrogen oxide emissions, and scrubbers and diesel particulate filters can be used to reduce sulphur dioxide and particle emissions.

M. Wilewska-Bien (✉) · J.F. Lindgren · M. Zetterdahl · K. Salo
C. Gabrielii · L. Granhag · S. Brynolf
Shipping and Marine Technology, Chalmers University of Technology, Gothenburg, Sweden
e-mail: magda.w.bien@chalmers.se

J.F. Lindgren
e-mail: f.lindgren@chalmers.se

M. Zetterdahl
e-mail: maria.zetterdahl@chalmers.se

K. Salo
e-mail: kent.salo@chalmers.se

C. Gabrielii
e-mail: cecilia.gabrielii@chalmers.se

L. Granhag
e-mail: lena.granhag@chalmers.se

S. Brynolf
e-mail: selma.brynolf@chalmers.se

M. Magnusson
Volvo Group Trucks Technology AB Volvo, Gothenburg, Sweden
e-mail: mathias.magnusson.2@volvo.com

K. Andersson et al. (eds.), *Shipping and the Environment*,
DOI 10.1007/978-3-662-49045-7_11

The restoration or remediation of natural environments may also be required after large oil spills. Possible remediation methods include booms, mechanical techniques and dispersant chemicals. These and several additional technical measures to reduce discharges and emissions are described in this chapter, including measures to reduce the impact of the infrastructure related to the shipping industry.

Keywords
Abatement techniques • Emissions • Discharges • Marine spatial planning

Shipping contributes to emissions to air primarily due to the combustion of fuels which produces greenhouse gases (GHGs), nitrogen oxides (NO_x), and sulphur oxides (SO_x); other sources include the transportation of volatile cargo (volatile organic compounds, VOCs) and refrigeration units (halocarbons). These sources are discussed in detail in Chap. 5. This chapter focuses on solutions to reduce these emissions or remediate their impacts if reduction is not possible. This section separately addresses the different emissions and the possible solutions to reduce their impacts. Some solutions can be used to reduce multiple types of emissions; these solutions are described in detail for only one emission type and are referenced for the other emissions. When reducing or remediating various discharges to the sea, proactive methods to minimise discharges are generally applied via regulation. However, discharges still occur through both accidents and legally approved discharges. In such cases, the issue must be addressed by alternative three or four in Fig. 8.2 in Chap. 8. End-of-pipe techniques are used on ballast water to hinder invasive species and on wastewater to remove contaminants so that the resulting discharge is not harmful to the marine environment. Restoration or remediation of natural environments is usually used after large oil spill events.

11.1 Remediation of Oil Spills

Several marine governance approaches have been implemented to limit the number of oil spills and the spilled volume (see Sect. 4.1.1.2). However, spills still occur. When this happens, options are available to limit the environmental impacts. Different techniques can be used to remediate oil after a spill depending on the size of the discharge, viscosity of the oil, and geographic location. The remediation of a spill at sea is preferable because it is more difficult and more costly to remediate oil after it has reached shores or near-shore areas [1]. Smaller spills are typically not remediated because it is often not economically feasible.

11.1.1 Techniques Used at Sea

Four main oil spill removal techniques are used at sea: booms that contain the spill; mechanical techniques that remove the oil; in situ burning, in which the oil slick is ignited; and dispersants, which reduce the tension between oil and water, causing the oil to be broken into smaller droplets that mix with the surrounding water, removing the oil from the ocean surface.

11.1.1.1 Booms

Booms are inflatable floating barriers used to contain oil on the surface of the water, preferably encircling the entire spill, and to protect biologically sensitive areas or to divert the spill to areas that are more suitable for recovery. Booms prevent spills from spreading and concentrate the oil. After a spill, booms are often the first technique that is employed [2, 3]. Different types of booms are deployed in different areas: offshore, inshore, in a harbour or in rivers. A boom generally comprises four different components: a means of floatation, a freeboard section that prevents oil from flowing over the boom, a skirt that prevents oil from slipping under the boom, and tension members to support the entire construction. Booms are generally constructed in 15–30 m long sections, with the possibility to tow, anchor or connect several booms in a row; they are generally composed of materials that include polyvinyl chloride (PVC), polyester, nylon or aramid [3]. A drawback of using booms is that a portion of the oil will inevitably sink to the bottom of the ocean, affecting the marine benthic life. At the sea floor, the oil might form tar balls that could wash up on shore [2]. Another drawback of booms is that they can fail to contain the oil when affected by water currents, high waves and/or winds [3].

11.1.1.2 Mechanical Techniques

Mechanical techniques to remove oil can include mechanical grabs, such as scoops that are used when environmental conditions, e.g., cold weather, or the type of petroleum product, e.g., heavy fuel oil (HFO), yield highly viscous oil that is not suitable for other remediation actions. More commonly, skimmers are used to mechanically remediate oil that is enclosed by booms. Skimmers work by scratching, sucking or brushing the oil from the surface, without removing too much seawater in the process, and depositing the collected oil in storage tanks in an oil response vessel. There are many types of skimmers, including elevating, vacuum, weir and oleophilic skimmers. The optimal conditions for skimmers are calm weather, weak currents (<1 knot), low wave heights (<1 m) and a thick oil film [3]. Highly vicious oil types, debris and ice can hinder skimmer uptake and require the apparatus to be frequently cleaned by hand. These characteristics of skimmers cause them to often be an inefficient remediation technique [2].

In many circumstances, various types of sorbent materials are used to facilitate the successful removal of oil from the ocean or near-shore areas. These materials work either by adsorption, in which the oil is distributed over the surface of the adsorbent, or by absorption, in which the oil is distributed throughout the body of the absorbent. An ideal sorbent material should have high oleophilicity (affinity for oil), hydrophobicity (not taking up water), uptake capacity, buoyancy and rate of uptake [4, 5]. These products are typically composed of inorganic minerals, organic synthetic products or organic vegetable products, e.g., volcanic ash, polypropylene and wood fibres. Organic synthetic products typically have the highest oil removal ability, although these products also have a very slow degradation rate [2].

Although herder chemicals have existed since the 1970s, they have recently been further developed with silicone- and fluor-based surfactants [6]. Instead of acting through dispersion, oil herder chemicals are designed to increase the surface tension between the oil and water, thereby reducing the dispersion of the spilled oil [2], allowing easier manual removal of the oil from the water surface or in situ burning of the slick; however, oil herders also increase the rate at which the oil sinks to the ocean floor [3, 6].

11.1.1.3 In Situ Burning

In situ burning requires that an oil spill be ignited on site. The oil is contained within fire resistant booms and ignited in a controlled manner. Certain basic conditions must be fulfilled for spilled oil to burn: the presence of fuel, oxygen and an ignition source. Sufficient vaporisation of the oil must be ensured for the fire to ignite and continually burn [3]. The minimum required thickness of weathered crude oil required for ignition is three millimetres; oil layers that are too thin act to cool the oil, preventing it from producing enough vapours to ignite [6]. If the conditions are sufficient to begin a controlled burn, in situ burning rapidly removes large amounts of oil from the water surface and is highly effective as an oil removal technique, requiring little equipment and few labour personnel. Under certain conditions, e.g., oil-in-ice conditions, in situ burning may be the only possible clean-up technique. Negative aspects of in situ burning limit the use of this technique. For example, the creation of a large smoke plume with toxic emissions (e.g., soot particles and polycyclic aromatic hydrocarbons [PAHs]) can be problematic when burns are performed in near-shore areas, and the brief time frame and the required oil thickness for ignition limit broad application [3].

11.1.1.4 Dispersant Chemicals

Dispersant chemicals consist of solvents that dissolve oil in water, surfactants that lower the surface tension between oil and water, and other additives with the general purpose of reducing the surface tension between oil and water. This remediation technique is performed by adding chemicals with hydrophilic and lipophilic ends, causing the oil and water to bond. An oil spill can subsequently be broken apart by wave and wind power, causing the oil to form smaller droplets that eventually become denser than water, causing them to sink into the water column or even down to the sea floor. This technique does not alter the amount of leaked oil but alters its transport, fate and possible environmental effects. Instead of ultimately washing up on biologically sensitive or economically valuable shores or attracting and affecting, e.g., sea birds, the spilled oil is translocated to the water column and sea floor where it will have toxic effects on the marine life in these habitats. Therefore, the use of dispersants is a trade-off between possibly decreasing the effects of the spilled oil on the water surface and increasing the effects in the water column and at the sea floor [3]. Dispersants are less toxic today than when first introduced in the late 1960s. In many cases, the old dispersant chemicals were more toxic than the oil itself. Examples of this include the *Torrey Canyon* oil spill in 1967, in which more than 10,000 tonnes of dispersants were used, and the *Sea Empress* spill in 1996. On both occasions, the applied dispersants had lethal effects on numerous animals that survived the effects of the spilled oil [7].

11.1.2 Techniques Used on Shores

Natural recovery is the first alternative that must be considered when discussing clean-up strategies for shores affected by oil spills. The amount of oil that reaches a beach may be limited, and the available clean-up methods may not improve the environmental status and could actually increase the damage to the environment. Moreover, the safety of the clean-up response personnel might not be ensured due to the environmental conditions or the risks associated with the oil itself [3]. If techniques other than natural recovery are desirable, two main options exist.

11.1.2.1 Mechanical Techniques

Machinery, such as excavators, can rapidly clean a beach with minimum labour. However, this process produces up to ten times the amount of oiled waste produced by manual removal and can only be used on flat shores. If oil has penetrated into or been buried deep in the sand or sediment, this clean-up method will lead to even larger volumes of oil-contaminated waste [3]. When the necessary conditions are not fulfilled, e.g., on steep shores or where the coastline is littered with boulders, manual labour using rakes and shuffles is necessary to remove the oil from the beach. In many oil spills, humans with shovels are the only precise approach to remove the oil without damaging the underlying habitat [7]. On shores consisting of gravel or boulders or on anthropogenic structures, water hoses with cold water can be used to clean, e.g., rocks and piers, pushing the oil back into the water where it can be removed using other clean-up techniques.

11.1.2.2 Biological Degradation

Two variants of biological degradation exist: (1) oil-degrading microorganisms (bacteria, fungi, and algae) [8] can be added to the affected shore or (2) nutrients, such as nitrogen and phosphorus, can be added to already existing oil-degrading microbial communities. Several hundred species of microorganisms are capable of degrading petroleum [9]. This remediation option is more commonly used when the cleaning operation is in its final stage or on habitats affected by oil trapped in anoxic habitats. However, biological degradation has limitations. For example, the affected area must be sheltered, preventing oil and microorganisms from being washed away. Oil trapped in sediments, especially under anoxic conditions, is more difficult to remediate because microorganisms require oxygen for their metabolisms [2]. Low water temperatures also slow the metabolisms of microorganisms, and high concentrations of PAHs can be toxic even to the oil-degrading microorganisms [8]. Furthermore, heavier types of oil are more difficult for bacteria to degrade because they contain non-biodegradable compounds [7].

11.1.3 Treatment of Bilge Water

For ships over 400 GT, regulations contained in Annex I of MARPOL 73/78 state that the hydrocarbon content cannot exceed 15 ppm in treated bilge water that is to be discharged overboard [10]. Oily water exhibits a natural tendency to separate. The rate of this separation depends on the oil dispersion, although it generally occurs within 12–24 h. However, because detergents are typically contained in bilge water, emulsions form, which makes it challenging to treat the mixture. In general, there are two types of emulsions: coarsely dispersed, with a droplet size greater than 50 μm, and finely dispersed, with a droplet size of 0.2–50 μm. These differences in oil droplet sizes have implications for the applied oil pollution abatement technique because smaller droplets require more time to rise to the surface and enable gravity separation [11]. Gravity oily water separators (OWS), which use the density difference between water and hydrocarbon phases, have historically been used to treat bilge water because the onshore treatment of oily water is expensive (100 €/m^3). However, when stable oily liquid emulsions are present, the oil droplets are small, 0.2–50 μm. Thus, a standard OWS can only reduce the oil content in the bilge water to 20–100 ppm, which does not meet the

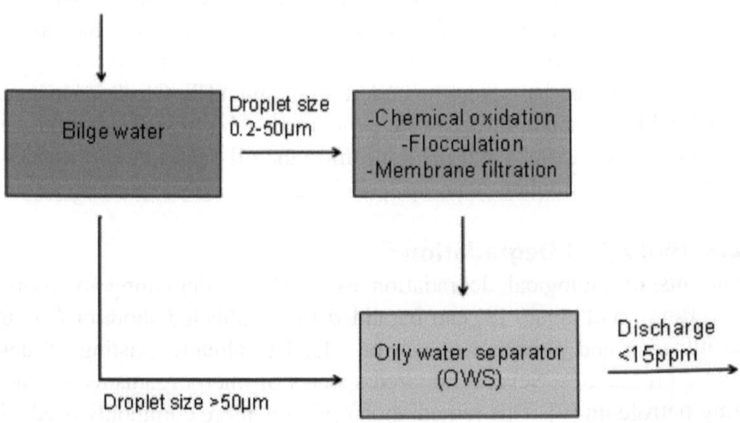

Fig. 11.1 An overview of a modern bilge water treatment procedure. For highly emulsified bilge water, an additional method must be used in conjunction with an OWS. For bilge water with oil droplet diameters exceeding 50 μm, only OWS treatment is necessary to comply with MARPOL 73/78 regulations

MARPOL 73/78 requirement (e.g., 15 ppm). Therefore, new methods have been developed in which micron-scale oil droplets are merged into larger droplets that are easily separated by gravity; these methods can be used in combination with a standard OWS. Examples of these additional methods include membrane technologies (ultrafiltration), chemical oxidation, flocculation and biological treatment (Fig. 11.1) [12, 13].

11.2 Antifouling

When submerged in water, ship hulls are quickly colonised by marine plants and animals, resulting in increased drag (see Sect. 4.3). To prevent the growth of marine organisms (biofouling) on ship hulls, antifouling techniques are needed. Until the enforcement of the Antifouling Systems Convention by IMO in 2008, antifouling systems with tri-butyl-tin as the active component were primarily used. This biocide has been replaced with copper, which is non-degradable and is also toxic to organisms. Following this development, demand has increased for new non-toxic antifouling strategies.

11.2.1 Non-toxic Antifouling Technologies

The development of non-toxic antifouling paints that are as effective as current biocide technologies has been challenging. However, one non-toxic alternative deserves special attention, i.e., foul release/non-stick fouling release coatings. This technique was developed simultaneously with self-polishing copolymer systems; however, because self-polishing systems were less expensive and more effective, the development of the non-stick fouling technique remained in the early developmental stages until TBT-based products were banned. The non-stick fouling release system is based on a combination of low surface energy, glass transition temperature and modulus [14] that provides an ultra-smooth surface. These features do not impair the ability of the fouling organisms to attach to a hull [15]; instead, they lower the attachment strength of the fouling organisms, allowing their removal at specified critical speeds [16]. Depending on the fouling community that is attached to a hull, these speeds are typically between 10 and 20 knots [17]. Because these coatings do not leach toxic substances, their environmental impact is very low. Two types of foul release approaches exist: silicone-based and fluoropolymer coatings. Silicone elastomers are applied to a hull in thick layers (6 mm) [18]. Drawbacks of the silicone-based coatings include their high cost, low resistance to mechanical damage and difficulties related to paint application [15, 19]. Fluoropolymer coatings also form non-porous, low surface energy surfaces with good non-stick characteristics. Drawbacks of these coatings include the limited mobility around the fluorine atoms and the requirement of higher stress, i.e., ship speed, to break the adhesion of foulers to the hull substrate [18].

A ship hull is typically painted at intervals of 3–5 year. Between repaintings, many ships are cleaned. Hull cleaning is most commonly performed by divers using different brushes or water jet systems. Unmanned remotely operated vehicles (ROVs) are also used for in-water cleaning. To avoid damage to the antifouling paint during cleaning, specific practices, such as a seawater jet with adjustable pressure, can be used. Devices for collecting paint fragments and biota (including possible invasive species) during cleaning are available. If using hull cleaning as an antifouling technique, use of a hard paint that is not affected by cleaning devices is recommended.

11.2.2 Areas of Research

To develop new non-toxic antifouling strategies, research is simultaneously needed in several areas. One of these areas is in biomimetics, in which the natural world is a model for possible solutions to the fouling problem. Marine organisms have both physical and chemical methods of protecting themselves from fouling by other organisms. Chemical antifouling methods often use secondary metabolites (natural products) that are produced by the organisms. Groups of organisms that have been investigated for potential effective secondary metabolites include bacteria, echinoderms, cnidarians, algae, bryozoans, sea squirts and sea sponges [17]. Some examples of secondary metabolites include barretin from the sponge *Geodia barretti* [20] and 1,1,3,3-tetrabromo-2-heptanone from the red algae *Bonnemaisonia hamifera* [21]. However, although research on natural products has been conducted for more than 20 years, no product has yet been successfully incorporated into an antifouling system.

Physical antifouling methods by which organisms protect themselves from fouling range from micro to macro scales. At the micro scale, microtextures and nanostructured surfaces, e.g., the micro-sized spicules used by echinoderms, tend to interfere with the sensitive settlement behaviour of many fouling species, e.g., barnacles and algal spores. At the macro scale, the skin properties of whales, sharks and dolphins also prevent fouling [22, 23]. However, no successful product has reached the market [17]. New antifouling techniques that are not based on biomimetics are also being considered, e.g., the potent neurological blocking agent medetomidine ((+)-4-[1-(2,3-dimethylphenyl)ethyl]-1H-imidazole), which is commonly used in veterinary medicine and has an inhibiting effect on barnacle settlement at very low concentrations [24]. Medetomidine is approved for use in South Korea, Japan, and China; approval is expected in the EU and the USA in 2017. However, this substance only affects barnacles. Thus, to create a successful antifouling system, medetomidine must be used together with other biocides [25]. Another technique that has been developed and commercialised for the leisure boat market is the low oxygen technique, in which proteins are added to the paint matrix to provide nutrients for the biofilm on the hull surface, which subsequently uses all of the oxygen near the hull. This condition inhibits the settlement of several types of macrofoulers, e.g., barnacles, mussels and bryozoans [16]. A third example of potential future antifouling agents is the anti-parasitic agent Ivermectin, which is commonly used in veterinary medicine as a pesticide and in fish farming; Ivermectin has a strong biocidal effect against barnacles [26].

11.3 Ballast Water

The spread of invasive marine species with ship ballast water (see Sect. 4.4) is a well-known problem, and the Ballast Water Management Convention (BWMC) under IMO requires that ships over 400 GT take measures to not discharge ballast water, including marine organisms, over set thresholds.

Presently (May 2015), approximately 60 ballast water treatment systems (BWTSs) have been approved by IMO. Technologies used for ballast water treatment have generally been derived from existing industrial applications of wastewater treatment systems in which solid-liquid separation and disinfection are applied. In separation, solid suspended material is removed via either sedimentation or filtration. Disinfection can be achieved in several ways, e.g., through the use of biocides that kill aquatic organisms. Treatment systems commonly include a combination of fundamental technologies, i.e., mechanical, physical and/or chemical treatment. Mechanical treatments include filtration and cyclonic separation. Physical disinfection could include UV irradiation, deoxygenation, ultrasound, cavitation and heat, whereas chemical treatment and biocides could involve chlorination or ozonation. A two-stage approach is used in most systems, in which mechanical separation is used for the first step, and physical/chemical treatment is applied in the second step.

Each system has advantages and disadvantages related to cost, efficiency and capacity. Running a ballast water treatment system can have operational implications, such as extended ballasting time and energy requirements.

11.3.1 Mechanical and Physical Treatments

The most established and developed physical treatment of ballast water is a filtration method combined with UV irradiation. This system uses UV light at different wavelengths and intensities to destroy microorganism cell membranes. The use of UV light, which is recognised as being highly effective against many microorganisms, depends on the clarity of the water to function properly. Hence, UV treatment is adopted as a secondary step to filtration, where larger (>50 μm) organisms and particles are first removed by filtration, increasing the light transmission of the UV treatment [27] (Fig. 11.2).

In the deoxygenation method, oxygen is removed by injecting an inert gas, such as nitrogen, or by inducing a vacuum. The removal of oxygen is fatal for many organisms, although it may be less effective for some taxa that are adapted to low-oxygen environments [28]. Oxygen removal as a treatment for ballast water

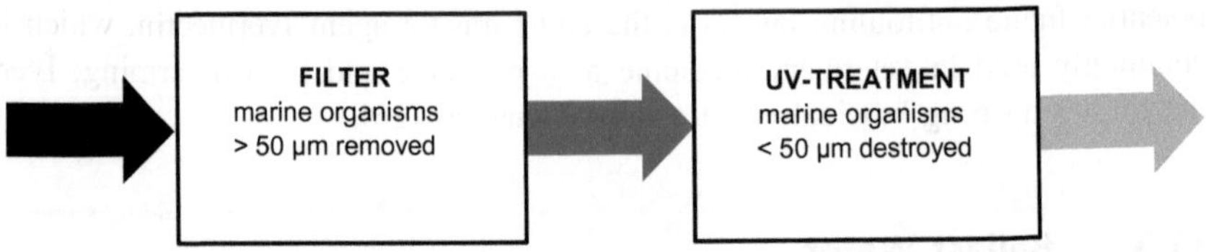

Fig. 11.2 An overview of an example of a common ballast water treatment method

can have the additional positive effect of reduced corrosion in the tanks. Ultrasound and cavitation systems use high-amplitude and high-frequency energy to cause mechanical stress, destroying the cell membranes of organisms, and are dependent on the wave amplitude and residence time for efficiency. For heat treatment of ballast water, the mortality of marine organisms has been demonstrated when temperatures are sustained for extended periods [29]. An advantage of heat treatment is the low or zero operating costs due to the utilisation of a ship's waste heat.

11.3.2 Treatment Methods Using Chemicals

Chemicals are effective for killing marine organisms; however, the potential for harm to the environment by the chemicals and/or their by-products is the reason for extensive testing before being approved for use in BWTSs. The effectiveness of biocides is related to changes in water conditions, salinity, temperature, pH and, most importantly, the various types of organisms in the water [27]. The BWTSs that use chemicals (in contrast to systems without chemicals) require extra tests for basic approval to ensure that the chemicals do not cause unwanted side effects. The basic approval is granted by IMO/MEPC, and only after achieving this approval can a system continue the process towards final approval.

Biocides used in BWTSs include chlorine, ozone, chlorine dioxide, hydrogen peroxide, peracetic acid and menadione/vitamin K. Most biocides are oxidisers, which destroy the cell walls of biota, and each biocide exhibits specific characteristics. Chlorine diluted in water destroys the cell walls of organisms; this method is well established for municipal and industrial applications. However, chlorine is not effective against the dormant stages of microalgae with stronger shells. In a study of five BWTSs based on chlorination treatment, 22 disinfection by-products were identified, of which four compounds were found at concentrations that may pose risks to the local aquatic environment [30]. Ozone bubbled through water as a gas is effective at killing microorganisms, although ozone requires expensive and complex machinery for on-board production and produces bromate by-products [31]. Chlorine dioxide is also effective in high-turbidity waters (where UV treatment is less effective due to the particles in the water) and has a half-life of 6–12 h. Hydrogen peroxide and peracetic acid are also effective and have few harmful by-products; however, both require high-dose applications, are expensive, and demand large on-board storage capacity [27]. Menadione/vitamin K is a non-oxidising biocide that is toxic to invertebrates. It is a natural product that can be synthetically produced for commercial use. Moreover, Menadione/vitamin K is safe to handle, although treated water will require neutralisation before discharge. All chemical treatments require the pre-treatment of ballast water, i.e., solid-liquid filtration, to perform satisfactorily. Post-treatment procedures may also be required when unwanted by-products must be removed before discharge.

The price estimate for a ballast water treatment system is approximately 1 million US dollars, and the number of treatment units required is dependent on the size of a ship [32]. The system requirements, such as pump capacity, also vary with ship type and size. In general, the installation of BWTSs in new ships is not a challenge; however, when retrofitting systems in old ships and in already existing engine rooms, the available space is often limited. The running costs of BWTSs vary with the type of equipment and use of consumables, such as UV lamps and chemicals. A challenge in the development of BWTSs is that they need to work properly in all possible combinations of salinity and temperature conditions that ships endure along their routes.

11.3.3 Other Alternatives

There are also alternatives to installing treatment systems according to the BWMC. These methods are currently only at the concept stage or in pilot programmes. Alternatives could include a flow-through system that does not treat water; instead it constantly overflows the tanks, resulting in continual exchange of water within the ballast tank. Furthermore, a future project will design ships that do not need ballast.

An interim solution until the Ballast Water Management Convention enters into force is to perform ballast water exchange (BWE) when a ship enters oceanic water. Because oceanic water contains many fewer plankton and marine organisms than coastal water, the exchange should decrease the concentration of organisms in the ballast water tanks. Oceanic water is at least 200 nautical miles from shore and at least 200 m deep. For traffic moving only within coastal areas, finding oceanic water is sometimes not possible, which limits this option. BWE can be used voluntarily but is not approved by the convention to be used as the solitary management method. However, in some areas, BWE is required due to national regulations, such as for ships bound for the Great Lakes in Canada and the US or Australia. BWE under poor weather conditions can place the ship at risk due to lost stability.

11.4 Wastewater

Sewage (black water) and grey water are wastewater streams generated on ships (described in Sect. 4.2) that require considerable on-board storage space and should be properly managed if discharged into the sea. Wastewater may contain organic matter, pathogens, and nutrients that can negatively impact the marine environment when discharged overboard, creating health hazards, local eutrophication effects and oxygen deficiencies. By reducing the volume of water used at the source, one can reduce the volume of wastewater that must be handled. An example of this is a common technique to decrease the volume of sewage by using vacuum toilets, which can reduce the volume from ca.70 to 25l per person per day [33]. The overboard discharge of sewage is not an environmentally sound solution, although it is permitted under specific conditions according to IMO regulations. The MARPOL 73/78 Annex IV [34] regulation permits the discharge of untreated sewage at a distance of more than 12 nautical miles from the nearest land. The discharge of sewage that has been stored in holding tanks should not be instantaneously performed. Instead, a moderate discharge rate should be used when a ship is en route and proceeding at a speed of at least 4 knots. Concern has been raised regarding the discharge of sewage into the marine environment, and in some areas, a ban has been introduced (i.e., in the Baltic Sea, not yet in force). No global regulations regarding the discharge of grey water exist, although local restrictions may be placed on territorial waters.

Sewage can be stored on ships in holding tanks with sufficient volume for discharge in a port reception facility (PRF). The storage of sewage should be controlled due to the risk of hydrogen sulphide generation, which has an offensive odour and possesses health and corrosion risks. Aeration (adding oxygen to the tank) and minimising the storage time are examples of methods by which the formation of hydrogen sulphide can be greatly reduced [35]. While in port, sewage from ships is typically pumped into the PRF, and the discharge should be completed in a reasonable amount time, which is of interest to both the port and the ship. Sewage from ships can be received via fixed sewage pipelines and hoses or mobile equipment, such as barges or trucks. The sewage can later be treated in a municipal treatment plant or in a treatment plant located at the port, if possible. Because the composition of sewage from ships may differ from the composition of wastewater that is usually treated at the municipal treatment plant, it may be regarded as industrial wastewater, which is associated with higher treatment costs [35].

Both sewage and grey water can be treated on ships. Grey water from the galley typically passes through grease traps to remove grease and oil from the wastewater to prevent clogging. Some systems treat grey water by reverse osmosis filtration and UV disinfection [36, 37]; direct nanofiltration has also been tested [37]. If the treated water is of satisfactory quality, it can be reused for firefighting, laundry or other appropriate, non-potable purposes on ships. Sewage treatment plants are primarily designed to treat sewage, although grey water is also treated in these systems in some cases. The goal may be to purify the grey water or to mix it with the black water before treatment to improve the functionality of the treatment plant. Sewage, especially that resulting from vacuum flush systems used on a ship, can be concentrated in solids. By mixing grey water and sewage, the volume of wastewater requiring treatment is greater, while the solids are less concentrated [38].

The same principles used for municipal wastewater purification can also be used on ships, although they must be adapted for minimal equipment size, ease of operation, and resistance to ship movements. Additionally, wastewater treatment plants are exposed to fluctuations in plant feed [39], and the treatment process typically includes several steps: mechanical screening, oxidation, clarification and disinfection [36]. Generally, sewage plants may be biological or chemical plants because the organic load can be oxidised either by biological or chemical means. Typically, the mechanical pre-treatment of sewage (screening, sieving, or sedimentation) is applied first. The pre-treatment may achieve up to a 50 % reduction in the organic load [36]. The biochemical oxygen demand (BOD) quantifies the amount of oxygen required by microorganisms to decompose organic matter in wastewater [40], i.e., a higher BOD is indicative of more organic impurities in the water. Biological treatment plants are often installed on cruise ships. These plants use microorganisms to consume the organic material in the sewage. The conventional solution is activated sludge, in which sewage enters the bioreactor and is mixed with microorganisms (bacteria and protozoa). Oxygen is continuously supplied to the bioreactor to enhance the digestion process. The organic material is biologically oxidised and converted to biomass (new cells are formed), carbon dioxide and water. After a predetermined time, the microorganisms are removed from the mixture in a clarifier (settling tank), and a portion of them is returned as activated sludge to the bioreactor. The excess sludge is regarded as waste. Biological plants require a relatively long start-up time, i.e., approximately two weeks [38]. A high reduction of BOD (80–95 %) and phosphorous (20–40 %) can be obtained in these biological processes [36]. Some biological plants also have a membrane installed to filter solids. The membrane acts as a barrier to separate biological and inorganic solids from the treated wastewater; such systems are called membrane bioreactors (MBR) [41]. Solutions of this type are referred to as advanced wastewater treatment systems (AWTs) [39]. The application of a membrane helps to eliminate the need for the conventional clarification process that is used in standard activated sludge systems. The final step is disinfection, which is performed via either chlorination or UV light. The effectiveness of disinfection depends partly on the effectiveness of the previous steps (removal of suspended solids and decomposition of organic matter). Chemical plants macerate the sewage to break the solids into smaller pieces. A chemical oxidant, i.e., sodium hypochlorite, is added to oxidise organic impurities [42], which is similar to the treatment of marine organisms in BWTSs. These plants are typically smaller than biological plants and are more common on small or medium ships [38]. The applied chemicals should be handled with care. In some systems, chlorine does not have to be added as hypochlorite because it is produced in situ by electrolysis using seawater [42]. If chlorine is used as an oxidant, an effluent de-chlorination unit may be required to meet the effluent criteria. In addition, trials are being conducted using advanced oxidation, which is a process involving the generation and use of hydroxyl free radicals as a strong oxidant to destroy compounds in sewage. This process is induced through the reaction of ozone-rich water with UV light [38].

The excess sludge that is generated during sewage treatment should be managed. All vessels that have on-board biological wastewater treatment collect sludge that consists primarily of excess biological mass [39]. This sludge can be either discharged overboard 12 nautical miles from the nearest land area [43] or discharged to port reception facilities while in port. Up to 270 m^3 of sludge from one cruise ship may be discharged to a PRF during a port call [39]. This sludge is characterised by a high water content and the presence of organic matter and pathogens [42]. In general, the sludge can first be centrifuged to increase the dry substance content and either stored in a holding tank for land disposal or prepared (by steam drying or filter pressing) for incineration [35].

The treated water (effluent) should not produce visible floating solids or discolour the surrounding water. The effluent should also comply with the effluent standards stated in the amended Annex IV of MARPOL 73/78. The limits of resolutions have become stricter with time, and new parameters have been added (Table 11.1).

Table 11.1 Effluent criteria in the amended MEPC resolutions (adapted from Hedberg and Lengquist [44])

Effluent criteria	MEPC.2(VI)	MEPC.159(55)	MEPC.227(64)
	STP installed before 1 January 2010	STP installed between 1 January 2010–31 December 2015	STP installed on 1 January 2016 or later
Faecal coliform standard (FCS)	250 FC/100 ml	Not applicable	Not applicable
Thermotolerant coliform standard (TCS)	Not applicable	100 TC/100 ml	100 TC/100 ml
Total suspended solids (TSS)	50 mg/l	35 mg/l	35x(Qi/Qe)[1] mg/l
Biochemical oxygen demand (BOD_5)	50 mg/l	25 mg/l	25x(Qi/Qe)[1] mg/l
Chemical oxygen demand (COD)	Not applicable	125 mg/l	125x(Qi/Qe)[1] mg/l
pH	Not applicable	6–8.5	6–8.5
Chlorine	Not applicable	0.5 mg/l	0.5 mg/l
Nitrogen[2]	Not applicable	Not applicable	20x(Qi/Qe)[1] mg/l or minimum 70 % reduction
Phosphorus[2]	Not applicable	Not applicable	1.0x(Qi/Qe)[1] mg/l or minimum 80 % reduction

[1]Qi/Qe dilution compensation factor
[2]Nitrogen and phosphorous removal standards are for passenger ships intending to discharge sewage effluent in special areas

The regulations have become stricter over time, although areas for improvement remain. Sewage treatment plants currently installed on ships sometimes have difficulties in meeting the standards of resolution MEPC.2(VI) and MEPC.159(55), which may be due to insufficient maintenance, improper use of cleaning additives or incorrect dimensioning [39, 45].

11.5 Solid Waste

As described in Sect. 4.5, marine litter is a global problem that negatively affects the marine environment. Plastic waste in different sizes is especially important because it is consumed by organisms at the base of the food web and by larger animals.

Various measures exist to reduce the amount of generated solid waste and to minimise the risk of these wastes being discharged into the sea. In general, waste management can be divided into reduction at the source, reuse, recycling, and reduction of the waste volume and mass through technological measures and disposal. Different types of solid waste can be generated on-board ships; Box 11.1 describes a case involving passenger ships. The prevention of environmental impacts should always be proactive. The first priority should be to reduce on-board waste generation. Developing and improving purchasing strategies and management yields resource conservation and waste minimisation. The management of ship-generated waste does not end at the PRF, because ports typically do not treat ship-generated waste. Third parties are engaged in the treatment of this waste, its recycling, and its delivery for incineration or to a landfill. Sustainable waste management on a ship is naturally a combination of environmental policies, commitment from the ship's operating company and the engagement of the crew that must manage waste on a daily basis. Some instruments can assist with more sustainable waste management practices, such as garbage management plans, which are written procedures that define how garbage should be collected and managed on ships. According to MARPOL73/78 Annex V Reg. 10, these plans are mandatory for all ships greater than 100 GT and on ships certified to carry a minimum of 15 persons. Every ship greater than 400 GT and every ship certified to carry a minimum of 15 persons is required to keep a garbage record book, where all discharges and incinerations must be recorded [46]. This book should be available for inspection and should be preserved for at least two years from the date of the last recorded entry.

Box 11.1 Waste generation on cruise ships

A typical cruise ship with 3000 passengers and crew generates considerable number of tonnes of solid waste per week. One passenger on a cruise ship is estimated to generate 1 kg solid waste, two bottles and two cans per day [47].

Food waste is often the single largest component of the solid waste that is generated on ships [48]. On one cruise ship, approximately 12 m^3 of food waste can be generated per week [42]. Within special areas, food waste can be discharged only if it is ground into pieces less than 25 mm and the ship is en route and more than 12 nautical miles from the nearest land. The treatment of food waste on ships can vary; it can be pulped, compressed and/or incinerated [42]. Disposers (grinders or macerators) are used to grind food waste. To minimise problems with odours and hygiene, temperature-regulated storage rooms may be used.

When a ship arrives at a port of call, the time for waste disposal is often very limited, and thus it is important that it is performed smoothly. Therefore, the port of call should be notified in advance of the types and estimated amounts of wastes that must be left at the port reception facilities. Differences may exist in the system for labelling and sorting garbage on different ships and in different ports; therefore, harmonisation of these systems must be sought. For example, two ISO standards have been developed (for ships and for ports) that aim to homogenise the management of solid waste generated on ships. ISO 21070:2011 [49] applies to "the management and handling of garbage generated on ships while it remains on board according to the definitions stated in Annex V of MARPOL 73/78". Additionally, "it addresses the vessel-to-shore interface and the transfer of garbage from the vessel to the PRF". ISO 16304:2013 [50] applies to "the management of ship-generated waste regulated by MARPOL 73/78 that is discharged at ports and terminals". It also covers "principles and issues that should be considered in the development and implementation of a port waste management plan and the operation of port reception facilities". Other standards have also been developed in the shipping sector, such as a cruise industry standard for "Waste Management Practices and Procedures" that can be incorporated into ship management systems, which is part of the ISM Code [47].

The Port Reception Facility Database is a module of the Global Integrated Shipping Information System (GISIS) and provides data on facilities for the reception of all categories of ship-generated waste in ports. This information is updated by port states. This module also enables shipmasters that encounter difficulties in discharging waste to reception facilities to report cases of alleged inadequacies.

In the past, the discharge of garbage from ships was not as strictly regulated as it is today. The objective of the MARPOL 73/78 is to eliminate and reduce the discharge of garbage from ships into the sea; it prohibits the disposal of almost all types of garbage at sea and recommends discharge ashore in an appropriate PRF. The amended Annex V of MARPOL 73/78 prohibits the disposal of nearly all types of garbage at sea except for food waste, animal carcasses, cargo residues contained in wash water and cleaning agents and additives under specific requirements.

The volume of produced garbage can be reduced by various methods. Reducing the volume is important for several reasons, including the limited on-board storage space and for easier packing and transportation [51]. Economical savings can also occur when onshore fees for the delivery of waste are charged per unit volume (m^3).

In general, the equipment used for mechanical methods includes, shredders, compactors, crushers, pulpers and other processors [51]. A shredder can be described as a chamber with cutters attached to counter-rotating shafts. Shredders can crush and tear various materials, such as metal, paper and plastic, and can be applied to mixed waste or to individual waste types. After shredding, the small pieces of material can be stored or fed into a compactor for further processing [51].

Another possible treatment option for garbage is incineration, which also reduces the volume. This option is typically applied to waste that cannot be recycled. The incinerators used on ships are often automated, including feeding, controlling combustion sequences, and ash handling [51]. Incineration of garbage containing more than traces of heavy metals is not permitted according to MARPOL 73/78 Annex VI [52]. Incinerators must be approved in accordance with MEPC.76 (40)—The Standard Specification for Shipboard Incinerators. Incinerators are typically large and heavy, and their installation on existing ships requires major modifications. Incineration of garbage may cause emissions of unwanted by-products, such as dioxins, furans, PAHs, heavy metals, hydrochloric acid and particulate matter [53]. The generation of dioxins can be controlled by reducing the extent of incomplete combustion, assuring a specific residence time and temperature span [51]. The gaseous emissions from incinerator are vented through a stack along with engine exhaust, although some portion can be emitted to the incinerator room; therefore, operators should be trained and protected with regard to occupational health [51]. The ash that remains after combustion may contain heavy metals and other hazardous substances and should be disposed in port. The fly ash, which is a fraction of smaller particles, typically follows the flue gas and can be partially removed by exhaust gas cleaning systems. The incineration of ship-generated waste is prohibited in some areas, e.g., in the territorial waters of the Baltic Sea.

In addition to incineration, other technologies may be considered, such as plasma arc destruction, which has been used in trials on a passenger ship [54]. In this process, solid wastes, including paper, cardboard, plastics, textiles, wood and food, are shredded and milled before being rapidly gasified in a plasma-fired educator. Via this process, waste is exposed to extreme heat, and a synthesis gas is formed that consists primarily of carbon monoxide and hydrogen. In the next step, the synthesis gas is combusted with excess air, resulting in CO_2 and H_2O [54].

11.6 Greenhouse Gases (GHGs)

The most significant GHGs from ship operations are CO_2, methane, nitrous oxide and halocarbons (see Sect. 5.2). The CO_2 emissions from ship operations are related to the burning of fuels containing carbon and can be reduced by (1) transporting less, which reduces the need for fuel, (2) using less fuel per unit of transported cargo or passenger, i.e., increased energy efficiency, (3) changing to a fuel that does not contain carbon or utilises modern (not fossil) carbon, or (4) capturing CO_2 from the exhaust gas and storing it instead of emitting it into the atmosphere. The first two alternatives are in accordance with the first step in Fig. 8.2. The third alternative is related to the second step, whereas the fourth alternative is related to the third step. Possible measures to reduce fuel consumption through increased energy efficiency are discussed in Sects. 10.2 and 10.3, and potential alternative fuels are discussed in Sect. 10.4.

Carbon dioxide capture and sequestration (CCS) is a group of technologies for reducing CO_2 emissions from power plants and industrial processes. In general, CCS includes capturing, transporting, and storing CO_2 underground (either onshore or offshore) [55]. If an offshore storage option is chosen, the CO_2 captured from the exhaust gas can be delivered either via pipelines or ships [55]. Uncertainties remain regarding the environmental and security issues associated with the storage of CO_2 [56].

Different methods can be used to capture CO_2 from gas streams, including solvents, membranes, adsorption or cryogenic/condensation systems [56]. The current CCS methods designed for use on land are not suitable for direct application on ships because they would impact shipping performance [57]. Laboratory tests to develop methods that meet the requirements and conditions present on ships are on-going. One example is a process in which CO_2 is absorbed from the exhaust gas and stored in solid form ($CaCO_3$) on a ship after a sequence of chemical reactions are performed [57].

Methane emissions from modern diesel engines are not a large problem and do not result in a large contribution to the total greenhouse gas emissions from shipping. However, if the use of liquefied natural gas (LNG) as ship fuel increases, methane emissions might also increase. Engine methane slip has three primary causes: (1) gas in the intake port combined with scavenging,[1] (2) incomplete combustion and (3) crevices in the combustion chamber [58]. The methane slips reported for lean-burn SI engines and four-stroke lean-burn DF engines are 1.4–4.1 and 2.4–7 %, respectively, at full load. Methane emissions can be very high at low engine loads. Engine modifications can reduce methane slip from gas engines. Moreover, after-treatment methods can also be applied, e.g., oxidation catalysts, to further reduce emissions [59].

11.7 Nitrogen Oxides

Under high-temperature combustion, oxygen and nitrogen react to form nitrogen oxides (NO_x). Thus, NO_x formation depends on the combustion temperature, the duration of the combustion and the concentration of oxygen and nitrogen in the combustion chamber. These areas could therefore be addressed to decrease NO_x emissions from marine engines. In principle, three different methods exist to decrease NO_x emissions: fuel switching, combustion-related techniques and the after-treatment of exhaust gases. A decrease in NO_x may also be achieved by

[1]Removal of spent gases from an internal combustion engine cylinder and replacement with a fresh charge or air.

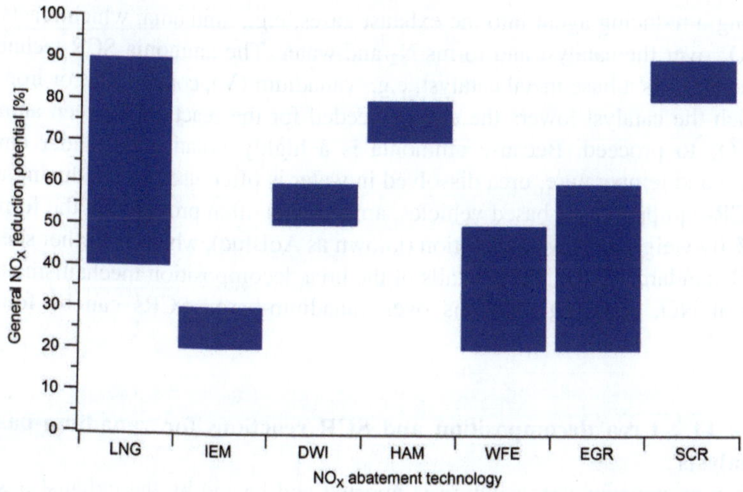

Fig. 11.3 General NO_x reduction potential for liquefied natural gas (*LNG*), internal engine modifications (*IEMs*), direct water injection (*DWI*), humid air motor (*HAM*), water fuel emulsion (*WFE*), exhaust gas recirculation (*EGR*) and selective catalytic reduction (*SCR*) [63]

changing the propulsion technology, e.g., switching from diesel engines to fuel cells or batteries (see Sect. 10.1.2 for more information regarding fuel cells and Sect. 12.3.3 for a discussion of the requirements for a future zero-emission ship).

Several different NO_x-abatement technologies for marine diesel engines are currently on the market [60], such as basic and advanced internal engine modifications (IEMs), exhaust gas recirculation (EGR), direct water injection (DWI), humid air motors (HAMs) and selective catalytic reduction (SCR). Not all of these technologies can independently reduce NO_x emissions to comply with the Tier III regulation (see Chap. 5), although the possibility exists for combining two or more abatement technologies to achieve the Tier III standard. However, SCR or the use of LNG in lean premixed combustion are considered two alternatives that can each independently comply with the Tier III regulation [61]. In a 2013 IMO review of the current status of possible NO_x Tier III abatement technologies, SCR, LNG and possibly EGR were suggested as being capable of achieving Tier III compliance [62]. The general NO_x reduction potential of some abatement technologies in relation to a marine diesel engine without any abatement technology is illustrated in Fig. 11.3.

11.7.1 Selective Catalytic Reduction (SCR)

SCR technology is the current technology with the greatest potential to reduce nitrogen oxide emissions (see Fig. 11.3). In general, SCR is a catalytic exhaust gas after-treatment system that is installed after the engine and selectively reduces the NO_x concentration in the exhaust gases. The NO_x reduction is performed by injecting a reducing agent into the exhaust gases, e.g., ammonia, which reacts with the NO_x over the catalyst and forms N_2 and water. The ammonia-SCR technology commonly uses a base metal catalyst, e.g., vanadium (V), copper (Cu) or iron (Fe), in which the catalyst lowers the energy needed for the reaction between ammonia and NO_x to proceed. Because ammonia is a highly volatile gas under ambient pressure and temperature, urea dissolved in water is often used as a reducing agent. For SCR-equipped land-based vehicles, ammonia is often provided in the form of a 32.5 % by weight urea water solution (known as AdBlue), which is further specified by ISO standard 22241. More details of the urea decomposition mechanism and the chemical NO_x reducing reactions over vanadium-based SCRs can be found in Box 11.2.

Box 11.2 Urea decomposition and SCR reactions for vanadium-based catalysts

When an aqueous urea solution is injected and heated by the exhaust gases, the water evaporates, and the urea decomposes into ammonia and isocyanic acid (thermolysis). In the presence of water, the isocyanic acid (HNCO) is further decomposed (hydrolysis) over the SCR catalyst into ammonia and carbon dioxide [64]. In the ideal decomposition process, one mole of urea produces two moles of ammonia and one mole of carbon dioxide according to the following reactions:

$$NH_2{-}CO{-}NH_{2(s)} \rightarrow NH_{3(g)} + HNCO_{(g)} \quad (11.1)$$

$$HNCO_{(g)} + H_2O_{(g)} \rightarrow NH_{3(g)} + CO_{2(g)} \quad (11.2)$$

However, side reactions may occur that can interrupt/disturb the ideal urea decomposition. The NO_x-reducing reactions that occur over vanadium-based SCR catalysts are dependent on the composition of NO_x, i.e., the NO_2/NO_x ratio, in the gas feed [65–69]. For a gas feed consisting of pure NO in the presence of O_2, the following suggested reaction is denoted as *standard SCR*:

$$4NH_3 + 4NO + O_2 \rightarrow 4N_2 + 6H_2O \quad (11.3)$$

In a gas feed with both NO and NO_2, the following reaction is called *fast SCR*:

$$4NH_3 + 2NO + 2NO_2 \rightarrow 4N_2 + 6H_2O \quad (11.4)$$

For pure NO_2, an additional reduction pathway is possible (called *NO_2 SCR*):

$$4NH_3 + 3NO_2 \rightarrow 3.5N_2 + 6H_2O \quad (11.5)$$

Additionally, for V_2O_5–WO_3–TiO_2 catalysts at high temperatures ($T > 400$ °C), ammonia, NO and oxygen may react and form nitrous oxide (N_2O). Moreover, at even higher temperatures, ammonia may begin to oxidise into NO [70]. Hence, at high temperatures, the NO_x reduction performance over a V_2O_5–WO_3–TiO_2 catalyst decreases. The following order for SCR reaction rates over V_2O_5–WO_3–TiO_2 catalysts is suggested: fast SCR ≫ standard SCR > NO_2 SCR [65].

Ammonia-SCR technology requires the urea dosage to be adjusted to correspond to the NO_x concentration in the exhaust gases. This is a complicated task due to dynamic NO_x emissions, engine temperatures from engines, temperature-related catalytic activity, and ammonia storage capacity in the catalyst. Ammonia-SCR technology has been used for several decades in different industrial sectors, such as power plants and land-based transportation (trucks). During this time, the technology has been continuously improved (see Box 11.3). NO_x reduction efficiencies of 80–90 % and even higher have been demonstrated [71].

Box 11.3 Development of ammonia-SCR catalysts for land-based applications

One of the most commonly used ammonia-SCR catalysts in the industry is a vanadium-based (V_2O_5) catalyst supported by titanium dioxide (TiO_2). However, problems remain with vanadium-based catalysts regarding the oxidation of SO_2 to SO_3, which can react with ammonia and form ammonia sulphates, the decrease in activity and selectivity at high temperatures and the toxicity of vanadium species, which begin to volatilise at temperatures exceeding 650 °C. In addition, during active regeneration of diesel particulate filters (DPFs), the exhaust gas temperature over the SCR catalysts can exceed 500 °C if placed downstream of the filter [72–74]. In this case, the use of vanadium-based SCR catalysts is less preferable. Thus, research has been directed towards the development of new types of SCR catalysts, such as metal-exchanged zeolites[71], with a focus on copper- (Cu) and iron- (Fe) exchanged zeolite catalysts. Several studies have been completed regarding catalytic activity [74–79] and thermal stability [80–84]. In general, Cu-zeolites are active at lower temperatures (e.g., less than 300 °C), whereas Fe-zeolites are more active at higher temperatures. However, Cu-based catalysts are more sensitive to sulphur poisoning than Fe, and the use of either Fe- or Cu-exchanged zeolites may require ultra-low sulphur fuels with less than 15 ppm sulphur [72, 85–87]. Recent research has investigated the possibility of combining Cu- and Fe-zeolites to achieve a wider operational window [88, 89].

V_2O_5-based catalysts may function well at temperatures of 260–450 °C [90] and are less sensitive to sulphur poisoning than both Fe- and

Cu-exchanged zeolites [91]. The support material is also an important factor that should be considered; both TiO_2 and Al_2O_3 have been reported to provide improved catalytic activity for vanadium-based catalysts, with TiO_2 being the superior material [92]. TiO_2 is also less sensitive to sulphur poisoning than Al_2O_3 [93, 94]. In addition, commercial vanadium-based catalysts can be doped with WO_3, which increases their activity, widens the temperature range for the SCR reaction, increases the poison resistance for both alkali metal oxides and arsenic oxide and reduces the oxidation of ammonia and SO_2 [95]. Hence, due to the relatively high sulphur levels in marine fuels, vanadium-based catalysts supported on TiO_2 and possibly doped with WO_3 are commonly being used for marine applications (ships).

Several different forms of catalytic deactivation occur, including chemical, mechanical and thermal deactivation. An important deactivation mechanism for an ammonia-SCR catalyst operating downstream of a diesel engine in the presence of ammonium, sulphur and water is the formation of ammonium sulphates, which may result in pore blocking and related catalytic deactivation [96–99] The vanadium content in an SCR catalyst has a positive effect on NO_x reduction [100] and simultaneously increases the less preferable oxidation of SO_2 to SO_3 [71, 101, 102], which later can react with ammonia and water to form ammonium bisulphate (NH_4HSO_4) and/or ammonia sulphate ($(NH_4)_2SO_4$) [94]. Hence, vanadium-based NH_3-SCR for mobile diesel applications usually contains 1.5–2.0 wt% V_2O_5 and approximately 8 wt% WO_3 on a TiO_2 support [103]. Additionally, ammonium nitrates may form and deposit on surfaces at temperatures less than 170 °C in the presence of NH_3 and NO_2 and begin to decompose at higher temperatures [66, 67, 69, 104]. Thus, the sulphur, phosphorous and alkali metal contents in marine fuels, lube oils and urea solutions should be minimised to avoid the catalytic deactivation of marine vanadium-based SCR catalysts.

An example of an SCR arrangement on a ship is illustrated in Fig. 11.4 [105]. In this figure, a diesel engine is illustrated in the lower left corner, with the exhaust gas pipe extending to the upper right corner. Figure 11.4 illustrates the urea injection position upstream of the catalyst (denoted as the converter/silencer), followed by mixers, which allow the urea to decompose into ammonia and mix properly with the exhaust gases before reaching the catalyst. A dust blower system is included in the converter and utilises pressurised air to remove dust, ash and soot from the catalytic material. Additionally, the emission measurement port and the NO analyser can be used to control and regulate the urea dosage.

SCR catalysts have been installed and operated on many ships (possibly more than 1000 vessels) [62]. Most of the installations have been on four-stroke engines, although an increasing number of installations have occurred on two-stroke engines. However, in the latter case, the SCR unit may require installation upstream of the turbine for increased exhaust gas temperatures. SCR will be able to meet the Tier III requirement for a majority of marine engines, while three general technical

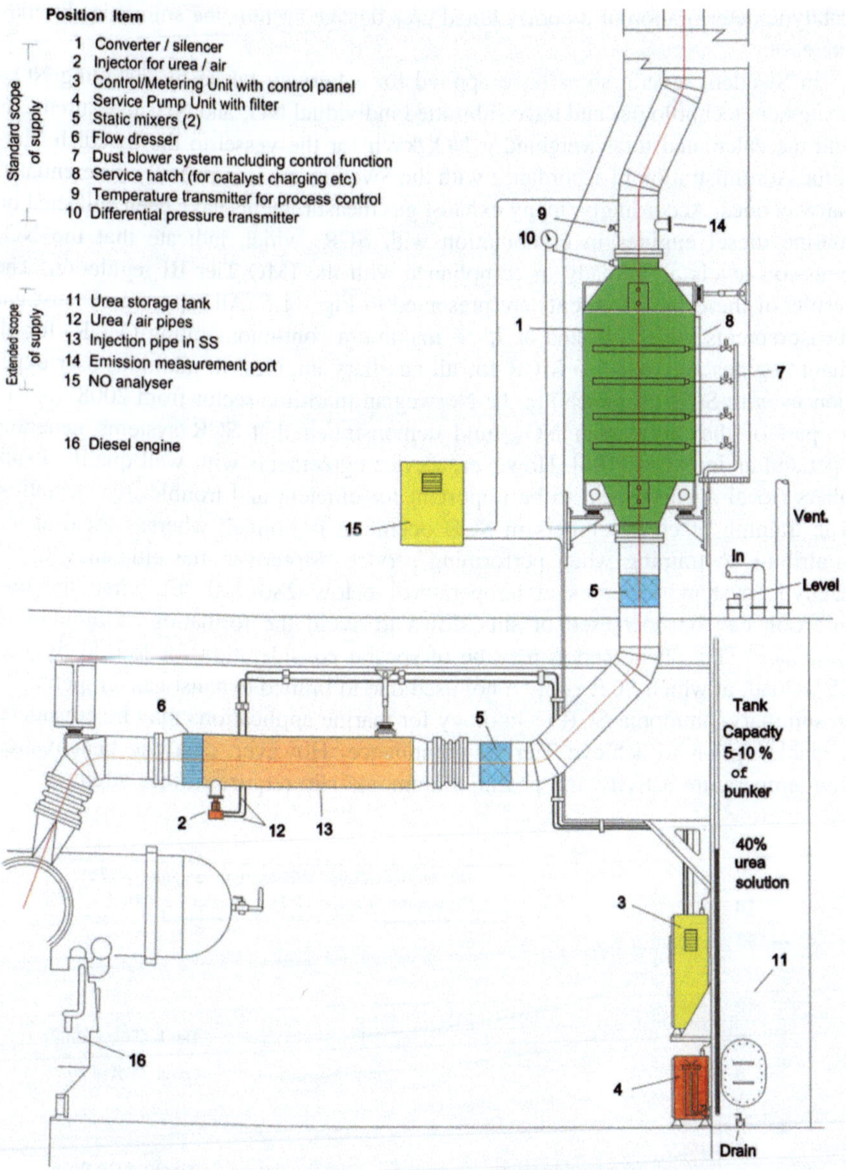

Fig. 11.4 General SCR arrangement. Printed with permission from D.E.C. Marine AB [105]

issues with SCR have been raised [62]. First, low-temperature operation, e.g., at engine loads less than 25 %, results in insufficient NO_x reduction and the potential formation of ammonium sulphates. Second, catalyst deterioration may occur by poisoning and fouling by soot, ash and ammonium sulphates. Third, as a result of catalytic deterioration or a poorly tuned urea dosage system, the ammonia slip may increase.

In Sweden, several ships have applied for a fairway rebate by installing NO_x-abatement technologies and have submitted individual NO_x and NH_3 measurements and the calculated total weighted g NO_x/kWh for the vessel to the Swedish Maritime Administration in accordance with the Swedish environmentally differentiated fairway dues. Accordingly, many exhaust gas measurements have been collected on marine diesel engines in combination with SCR, which indicate that the NO_x emission levels are already in compliance with the IMO Tier III regulation. The results of these measurements are presented in Fig. 11.5. All reported exhaust gas measurements were collected at 75 % maximum continuous rating (MCR) for all main engines and at 50 % MCR for all auxiliary engines. In addition, user experiences with SCR (8 vessels) in the Norwegian maritime sector from 2008 to 2011 as part of the Norwegian NO_x fund demonstrated that SCR systems generally operated as intended [106]. However, service agreements with well-qualified suppliers (local agents) seem to be important for efficient and trouble-free operation. The training of crewmembers in SCR operation is limited, whereas local agents contribute to training when performing service. Moreover, the efficiency of the catalytic system decreases at temperatures below 280–320 °C, when the urea injection can be decreased or shut down to avoid the formation of ammonium sulphates. This phenomenon may be of special consideration for low loads, e.g., 25 % load, at which SCR is often not used due to limited exhaust gas temperatures. In summary, ammonia-SCR technology for marine applications may be considered a viable option to achieve Tier III compliance. However, catalytic deactivation, low-temperature activity and potential ammonia slip require further attention.

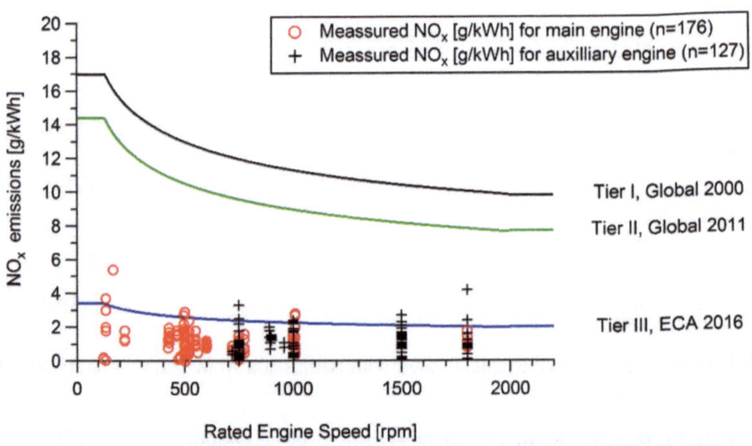

Fig. 11.5 IMO NO_x emission regulations and measured NO_x emissions for SCR installations in accordance with Swedish environmental differentiated fairway dues. Adapted from Brynolf et al. [107]

11.7.2 Alternative Fuels

The potential NO_x reduction from switching fuels to LNG is dependent on the engine type. Premixed gas-fuelled lean-burn gas engines, which have a pre-chamber and spark plug for ignition, and dual fuel engines, which use a 1–5 % pilot injection of diesel fuel for ignition, are both operated as an Otto cycle in lean conditions. The Otto cycle and excess air (lean burn) contribute to reduced peak temperatures compared with the Diesel cycle and may reduce NO_x emissions by approximately 75–90 % [61, 108, 109]. The reduction can be explained by the reduced compression ratio required by the Otto cycle, which generates lower temperatures that are further reduced by the over-stoichiometric air inflow. However, high-pressure gas injection engines are operated as Diesel cycle engines, which use a 1–5 % pilot injection of diesel for improved ignition and can reduce NO_x formation by approximately 40–50 % [109]. Both lean-burn gas engines and gas-fuelled dual-fuel engines are sensitive to knock, i.e., auto-ignition, whereas high-pressure gas injection engines are much less sensitive to knock [110].

Methanol could also contribute to meeting the Emission Control Area (ECA) SO_x and NO_x requirements [119]. The fuel heating value of methanol is approximately half that of diesel fuel (approximately 20 and 43 MJ/kg, respectively), requiring increased mass flow and evaporation of fuel to maintain the same power output from the engine. Hence, this fuel change would lead to suppressed combustion temperatures and reduced NO_x formation. However, pure methanol may not reduce NO_x formation by more than 60–70 %, whereas methanol in combination with 10 % water (i.e., a 90/10 methanol/water mixture) may reach Tier III compliance. However, these results describe a diesel engine operating on methanol, e.g., the high-pressure injection of methanol with 5–10 % diesel as a pilot fuel. Methanol is further discussed in Sect. 10.4.2.3.

11.7.3 Basic and Advanced Internal Engine Modification (IEM)

IEMs are considered to be combustion adjustment methods because they attempt to reduce the formation of NO_x in the combustion process. Several different techniques have been applied in this field, such as reduced scavenging air temperature, the Miller process, delayed fuel injection, compression ratio modification and injector-nozzle design [112]. Only the slide-valve technique, which is a change in the injector-nozzle design, is considered to be a basic IEM. This technique uses improved fuel spray to delay fuel injection and subsequently reduces NO_x formation while maintaining fuel efficiency [113]. The other technologies are advanced IEMs and are under further development by their manufacturers. Hence, their individual NO_x reduction potential may not yet be fully demonstrated, although a total overall NO_x reduction performance of 30 % may be assumed for all advanced IEM measures [60].

11.7.3.1 Scavenging Air Temperature

Reducing the scavenging air temperature can reduce NO_x formation during the combustion process. However, the air temperature must not be reduced below the ambient dew point because excessive water condensation may become an issue. The charge air temperature should not be reduced below 45 °C for modern diesel engines to avoid precipitation of water in the charge air system [110]. Moreover, based on empirical data, the temperature should be maintained above 50 °C when using HFO.

11.7.3.2 Miller Process

The Miller process adjusts the valve timing, i.e., closing the inlet valve before the piston has reached the bottom dead centre, which results in the expansion of the air contained in the cylinder and a subsequent temperature decrease. Thereafter, the compression temperature decreases, and NO_x formation is reduced. Miller timing may reduce NO_x formation by 15–30 %; however, possible drawbacks with Miller timing may include difficulties with starting the engine and increased soot formation with partial loads because the turbo blower pressure is too low to provide sufficient air to the cylinders. This problem could be counteracted by introducing flexible camshaft technology, which allows the valve closing and fuel injection to be adjusted at start-up and for partial loads [112]. High-efficiency turbochargers or two-stage turbochargers could be used to compensate for the shorter valve opening times in Miller timing, which provides sufficient air to the combustion process and allows the related engine efficiency to remain unaffected [108, 112]. The combination of extreme Miller timing and two-stage turbo charging with EGR could potentially decrease NO_x formation by 60 % [110].

11.7.3.3 Delayed Fuel Injection

A substantial amount of NO_x formation is related to non-uniform combustion processes in which early burned gases are further compressed and higher combustion temperatures are produced until the piston reaches the top dead centre; this process results in increased NO_x formation [114]. By delaying the fuel injection, the post-compression of burned gases can be reduced, which results in less NO_x formation. However, the reduced peak combustion pressure may imply slightly lower engine efficiency. A delay in the fuel injection of two crank angles on a two-stroke engine can produce a reduction in the peak combustion pressure of 10 bar, which reduces NO_x formation by 10 % with a fuel penalty of 3 g/kWh. The fuel delay technique may achieve a NO_x reduction of up to 25 % [112].

11.7.3.4 Compression Ratio

The compression ratio and combustion temperature are interrelated, i.e., higher compression ratios are related to higher temperatures. By using an increased compression ratio and a reduced fuel ignition delay, the fuel injection can be partially delayed, causing a modest NO_x reduction without a fuel penalty [112]. However, NO_x formation can be reduced by up to 50 % for medium-speed engines that are tuned for increased compression ratios in combination with fuel injection systems that are optimised for late injection and short duration [108]. Additionally, a reduced compression ratio also results in less NO_x formation due to reduced combustion temperatures; however, the reduced compression ratio results in a fuel penalty.

11.7.3.5 Injector Nozzle Design

The design of the fuel injector nozzle is important for the performance and emissions of diesel engines and can lower the NO_x emissions via delayed fuel injection. The fuel injector design affects the evaporation of fuel, air-to-fuel mixing and ultimately the entire combustion process [112, 114]. Two different principal fuel injection designs are typically used depending on the engine type. Four-stroke engines typically have a centrally positioned fuel injector at the cylinder head, whereas two-stroke engines have multiple injectors configured in a ring in the top cylinder region [112].

The most widespread IEM is the replacement of ordinary fuel valves with low-NO_x slide-type fuel valves (slide valves). However, this technique only applies to two-stroke engines, and nearly all new, slow-speed two-stroke engines delivered after 2000 are equipped with slide valves to meet the IMO NO_x standards [60]. Slide valves are designed to optimise the spray distribution, which allows for delayed fuel injection and a subsequent reduction in NO_x formation while maintaining fuel efficiency [113]. The initial objective was to avoid engine fouling by reducing the amount of fuel that was released from the SAC volume, i.e., the volume between the injector needle and the nozzle orifice(s), in the fuel injectors during the scavenging interval, thereby reducing the formation of soot, VOCs and NO_x. The assumed reduction in NO_x through slide valves is 20 %. Depending on the fuel type, slide valves may also lead to a reduction in PM and VOC formation of up to 50 % [60]. Additionally, "low-NO_x" nozzles may reduce NO_x formation by 10–25 % depending on the engine type [115].

11.7.4 Addition of Water to the Combustion Process

The addition of water to the combustion process has been proven to substantially reduce NO_x formation by decreasing the combustion temperature via the increased heat absorption from the added mass (fuel + water) and the increased heat capacity and evaporation. Generally, the addition of water can occur in three ways: DWI into the combustion chamber, adding atomised water[2] or water vapour (saturation) to the scavenging air, and producing water-fuel emulsions. The water quality must be acceptable for these processes to be effective; thus, the water used in these processes is generally produced by evaporator systems [112]. The use of water addition

[2] In this context, the phrase atomised water is used to describe water droplets with a very small droplet diameter, e.g., a fine water mist.

techniques may have an effect on the lubrication oil film in the cylinder liner, which may increase cylinder wear [116].

11.7.4.1 Direct Water Injection (DWI)

DWI is achieved via an independent injection system and monitored by an electronic control system, thereby allowing water injection to occur without compromising fuel injection. The water injection timing is not required to be identical to the fuel injection timing, which allows the water to be injected directly into the combustion chamber at the necessary place and time to achieve satisfactory NO_x reductions [112]. DWI may not be feasible for low engine loads, i.e., less than 30 %, because the combustion temperature may become too low for complete combustion [113]. Typical water-to-fuel ratios of 40–70 % achieve NO_x reductions of 50–60 % [60, 115, 117]. In general, a NO_x reduction of 50 % is achieved at a 40 % water-to-fuel ratio [60]. DWI can result in a slight increase in smoke/soot and PM emissions in combination with a fuel penalty of approximately 0–5 % [115, 117]. However, when using DWI, the fuel's sulphur content must be considered to avoid corrosion by sulphuric acid. The upper limit is approximately 1.5 wt% [110, 113].

11.7.4.2 Atomised Water or Water Vapour Techniques

The addition of atomised water or water vapour to the scavenging air involves similar techniques, although there is one major difference with respect to the water supply. For a water atomising system, the water is injected into the scavenging air and must be pure, i.e., no impurities are permitted. Therefore, an on-board freshwater production and/or storage system is required. However, a water vapour system can directly use ordinary seawater in the vapour process. The SAM system and the WetPack method are two techniques using technically pure water, i.e., deionised or evaporated, whereas the HAM concept is an evaporation technique [116]. The WetPack system can reduce NO_x formation by 30–60 % in four-stroke engines using a water/fuel ratio of approximately 40 % [108]. The HAM approach enriches (saturates) the charge air with evaporated seawater using waste heat from the engines. Approximately three times as much water vapour as fuel should be introduced into the engine to achieve a NO_x reduction of 70–80 % using HAM technology [60]. However, the saturation point of water in air depends on the surrounding absolute pressure and temperature, and increased temperatures and/or reduced pressures increase water evaporation, i.e., the temperature and pressure of the charge air determine the amount of water vapour that can be transferred into the combustion chamber.

11.7.4.3 Water-Fuel Emulsion (WFE)

The third water technique is WFE, which can achieve NO_x reductions of up to 50 % [112]. By adding water to the fuel, the total required fluid volume to be injected increases, i.e., the same amount of fuel may be used plus additional water, which may require portions of the fuel system, such as fuel pumps and injectors, to be redesigned. The WFE must also retain a high mixing quality to prevent water particles from merging and possibly displacing the fuel, which would result in heterogeneous water-fuel injection. Such heterogeneity may reduce the engine power output. A water addition of approximately 20 % is considered to be the maximum for standard engines due to the volumetric capacity of the injection pumps at full load [115]. However, the upper limit for water addition may be 70 %. In general, a 10 % NO_x reduction can be achieved using a 10 % water addition. However, this method may cause a fuel penalty.

11.7.5 Exhaust Gas Recirculation (EGR)

EGR systems filter and possibly cool a fraction of the exhaust gases before being re-circulated into the engine charge air. EGR systems can be separated into three different categories: (1) external high-pressure EGR with recirculation on the engine side of the turbocharger, (2) external low-pressure EGR with recirculation after the turbocharger, and (3) internal EGR with advanced exhaust and inlet valve techniques that allow for some exhaust gas to return back into the cylinder from the exhaust gas receiver [110]. For four-stroke engines, EGR often involves external manifolds and possible recirculation on the engine side of the turbocharger. However, for two-stroke engines, EGR may be achieved by reducing the scavenging efficiency, which reduces the purity of the gas that is trapped in the cylinders when beginning compression [108].

By adding inert compounds, e.g., water and CO_2, to the intake air, the oxygen concentration is reduced, and the overall heat capacity of the combustion gases increases, which results in reductions in combustion temperature and related NO_x formation. However, a limited amount of EGR can be applied before the oxygen level becomes too low to allow complete combustion, yielding the formation of carbon monoxide, particulates and unburned hydrocarbons [118]. The dissociation[3] of CO_2, i.e., the separation of a CO_2 molecule into one carbon and two oxygen atoms, is an endothermic reaction which leads to a lower combustion temperature and has also been found to slightly reduce NO_x formation. In addition, similar arguments apply to water vapour due to its similar heat capacity and the possibility of dissociation [118].

A barrier to EGR is the difficulty related to removing all particulate matter from the exhaust gases before mixing them with the intake air, which may cause the deposition of particulates inside the engine, the contamination of the lubrication oil and the deterioration and wear of the combustion chamber [60]. Furthermore, EGR in combination with high-sulphur fuels may result in the corrosion of turbochargers, intercoolers and scavenging pipes [108]. The negative influence of EGR on the combustion efficiency can be partially reduced if an engine is designed with an EGR system from the beginning rather than a retrofit arrangement, e.g.,

[3]Dissociation is the separation or splitting of molecules into smaller particles such as atoms, ions or radicals.

significantly increased fuel injection pressure, earlier fuel injection and reduced compression ratio.[4] Thus, such an engine will not be optimised to run without EGR [119]. The reduction in NO_x formation using EGR is related to the recirculation fraction; however, the reduction potential may be assumed to be 30 % in combination with approximately 10–15 % EGR [60, 108, 115].

11.7.6 IMO Tier III Compliance Using Combined NO_x-Abatement Technologies

Among the previously described abatement technologies, only changing to LNG and the installation of SCR have demonstrated the ability to independently achieve Tier III compliance. However, IMO [62] concluded that the use of EGR also has the potential. Therefore, EGR will likely be combined with other NO_x-abatement technologies to achieve Tier III compliance. Furthermore, combinations of different technologies, such as water technologies, variable valve timing and lift and other alternative fuels, may also achieve Tier III compliance [62].

DWI has the potential to reduce NO_x formation by 50 % NO. However, in combination with Miller timing and 2-stage turbocharging, including the use of an intercooler between the compressor units, this combination may reach Tier III compliance. However, DWI may suffer a fuel penalty of approximately 2 % and works primarily at intermediate and high engine loads. For low loads, DWI reduces the temperature in the combustion chamber such that complete combustion may not possible. Additionally, DWI may require the use of marine fuels with a maximum sulphur content of 1 % by weight to avoid excessive corrosion. For retrofit installations, the increased mass flow over the turbine may require a re-matching of the turbine vanes to avoid over-spinning [113].

EGR can potentially reach a 40 % reduction in NO_x formation in combination with a fuel penalty of approximately 2 %, whereas EGR in combination with Miller timing and 2-stage turbocharging could attain Tier III compliance. However, at only 10 % EGR, the smoke emitted from the exhaust funnel may become visible; this effect must be considered [113]. Hence, as discussed above, EGR should be integrated directly into the engine design phase and not be installed as a retrofit solution. The negative impact of EGR on the combustion efficiency could be partially reduced if an engine is designed with an EGR system from the beginning rather than a retrofit arrangement, e.g., significantly increased fuel injection

[4]The compression ratio is reduced if EGR is operated such that the total amount of air plus the recirculated exhaust gases exceed the amount of air at 0 % EGR. This operation is performed to avoid conditions in which the combustion pressure in the cylinder is too high, which may mechanically damage the engine.

pressure, earlier fuel injection and a reduced compression ratio.[5] Thus, such an engine would not be optimised to run without EGR [119].

SCR technology or a fuel switch may currently be the best available options to achieve a substantial decrease in NO_x emissions from marine engines. These options are also used for land-based applications, such as trucks, and have been continuously developed over several decades to meet increasingly stringent emission regulations. However, when developing and adapting them to marine engines, the specific conditions of marine applications, such as fuel quality and exhaust gas temperatures, should be considered.

11.8 Sulphur Oxides

As discussed in Chap. 5, the sulphur in common marine fuels is oxidised to sulphur oxides, primarily SO_2, during combustion. Sulphur oxide emissions lead to the formation of acidic gases and particles that have adverse effects on the environment and human health.

11.8.1 Low-Sulphur Fuels

One strategy to reduce amount of sulphur oxide generated under combustion is to use fuels with lower sulphur content. The fuels under discussion in this section are derivatives of crude oil that are commonly used in the shipping sector and that can fulfil the requirements of stricter emission regulations, namely low-sulphur heavy fuel (LSHFO), marine diesel oil (MDO) and marine gas oil (MGO), which are suitable for ships operating in areas with sulphur limits of 0.1 mass% (SO_x ECA) and 0.5 mass% (global 2020/2025). Furthermore, other fuel alternatives exist that contain only a few ppm sulphur, including LNG and methanol (Sect. 10.4).

Although sulphur limits are becoming increasingly more stringent for ships, a large difference remains between them and the regulations for road vehicles (Fig. 11.6). The sulphur content in marine fuels used in the SO_x ECA from 2015 remains approximately 100 times greater than that in car fuel used in the EU.

As discussed in Chap. 2, crude oil is refined into different end products, such as jet fuel, gasoline, diesel, and HFO. HFO is the fraction that remains after the lighter fractions have been distilled in the refinery; due to its lower price, HFO is also commonly used on ships. This fraction has a relatively high sulphur content due to the sulphur content of crude oil. The global average sulphur content of HFO is 2.7 mass% [120]. LSHFO is generally produced by blending regular HFO with

[5]As discussed previously, the compression ratio is reduced if EGR is operated such that the total amount of air plus the recirculated exhaust gases exceeds the amount of air at 0 % EGR. This operation is performed to avoid conditions in which the combustion pressure in the cylinder is too high, which may mechanically damage the engine.

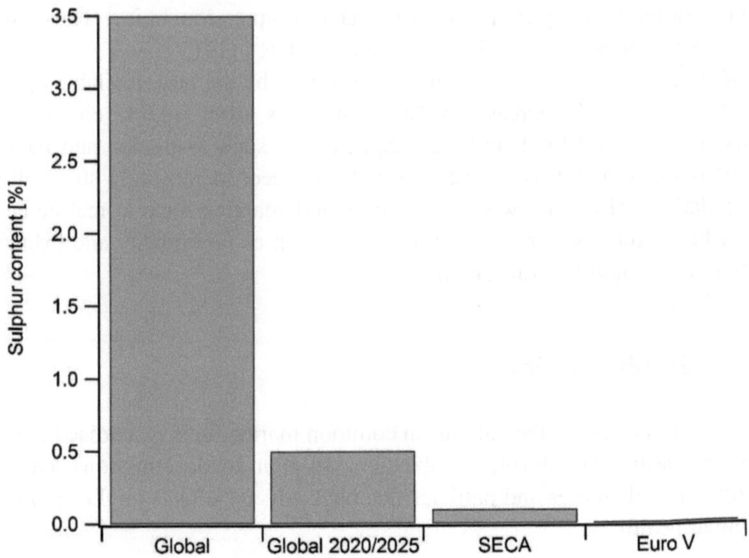

Fig. 11.6 Permitted sulphur content in marine fuels in different years and regions compared with the permitted content in European road diesel

low-sulphur products, including a variety of distillates with less than 1 mass% sulphur [121]. No engine modification is necessary for a ship to change from HFO to LSHFO. In addition to HFO, light fuel oil can also be used on ships, generally as MDO or MGO, which are composed of fractions that are separated in the refinery and contain less sulphur by weight than HFO [122]. MDO is a mixture of kerosene, light gas oil or heavy gas oil and contains up to 1.5 mass% sulphur in the EU [123]. Marine gas oil does not contain any residual components, which allows vessels to comply with the 0.1 mass% sulphur limits for marine fuels in the ECA. The distillates are typically more expensive than common bunker fuel and have worse lubrication properties [123]. As an alternative to distillates, hybrid fuels (also called ECA fuels) have been recently introduced. In general, they are blended products of different refinery streams and contain up to 0.1 mass% sulphur [124]. However, because they are blended products, they do not directly fit into traditional specifications set by ISO 8217:2012 [124], which specifies categories of fuels [125]. The standard categorises fuels according to several properties, including viscosity, flash point and sulphur content.

11.8.2 Scrubbers

An alternative to low-sulphur fuels is to continue using HFO and to clean the exhaust gases, i.e., "end-of-pipe" solutions. Scrubbing involves removing a component from a gas by absorption into liquid droplets. The term is occasionally also used for removal by reaction with a solid, i.e., *dry scrubbing*. Efficient scrubbing requires good contact between the gas and the scrubbing liquid or solid; thus, a large contact surface and long contact time are desired. The space required for the equipment is determined by this contact surface and time.

The technology for the removal of sulphur oxides by scrubbers from exhausts is well developed for land applications. Both dry scrubbers, using lime or other calcium-containing minerals to react with the sulphur, and wet scrubbers, using an alkaline liquid, are used. Although scrubbers have been applied on land for many years, their marine application is relatively young. Space and weight restrictions on equipment are relatively small for land-based applications, but they have become critical issues for shipping, especially for the retrofitting of existing units. Wet and dry scrubbers for ships exist, and, in general, high removal rates of sulphur oxides have been reported, i.e., up to 100 %.

The first prototype seawater scrubber on a vessel was installed in 1991 [126]. Early results have recently been published for retrofits achieving an effective reduction of SO_x of at least 98 %. Some examples that have been reported include an open-loop scrubber installed on a 1 MW auxiliary generator on a ferry in 2005 [127], a closed-loop scrubber installed on 680 kW auxiliary engine on a tanker in 2008 [128], a hybrid scrubber installed after a two-stroke 21 MW main engine on a ro-ro vessel, in 2009 [129], and a dry scrubber installed after a 3.8 MW main engine on a cargo ship in 2009 [127].

11.8.2.1 Wet Scrubbers

A wet scrubber uses a liquid medium that is either seawater or freshwater, in some cases with chemical additives such as sodium hydroxide (NaOH), to remove SO_x. As described elsewhere in this book (Sect. 11.10), wet scrubbers also provide the additional benefit of decreasing the amount of particulate matter from exhaust gas. Wet scrubbers can be divided into three subcategories: open-loop, closed-loop and hybrid systems; the latter is a combination of the first two types. The simplified chemistry behind scrubber processes can be found in Boxes 11.4 and 11.5.

In an open-loop scrubber, seawater is pumped onto a ship, interacts with and cleans the exhaust gases, is filtered and is subsequently discharged back into the sea (Fig. 11.7).

Large quantities of water are run through the system. The wash water flow rate is approximately 45 m^3/MWh [127], meaning that a 20 MW engine would require approximately 21,000 m^3 water per day, if operating at full load 24 h per day. The sulphur scrubbed from the exhaust is transferred to the sea along with the wash water.

This solution is based on the natural alkalinity of seawater, which is applied as an absorbent. Alkalinity is the natural buffering capacity of water and depends primarily on the concentration of bicarbonate $\left[HCO_3^-\right]$ and carbonate $\left[CO_3^{2-}\right]$ ions. The total alkalinity (A_T) of seawater is represented by the total concentrations of the following constituents [130]:

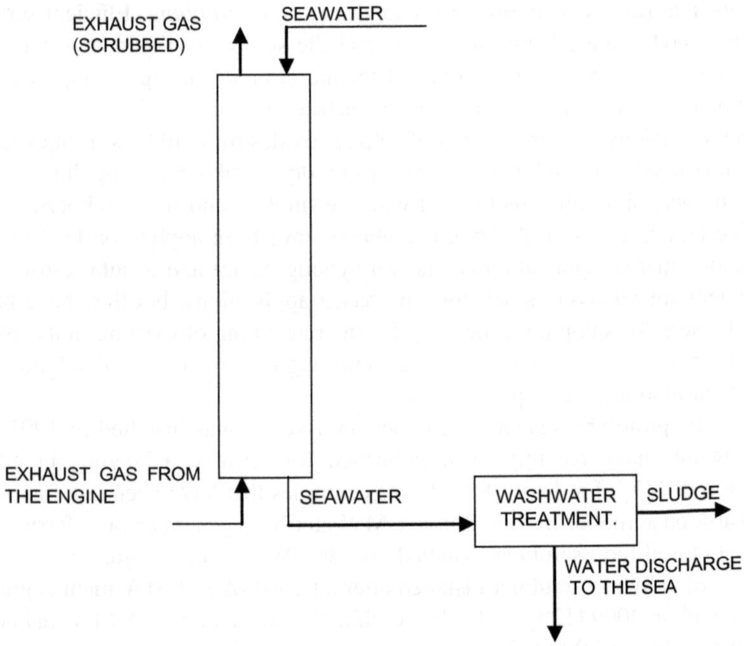

Fig. 11.7 Simplified view of an open-loop scrubber

$$\begin{aligned} A_T = & \left[HCO_3^-\right] + 2\left[CO_3^{2-}\right] + \left[B(OH)_4^-\right] + \left[OH^-\right] + \left[HPO_4^{2-}\right] \\ & + 2\left[PO_4^{3-}\right] + \left[SiO(OH)_3^-\right] + \left[NH_3\right] + \left[HS^-\right] - \left[H^+\right] - \left[HSO_4^-\right] \\ & - \left[HF\right] - \left[H_3PO_4\right] + \left[\text{minor bases} - \text{minor acids}\right] \end{aligned} \quad (11.6)$$

Because open-loop scrubbers rely on the natural alkalinity of seawater, this type of scrubber must be operated in waters with high alkalinity. In seawater with such an alkalinity, an open-loop scrubber has a removal efficiency of approximately 98 % [127]. The alkalinity in certain geographical areas, such as Bothnian Bay in the Baltic Sea or the Great Lakes in North America, can be insufficient to achieve the required SO_x reduction in the exhaust gas. Therefore, other solutions are recommended for these areas.

Box 11.4 Simplified chemistry of the open-loop scrubbing process
SO_x are formed during the combustion process due to the oxidation of naturally occurring sulphur in the fuel. In the exhaust gas, the major part of SO_x is present as SO_2, and a minor portion is SO_3 [131].

Reactions involving SO_2:

Gaseous SO_2 dissolves in seawater and subsequently is ionised, generating bisulphite and sulphite ions [127, 132, 133]:

$$SO_2 + H_2O \leftrightarrow H^+ + HSO_3^- \text{ (bisulphite)} \quad (11.7)$$

$$HSO_3^- \leftrightarrow H^+ + SO_3^{2-} \text{ (sulphite)} \quad (11.8)$$

Sulphite in seawater is readily oxidised to sulphate, in the presence of oxygen:

$$SO_3^{2-} + 0.5O_2 \leftrightarrow SO_4^{2-} \quad (11.9)$$

The generated hydrogen ions, are neutralised by the alkalinity of seawater (mainly due to its bicarbonate content), thereby consuming alkalinity.

Reactions involving SO_3:

Gaseous SO_3 dissolves in water, forming sulphuric acid, which dissociates to sulphate [132]:

$$SO_3 + H_2O \leftrightarrow H_2SO_4 \text{(sulphuric acid)} \quad (11.10)$$

$$H_2SO_4 \leftrightarrow H^+ + HSO_4^- \leftrightarrow 2H^+ + SO_4^{2-} \text{(sulphate)} \quad (11.11)$$

The generated hydrogen ions are neutralised by the alkalinity of seawater (mainly due to its bicarbonate content), thereby consuming alkalinity.

Before being discharged back to the sea, the wash water can be treated in a hydrocyclone, which separates suspended particles from water. Additionally, the wash water can be diluted with ambient seawater to raise the pH [134]. The wash water should meet the criteria for the pH, turbidity, temperature, and PAH and nitrate concentrations stated in the MARPOL 73/78 Annex VI resolution MEPC.259 (68) [135]. Monitoring conducted on discharge water from open-loop scrubbers show high concentrations of several contaminants, including metals such as copper and zinc [132]. Even though IMO has developed wash water discharge criteria with respect to some parameters there are currently no criteria with respect to metals. In addition, the IMO discharge criteria are not based on any scientific environmental risk assessment taking into account what impact the discharge may have on the marine environment on a local, regional or global scale. This is problematic, especially in sensitive coastal areas and in semi-enclosed environments such as harbours where a reduction in pH combined with elevated concentrations of contaminants may pose a risk to marine organisms. Due to this lack of knowledge, several countries have decided to adopt stricter regulations than required by IMO. For example, Belgium has prohibited discharges of wash water in its coastal waters (3 nautical miles off the coast) and California does not permit the use of scrubbers as an alternative to use of low sulphur fuel [136].

Another type of wet scrubber is a closed-loop scrubber (Fig. 11.8). In a closed-loop scrubber, water (freshwater or seawater) treated with sodium hydroxide (NaOH, caustic soda) is applied as the scrubbing medium. The NaOH solution should be handled with care. The dosage rate of the 50 % NaOH aqueous solution is approximately 15 l/MWh of scrubbed engine power (when a 2.7 mass% sulphur fuel is scrubbed to 0.1 %) [127]. In contrast to open-loop systems, the wash water in closed-loop scrubbers is recirculated with a wash water flow rate of approximately 20 m^3/MWh [127], which is less than a half that of the open-loop option [127]. Despite the differences, the exhaust gases are in direct contact with the wash water in both closed- and open-loop scrubbers. In closed-loop scrubbers, SO_x is removed from the exhaust gas as sodium sulphate (Na_2SO_4), sodium bisulphite or sodium sulphite (Box 11.5). During the wash water circulation, the concentration of Na_2SO_4 increases with time. To avoid saturation, small amounts of wash water must be removed, requiring the addition of freshwater. The rate at which this freshwater is added depends on the wash water discharge, losses due to evaporation and losses of liquid in the wash water treatment unit [127]. The consumption of freshwater can be reduced by capturing and reusing some of the water vapour within the exhaust [127]. Depending on the size of the wash water tank, this type of scrubber can operate without any discharge for a certain period of time [127].

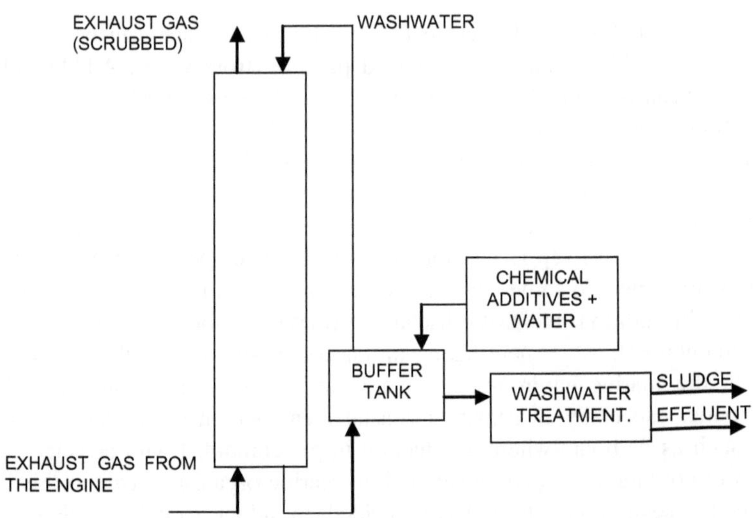

Fig. 11.8 Simplified view of a closed-loop scrubber

Box 11.5 Simplified chemistry of the closed-loop scrubbing process
In the aqueous solution, sodium hydroxide is present as ions [132]:

$$NaOH + H_2O \leftrightarrow Na^+ + OH^- + H_2O \quad (11.12)$$

Reactions regarding SO_2 [127]:

$$Na^+ + OH^- + SO_2 \rightarrow NaHSO_3(\text{aq sodium bisulphite}) \quad (11.13)$$

$$2Na^+ + 2OH^- + SO_2 \rightarrow Na_2SO_3(\text{aq sodium sulphite}) + H_2O \quad (11.14)$$

$$2Na^+ + 2OH^- + SO_2 + 0.5O_2 \rightarrow Na_2SO_4(\text{aq sodium sulphate}) + H_2O \quad (11.15)$$

Reactions regarding SO_3 [127]:

$$SO_3 + H_2O \rightarrow H_2SO_4 \quad (11.16)$$

$$2NaOH + H_2SO_4 \rightarrow Na_2SO_4(\text{aq sodium sulphate}) + 2H_2O \quad (11.17)$$

The surplus of hydrogen ions is neutralised by hydroxide ions (OH^-).

Hybrid scrubbers are more complex than the previously discussed types of scrubbers and can be operated in either open- or closed-loop mode; the mode can be chosen based on the local requirements.

One important technical aspect of wet scrubbers is the corrosivity of the wash water, which requires the construction materials to be chosen with care. Scrubbers add extra weight to ships—the weight of a typical 20 MW wet scrubber varies between 30 and 55 tonnes, excluding the wash water and treatment system [127]. The extra weight comes also from the generated sludge, which must be stored for land discharge because it is regarded as hazardous waste. Approximately 500 kg of sludge can be generated per 100 tonnes of HFO, assuming 70 % efficiency of PM removal [127]. Incineration of sludge and discharging it to the sea are not permitted. Therefore, this waste must be disposed in port reception facilities.

11.8.2.2 Dry Scrubbers

Wet scrubbers currently dominate the marine market, although dry marine scrubbers are also available that do not use wash water; instead, calcium hydroxide ($Ca(OH)_2$, slaked lime) is used as the absorbent in the form of 2- or 8-mm granules [127]. A sufficient mass of granulate must be loaded on a ship and stored. When the scrubberis in use, the granules are transported on a conveyor to the "absorber" chamber to react with the passing exhaust gas [134]. Calcium hydroxide must be handled with care because it is a harmful substance. Sulphur oxides in the exhaust gas react with calcium hydroxide to form gypsum [127], which is disposed of on land (Box 11.6). The consumption of calcium hydroxide granules is approximately 40 kg/MWh, requiring approximately 19 tonnes of granules per day for a 20 MW marine engine running on 2.7 % S fuel to meet the emission standard of 0.1 mass% sulphur fuel [127]. The used granules can be recycled on land for other applications [127].

Box 11.6 Simplified chemistry of the dry scrubbing process

Reactions regarding SO_2 [127]:

In the reaction between SO_2 and $Ca(OH)_2$, calcium sulphide is formed:

$$SO_2 + Ca(OH)_2 \rightarrow CaSO_3(\text{calcium sulphite}) + H_2O \tag{11.18}$$

...which is further oxidised:

$$2CaSO_3 + O_2 \leftrightarrow 2CaSO_4 \tag{11.19}$$

...and hydrated to form calcium sulphate dihydrate (gypsum), a solid product:

$$CaSO_4 + 2H_2O \leftrightarrow CaSO_4 \cdot 2H_2O \tag{11.20}$$

Reactions regarding SO_3:

$$SO_3 + Ca(OH)_2 + H_2O \leftrightarrow CaSO_4 \cdot 2H_2O \tag{11.21}$$

In a dry scrubber, most of the weight is installed on the upper levels of a ship, which can impact ship stability [127], although this is compensated by the relatively low power consumption, approximately 10 % of that of a wet scrubber for a similar engine size [132], and the lack of generated wash water.

Ships have different emissions to air. Although scrubbers remove sulphur oxides, nitrogen oxide emissions are managed by SCR. In some cases, both a SCR system and a scrubber must be installed on the same vessel. A common feature of all wet scrubbers is that they cool the exhaust gas, decreasing the temperature to below 300 °C, which is below the necessary temperature for SCR systems to function effectively and reduce NO_x emissions. Therefore, a wet scrubber should not be installed before a SCR system unless a reheater is placed in between to increase the temperature [127].

11.9 Non-methane Volatile Organic Compounds

As discussed in Sect. 5.6, the main source of non-methane volatile organic compounds (NMVOCs) from shipping is the handling of different types of oils as cargo. Thus, as mentioned in Sect. 5.6.3, to reduce these emissions, tankers should be equipped with a vapour emission control system (VECS). Such systems utilise different technologies to reduce NMVOC emissions to the atmosphere and can reduce the emissions by up to 100 %. These systems are often categorised as active or passive systems [137]. A passive system is designed to reduce the amount of NMVOC vapours formed by modified handling of cargo, which could include increased pressure in pipes and tanks, modified tubing to prevent pressure drop during loading, or the use of inert gas in different ways. Active systems are designed to manage crude oil vapours formed when cargo is handled, which often include compression of vapours followed by collection. Collection methods include condensation by cooling or absorption, often using cooled liquids, e.g., methanol or crude oil. Here, the vapours are passed through a packed column, intersecting a counter flow of the cooled liquid where the vapours are dissolved. The vapour-containing air can also be passed through a bed of adsorbing material, e.g., activated charcoal, which can be regenerated by a vacuum or steam when saturated. The collected vapours can also be burned in an open or closed flare. With a closed flare, the heat produced by the heat recovery process can be used. It is possible to use heat recovery in more complex systems using catalytic oxidation. However, all methods involving combustion generate CO_2 emissions and possibly other air pollutants [138].

11.10 Particles

Particle emissions from combustion processes impact human health, climate and the environment (see Sect. 5.5.2). Despite this knowledge, particle emissions from operating ships are not yet regulated. Consequently, no accepted abatement technologies for particle emissions exist within the shipping sector. For automobiles, diesel particulate filters (DPFs) are commonly used to abate particle emissions to comply with regulations [139]. The different methods to reduce particle emissions from operating ships include using (1) a different fuel, (2) a scrubber, (3) NO_x-abatement technology and (4) DPFs.

A change from low-quality HFO with a high sulphur content to a fuel of higher quality and with a lower sulphur content will reduce the particle emissions based on both number and mass [140] (also discussed in Sect. 5.5.1 and Table 5.7). Changing to a completely different fuel type, such as LNG, will result in considerable reductions in both the number of particles emitted and the emitted particle mass. When LNG is used for ship propulsion, the total number of particles emitted is reduced by 90–95 % compared with MGO (<0.05 mass% sulphur), and the mass is reduced by 99 % [141]. The use of biogenic fuels, e.g., palm oil, animal fat, soybean oil or sunflower oil, may yield a reduced mass of particles emitted compared with HFO, although it may also enhance the formation of particles, increasing the number of particles emitted [111]. A change in fuel, either to high-quality fuels with low sulphur contents or new marine fuels, such as LNG, is a consequence of the implemented regulation of sulphur content in marine fuel oils (MARPOL 73/78 Annex VI Regulation 14) in SO_x ECAs and at the global level. This regulation may also increase the use of scrubbers on operating ships (see Sect. 11.8.2.) for further discussion of scrubbers). Measurements conducted on ships operating with a scrubber have indicated that scrubbers considerably reduce particle emissions in terms of both number and mass; the total number of particles emitted decreased by 92 % with a scrubber, and the solid particles decreased by 48 %. These measurements imply a reduction in particles of all measureable sizes, of which scrubbers primarily remove volatile particles, and that particle emissions mainly consist of solid particles when a scrubber is used [142]. The removal of black carbon (BC) particles in a scrubber varies with the sulphur content of the fuel, where fuels with high sulphur contents, e.g., HFO, produce hygroscopic particles, i.e., particles that can take up water, which may be associated with BC and can cause increased removal of BC compared with fuels possessing lower sulphur contents (50–75 % removal for a high sulphur content and 20–55 % removal for a low sulphur content) [143, 144]. The use of a scrubber may also cause a change in the morphology of the emitted particles [145].

The two NO_x abatement technologies discussed herein regarding the reduction of particle emissions from operating ships are SCR and EGR. These two method/techniques are further described in Sects. 11.7.1 and 11.7.5. The use of EGR in combination with an internal scrubber will reduce particle emissions, although EGR itself will not reduce these emissions. Instead, an increase in particle emissions occurs because the recirculation of exhaust gases generates soot particles [146, 147]. SCR is applied within the maritime sector [146], and measurements on an operating ship with SCR indicated that the mass of particles with diameters less than 2.5 μm increased after the exhaust gases passed through an active SCR with urea injection. This depends on an increase in particles containing sulphate (the H_2SO_4*6.5 H_2O fraction). The increased H_2SO_4*6.5 H_2O fraction was determined to be 68–78 % for HFO and 89–92 % for MDO. The increased H_2SO_4*6.5 H_2O fraction was related to the vanadium catalyst used in the SCR, which is known to encourage the oxidation of sulphur dioxide in exhaust gases to sulphur trioxide, which is subsequently converted to sulphate [148]. This phenomenon has also been observed in particle emissions from heavy-duty engines equipped with SCR using low-sulphur fuels (300 ppm) [149]. The formed sulphate can be deposited in the catalyst, causing deterioration, and low-sulphur fuel (<0.05 mass% sulphur) is needed for the catalyst to function properly [139]. In addition, the emissions of elemental carbon (EC) and organic carbon (OC) decreased after the exhaust gases passed through the SCR system by 17–63 and 77–91 %, respectively [148]. Other on-board measurements demonstrated that the SCR had no effect on either the number of emitted particles or the mass of those particles [150].

DPFs and diesel traps are used for automotive purposes to reduce particle emissions from diesel engines to comply with regulations for land-based vehicles [139]. The particle emissions from automotive diesel engines can be reduced by more than 90 % using DPFs [149]. These filters tolerate high temperatures and can remove particles from exhaust gases. The particles accumulate in the filters. When the filters are full, cleaning is required through the regeneration of the DPF, in which the particles accumulated in the filter are oxidised, i.e., soot particles are oxidised to CO_2. Different methods are typically used to regenerate DPFs, e.g., fuel additives or NO_2 and catalytic oxidation [139]. DPFs are generally constructed with alternating flow channels that are plugged at the ends, which force the flow of exhaust gases through the filter medium (Fig. 11.9). The filtration process captures particles that contain EC, OC and sulphate [149]. The use of DPF on operating ships is limited due to the high sulphur contents of marine fuel oils. The DPFs that exist today cannot be used if the sulphur content of the fuel is greater than 0.05 mass% sulphur (500 ppm S), and catalysed DPFs require even lower sulphur contents of 15–50 ppm to function properly. Developments are on-going to manufacture marine DPFs that can tolerate higher sulphur contents than the DPFs used for automotive applications. Marine DPFs cannot tolerate sulphur contents greater than 0.1 mass% sulphur (1000 ppm), with less than 0.05 mass% sulphur being ideal [144]. Therefore, marine DPFs are currently not available for use on operating ships because the effectiveness of DPFs is considerably reduced when the fuel contains a high amount of fuel impurities, such as the HFO that is commonly used for the propulsion of ships. A high sulphur content causes the filters to be less effective, which means that the DPF will begin to produce particles, especially particles that contain sulphate [144, 146].

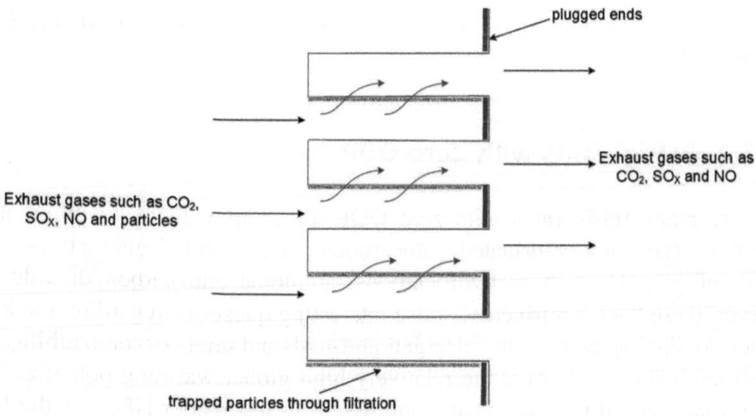

Fig. 11.9 Schematic of a DPF

11.11 Ozone-Depleting Substances (ODS)

Ozone-depleting substances (ODS), such as chlorofluorocarbons (CFCs) and hydrofluorochlorocarbons (HCFCs) have traditionally been used as refrigerants on merchant vessels. The presence of large refrigerant leakages from the refrigeration systems, together with the large ozone-depleting potential (ODP) of these refrigerants, has resulted in a series of international treaties demanding a gradual phase-out of these compounds. This is discussed further in Sect. 5.7.

There are two main options to prevent the emission of ODS [151]:

- Minimise all leakage from the refrigerating plant via technical and operational measures.
- Retrofit or replace the refrigeration plant to use refrigerants with zero ODP.

11.11.1 Minimising Leakage

As discussed in Sect. 5.7, yearly refrigerant leakage rates can be as high as 40 % of the initial refrigerant charge. Technical measures to reduce leaks include the following [61]:

- Using designs that are more resistant to corrosion, vibration and other stresses.
- Reducing the impact of leaks by decreasing the refrigerant charge (i.e., using indirect systems).
- Compartmentalising the piping system to isolate leaks.

Facilities must be available to allow safe and not unreasonably burdensome recovery of refrigerants during maintenance. Operational measures include planned maintenance and monitoring of the consumption of refrigerant to prevent and detect leaks [61].

11.11.2 Refrigerants with Zero ODP

There are many refrigerants with zero ODP. In addition to many different HFC refrigerants (i.e., non-chlorinated halocarbons), the potential alternatives include natural refrigerants, such as hydrocarbons, ammonia and carbon dioxide [152]. However, these "new" refrigerants raise interesting questions regarding the balance between conflicting environmental targets and goals and safety or compatibility [153].

International concern over the relatively high global warming potential (GWP) of HFCs has caused the European Commission to reconsider HFCs as the leading replacements [154]. For example, HFC-410A, which is one of the primary replacement refrigerants in new plants, increases the GWP by 16 % while lowering the attainable efficiency by 6 %, thereby increasing energy-related greenhouse gas emissions [153].

Using ammonia as a refrigerant requires certain design considerations, e.g., compatible materials, and the presence of additional safety equipment on ships due to its toxicity and flammability. Hydrocarbons, such as propane and butane, are extremely flammable, which is a serious obstacle to their widespread use as refrigerants. Carbon dioxide is one of the few natural refrigerants that is neither flammable nor toxic. However, its low critical temperature leads to disadvantages at very high operating pressures, i.e., reduced energy efficiency (especially at high ambient temperatures) and increased installation costs [155].

11.12 Measures to Reduce Noise

Concerns exist regarding the exposure of organisms to elevated levels of anthropogenic noise (see Chap. 6). Underwater noise emitted by ships can interfere with the detection of signals by marine organisms, which can hinder communication, mating calls, and searches for prey. In addition, noise generated in port areas impacts the surrounding environment.

The prevention of noise from active sonar, underwater explosions and airguns is under debate. Reducing noise from the source could include limiting its use; instead, finding viable alternatives and making technological improvements should be the aim. For example, alternatives and improvements to airgun technologies are in development. Improvements in the signal processing of the sound waves used to identify pockets of oil and gas in bedrock has reduced the need for low-frequency sounds. A new alternative to airguns is the marine vibrator, which uses less energy and produces noise that is lower in amplitude and distributed across a longer period of time [156]. In addition, the length of time in which the noise source is used can be limited, and its use can be restricted when sensitive organisms are in the area or during different sensitive seasons, e.g., spawning or mating.

Methods to reduce noise from ships have been studied for decades. In transit operating conditions, propeller cavitation is the main source of radiated noise, but, in low speed operations, machinery noise also becomes important. New and improved designs for propulsion systems have been further developed to reduce cavitation nuisance and meet expected regulations. Dampening the noise from engines using isolation mounts, dampening tiles and flexible hoses is also a possibility [156]. There are on-going international research projects in Europe to investigate and mitigate the effects of the underwater noise generated by ships, and the planned outcome is to disseminate guidelines for the regulation of underwater noise from commercial shipping [157, 158].

The units within a port that produce the most noise should be placed as far as possible from locations that require minimal noise levels [159]. Other solutions can also include placing them behind other port components or dampening building structures. These dampening structures (e.g., office buildings, terminals and storage areas) can be built to isolate adjacent neighbourhoods from noise sources. In addition, tunnels for railways and road infrastructure could be included in the port area to reduce noise production [159]. Further methods to reduce noise pollution include ensuring that equipment and other property in the port are well maintained and used in proper ways, such as roads or rail switches. Restrictions on engine operations and idling times can vary depending on the time of day, e.g., restrictions on the use of the noisiest machinery at night or replacing warning sounds with strobe lights [159]. Furthermore, the human factor should not be forgotten. Training of port employees and ship crews to reduce noise levels in the work place is essential. Cooperative actions of both port authorities and calling ships are very important in the process of reducing noise in port areas [159].

Docked ships should be able to plug into land-based electric utilities (shore power) instead of running auxiliary engines on fossil fuels, e.g., shore connection (cold ironing) [160], which reduces the negative health and environmental impacts from noise and vibrations.

11.13 Reducing the Impacts from Shipping Infrastructure

As described in Chap. 7, a broad infrastructure is connected to the activities of the shipping industry, e.g., transportation, building and recycling. This infrastructure services the industry, e.g., ports and the connected transport network of fairways and canals. This infrastructure must be created/built, operated and maintained, which creates environmental issues.

11.13.1 Measures Related to Ports

A port can reduce its emissions in several ways. Ports are often located in areas with windy conditions, which encourages investment in wind farms to reduce their CO_2 emissions. Solar power can also be used to complement electricity production based on fossil fuels.

Furthermore, engines in diesel trucks and tractors used by ports can be replaced or retrofitted with particle filters and NO_x control systems. The permitted sulphur content in the fuel can be decreased, e.g., current regulations are 15 ppm in the US and 10 ppm in the EU. The permitted idling time can be reduced. Electrification of port cranes, trucks and rail locomotives can be introduced. The availability of connections to the electricity grid for reefer trucks and containers can be introduced to avoid the use of diesel-powered refrigeration. Tugs operating in port areas should upgrade their engines to meet the standards for IMO Tier III NO_x reduction regulations. Ports are often already prepared for an oil spill accident, which allows for rapid containment and removal. Because the ballast water convention has not yet been ratified, ports can only urge vessels entering their waters to exchange ballast water at sea to minimise the risk of invasive species.

Ports can develop systems to provide ships at berth with shore-side electricity (cold ironing), allowing their auxiliary engines to be turned off. These engines are normally running while at berth to produce electricity. This action can reduce both air emissions and low-frequency noise and transfers the demanded power load of ships to the shore-side power supply without disrupting on-board services. Cold ironing provides electricity to emergency equipment, refrigeration, heating, lighting and other equipment that is used to load or unload cargo and reduces the negative health and environmental impacts that are generated by SO_x, NO_x, CO_2 and particle discharges, noise and vibrations. Countries with energy production based on nuclear and renewables will attain a net decrease in (air) emissions, whereas cold ironing in countries with energy production based on coal might attain a net increase in emissions using this approach.

The introduction of shore-side electricity is not without difficulties because a lack of electricity compatibility can occur, for example 220 or 110 V at 50 or 60 Hz. Connectors and cables are not standardised and can differ between ports. Ports must adapt the outgoing power to the demands of arriving ships, which could result in costly investments to ports. Ship owners must also retrofit their ships to accommodate shore-side power. Various organisations are working on developing and implementing standards for shore-side electricity. Currently, more than 20 ports offer shore-side electricity at one or several berths, including major ports such as the Port of Los Angeles and Rotterdam.

Strategies can be implemented to reduce the emissions from ships entering or leaving port areas. As of 2010, ships calling European ports or ships travelling inland are not permitted to use fuel with greater than 0.1 mass% sulphur content. The installation of NO_x-reducing technologies, especially for the auxiliary engines that are most frequently used when at berth, is another approach to reduce emissions.

For the transport networks serving ports, the use of railways rather than trucks to transport goods to and from the port areas can decrease both the emissions of air pollutants and congestion on the roads in major cities, especially when the locomotives are powered by electricity [161]. If they are not, regulations can be placed on locomotives operating in port areas, requiring that they use a fuel with a low sulphur content. The locomotives can furthermore be upgraded with new engines, and control systems for NO_x production can be introduced. In addition to decreasing the permitted idling times, regulations on non-electric locomotives can also be promoted for trucks.

11.13.2 Measures Related to Canals and Fairways

The mitigation measures to reduce the effects from shipping on canals and fairways can be divided into engineering and navigation measures. Engineered bank protections and the construction of side channels or pools can mitigate the negative effects on canal banks and beaches. Navigation rules can also mitigate these negative effects by limiting vessel speed, size, installed power and the proximity to the canal banks in protected areas [162].

11.13.3 Measures Related to Dredging

Many methods exist to reduce the environmental impacts of dredging activities, including encapsulated buckets or closed clamshells in dredging machinery, sub-suction that extracts sediment from deeper layers of the seafloor without disturbing the surface layer, and silt curtains to reduce the turbidity caused by dredging [163].

11.13.4 Measures Related to Shipbuilding and Scrapping

The shipbuilding industry must update its environmental agenda by introducing stricter environmental and technological requirements. Adopting requirements enforced in developed countries for newly ordered ships and ships to be scrapped may hasten the adoption of new technologies. Foremost, beaching should not be permitted; instead, ship recycling should be performed in an isolated and protected area. Oil, including sludge, inside ships should be cleaned or pumped before initiating the scrapping process. Barriers and booms to contain oil and other liquids when spilled should be sufficiently available. Reception facilities for hazardous waste material should be established, and fire and paramedic stations should be constructed near scrapping areas. Personal protective equipment should be provided to workers by both owner and contractors [164].

11.14 Shipwreck Remediation

There are a large number of shipwrecks around the world that are in need of remediation today and in the future; otherwise, large amounts of petroleum will be released into the marine environment. There are several remediation techniques available. After oil in a shipwreck has been identified and located, the suitable remediation technique needs to be selected. Locating the oil in a ship wreck is not always an easy task and can often be challenging because a wreck can be in various positions on the sea floor. Therefore, the oil might have travelled from the original tanks to other locations in the ship. The choice of remediation technique is mainly driven by, for example, the type of oil or the condition and location of the wreck. Four options for the oil remediation of ship wrecks exist: raising the whole ship to the surface, pumping the oil to the surface, biological degradation and using a tent construction.

For a wreck in moderately good condition in shallow waters, an alternative can be to raise the whole ship to the surface and salvage it. However, the right conditions for this alternative are rare, and other options are more commonly used.

Most commonly, various methods to pump the oil to the surface are used. These are controlled methods and can ensure that an oil tank is properly emptied. Oils with a low viscosity can be directly pumped to the surface using a vacuum pump and long hose. Centrifugal pumps are another alterative, but there is a risk of the oil becoming emulsified. In both cases, a flange is connected to the hull, and the oil is pumped to the surface. Problems can occur if the oil is of high viscosity or if there are impurities in the oil, which can clog the hose.

If the oil to be removed from the shipwreck is of high viscosity, pumping techniques are more difficult to employ. In this case, screw pumps are often needed. Another alternative is to reduce the viscosity of the oil by heating it. The existing piping in the ship that was used to heat the oil when in operation, if the ship was propelled using HFO, can then be used. If that possibility does not exist, steam can be directly inserted into the tank, with the negative effect of introducing condensed water in the tank. A third alternative is to mix the highly viscous oil with oil of low viscosity; however, this requires the ability to mix the contents of the oil tank.

Microbial degradation of the oil exploits the natural degradation of oil by microorganisms, e.g., bacteria and fungi. This remediation option is currently mostly in the developmental stage. Microorganisms capable of and specialised in oil degradation are added to the oil-containing tanks in the shipwreck, and, over time, the oil will be degraded to CO_2 and water. Limiting nutrients can also be added to enhance the degradation rate. Low water temperatures can be a limiting factor in using this option as a remediation alternative because the degradation rate can be significantly reduced [165].

A third alternative is to use a tent construction that covers the top of the ship and will catch the oil that escapes from the wreck. This alterative can be used when the condition of the wreck is poor, which prohibits drilling into the tanks or the attachment of flanges to the hull without a major risk of hull ruptures. When the tent is in place and collecting the escaping oil, additional holes in the hull can be created to enhance the leakage rate. Collected oil can then be pumped to vessels on the surface for collection and further treatment. The usage of a tent construction requires waters with low currents and a small tide. An example of a failed attempt using this remediation alternative is on the HMS *Royal Oak* in Scotland, where strong currents caused the remediation alternative to fail [166, 167].

11.15 Actions to Implement a Marine Spatial Plan

When a marine spatial plan has been prepared and approved, the required actions must be implemented. The implementation of a marine spatial plan can be performed through several actions whose usage must be specific to individual sites. In general, four management actions can be taken. Furthermore, in a particular area, one management action may be applied to meet the requirements of the marine spatial plan, whereas in other areas, several or all of these actions may be necessary.

- Input actions specify the input of human activities, e.g., limits on the maximum allowed horsepower of an engine or ship size, fishing capacity and/or activity, or the number of cruise ships permitted in a marine area.

- Process actions specify the nature of human activities, e.g., limits on the type of fishing gear or mesh size, requirements for minimum waste treatment technology, and specifications for the "best available technology" or "best environmental practice".
- Output actions specify the permitted output of human activities, e.g., limits on discharging pollutants to the marine environment, ballast water discharges, and sand or gravel excavation.
- Spatial and temporal actions specify where and when certain human activities are permitted, e.g., areas closed to fishing or energy development, requirements for low-sulphur bunker fuels in an area, the designation of marine protected areas, or the designation of areas of certain use, e.g., wind farms, waste disposal or sand and gravel excavation (see Box 11.7) [168].

Box 11.7 Altering shipping lanes: a spatial and temporal action
Over time, ships entering Boston Harbour (USA) have faced problems with whale collisions because the shipping lane in and out of the port takes them through an area with high concentrations of humpback and right whales. Data collected over 25 years have revealed that the shipping lanes are directly adjacent to an area where relatively few whales have been spotted. Subsequently, the shipping lane was moved 12° north, increasing ship travel by 10–22 min; however, the risk of hitting whales decreased by 58 to 81 % [168].

References

1. Etkin, D. S. *Estimating cleanup costs for oil spills.* in *International Oil Spill Conference.* 1999.
2. Bayat, A., Aghamiri, S. F., Moheb, A. & Vakili-Nezhaad, G. R., *Oil spill cleanup from sea water by sorbent materials.* Chem. Eng. Technol., 2005. **28**(12): p. 1525-1528.
3. Fingas, M., *Oil Spill Science and Technology, prevention, response and cleanup*. 2011, The Boulevard, Langford Lane, Kidlington, Oxford, OX5 1 GB, UK: Elsevier.
4. Chu, Y. & Pan, Q. M., *Three-Dimensionally Macroporous Fe/C Nanocomposites As Highly Selective Oil-Absorption Materials.* ACS Appl. Mater. Interfaces, 2012. **4**(5): p. 2420-2425.
5. Lin, J. *et al.*, *Nanoporous polystyrene fibers for oil spill cleanup.* Marine Pollution Bulletin, 2012. **64**: p. 347-352.
6. Buist, I., Potter, S., Nedwed, T. & Mullin, J., *Herding surfactants to contract and thicken oil spills in pack ice for in situ burning.* Cold Reg. Sci. Tech., 2011. **67**(1-2): p. 3-23.
7. Whitfield, J., *How to clean a beach.* Nature, 2003. **422**(6931): p. 464-466.
8. Juhasz, A. L. & Naidu, R., *Bioremediation of high molecular weight polycyclic aromatic hydrocarbons: a review of the microbial degradation of benzo a pyrene.* Int. Biodeterior. Biodegrad., 2000. **45**(1-2): p. 57-88.
9. Atlas, R. M. & Hazen, T. C., *Oil Biodegradation and Bioremediation: A Tale of the Two Worst Spills in US History.* Environ. Sci. Technol., 2011. **45**(16): p. 6709-6715.
10. IMO. *MARPOL - How to do it, Manual of the Practical Implication of Ratifying and Implementing MARPOL 73/78.* 1993: London.

11. Karakulski, K., Marawski, W. A. & Grzechulska, J., *Purification of bilge water by hybrid ultrafiltration and photocatalytic processes.* Separation and Purification Technology, 1998 (14): p. 163-173.
12. Ghidossi, R., Veyret, D., Scotto, J. L., Jalabert, T. & Moulin, P., *Ferry oil wastewater treatment.* Separation and Purification Technology, 2009(64): p. 296-303.
13. Sun, C., Leiknes, T., Weitzenböck, J. & Thorstensen, B., *The effect of bilge water on a Biofilm-MBR process in an integrated shipboard wastewater treatment system.* Desalination, 2009(236): p. 56-64.
14. Sun, Y. J., Guo, S. L., Walker, G. C., Kavanagh, C. J. & Swain, G. W., *Surface elastic modulus of barnacle adhesive and release characteristics from silicone surfaces.* Biofouling, 2004. **20**(6): p. 279-289.
15. Rath, S. K. *et al.*, *Two component silicone modified epoxy release coatings: Effect of modulus, surface energy and surface restructuring on pseudobarnacle and macrofouling behavior.* Applied Surface Science, 2010. **256**: p. 2440-2446.
16. Lindgren, J. F., Haeffner, M., Ericsson, C.T. & Jonsson, P. R., *Oxygen-depleated surfaces: a new antifouling technology.* Biofouling, 2009. **25**(5): p. 455-461.
17. Chambers, L. D., Stokes, K. R., Walsh, F. C. & Wood, R. J. K., *Modern approaches to marine biofouling coatings.* Surface & Coatings Technology, 2006. **201**: p. 3642-3652.
18. Yebra, D. M., Kiil, S. & Dam-Johansen, K., *Antifouling technology - past, present and future steps towards efficient and environmentally friendly antifouling coatings.* Progress In Organic Coatings, 2004. **50**(2): p. 75-104.
19. Beigbeder, A. *et al.*, *Preparation and characterisation of silicone-based coatings filled with carbon nanotubes and natural sepiolite and their application as marine fouling-release coatings.* Biofouling, 2008. **24**(4): p. 291-302.
20. Sjogren, M. *et al.*, *Antifouling activity of Brominated Cyclopeptides from the Marine Sponge Geodia barretti.* J. Nat. Prod., 2004. **67**: p. 368-372.
21. Nylund, G. M. *et al.*, *Seaweed defence against bacteria: a poly-brominated 2-heptanone from the red alga Bonnemaisonia hamifera inhibits bacterial colonisation.* Mar. Ecol.-Prog. Ser., 2008. **369**: p. 39-50.
22. Aldred, N. & Clare, A. S., *The adhesive strategies of cyprids and development of barnacle-resistant marine coatings.* Biofouling, 2008. **24**(5): p. 351-363.
23. Rosenhahn, A., Ederth, T. & Pettitt, M. E., *Advanced nanostructures for the control of biofouling: The FP6 EU Integrated Project AMBIO.* Biointerphases, 2008. **3**(1).
24. Dahlström, M., Mårtensson, L. G. E., Jonsson, P. R., Arnebrant, T. & Elwing, H., *Surface active adrenoceptor compounds prevent the settlement of cyprid larvae of Balanus improvisus.* Biofouling, 2000. **16**: p. 191-203.
25. MarinePaint. *Marine Paint Final Report.* 2012, Dept. of Biological and Environmental Sciences, University of Gothenburg.
26. Pinori, E. *et al.*, *Multi-seasonal barnacle (Balanus improvisus) protection achieved by trace amounts of a macrocyclic lactone (ivermectin) included in rosin-based coatings.* Biofouling, 2011. **27**(9): p. 941-953.
27. Armstrong G., *Ballast system design for flow-through exchange of ballast water.* Trans Inst Mar Eng, 1997. **109**: p. 257-269.
28. Tamburri, M. N., Wasson K, Matsuda M.,, *Ballast water deoxygenation can prevent aquatic introductions while reducing ship corrosion.* Biological Conservation 2002. **103**: p. 331-341.
29. Quilez-Badia, G. *et al.*, *On board short-time high temperature heat treatment of ballast water: A field trial under operational conditions.* Marine Pollution Bulletin, 2008. **56**(1): p. 127-135.
30. Delacroix, S., Vogelsang, C., Tobiesen, A. & Liltved, H., *Disinfection by-products and ecotoxicity of ballast water after oxidative treatment – Results and experiences from seven years of full-scale testing of ballast water management systems.* Marine Pollution Bulletin, 2013. **73**(1): p. 24-36.

31. Werschkun b, S. Y. a. b. S., *Disinfection by-products in ballast water treatment: an evaluation of regulatory data.* Water Res, 2012. **15**(46): p. 4884-4901.
32. King, D. M., Hagan, P. T., Riggio, M. & Wright, D. A., *Preview of global ballast water treatment markets.* Journal of Marine Engineering and Technology, 2012. **11**(1): p. 3-15.
33. HELCOM, *HELCOM recommendation 11/10. Guidelines for capacity calculation of sewage systems on board passenger ships.* 1990, Erweko Oy: Finland.
34. IMO, *Resolution MEPC.200(62), Amendments To The Annex Of The Protocol Of 1978 Relating To The International Convention For The Prevention Of Pollution From Ships, 1973, Special Area Provisions and the Designation of the Baltic Sea as a Special Area under MARPOL Annex IV, Adopted on 15 July 2011.* 2011, International Maritime Organisation, Marine Environment Protection Commitee.
35. HELCOM. *HELCOM interim guidance on technical and operational aspects of delivery of sewage by passenger ships to port reception facilities* 2013.
36. Hänninen S. & Sassi J. *Estimated nutrient load from waste waters originating from ships in the Baltic Sea area - Updated 2009.* 2009, VTT.
37. Guilbaud, J., Massé, A., Andrès, Y., Combe, F. & Jaouen, P., *Laundry water recycling in ship by direct nanofiltration with tubular membranes.* Resources, Conservation and Recycling, 2010. **55**(2): p. 148-154.
38. Markle, S. P., *Design and Prototype Development of Advanced Oxidation Black and Gray Water Treatment Systems.* Naval Engineers Journal, 2006. **118**(3): p. 51-64.
39. Kretschmann L., K. S., Westhof L., Kaiser A., Joswig M.,. *Wastewater Treatment on Cruise Liners: Current Situation and Solutions to Overcome Future Challenges.* in *Green ship technology conference 2015* 2015. Copenhagen.
40. Negulescu, M., *2 The Composition of Municipal Waste and of Stream Waters*, in *Developments in Water Science*, Negulescu Mircea, ed. 1985, Elsevier. p. 19-41.
41. Benson, J., Caplan, I. & Jacobs, R., *Blackwater and Graywater on U.S. Navy Ships: Technical Challenges and Solutions.* Naval Engineers Journal, 1999. **111**(3): p. 293-306.
42. U.S. EPA. *Cruise Ship Discharge Assesment Report.* 2008, U.S. Environmental Protection Agency: Washington, DC, USA.
43. Avellaneda, P. M. *et al.*, *Relative risk assessment of cruise ships biosolids disposal alternatives.* Marine Pollution Bulletin, 2011. **62**(10): p. 2157-2169.
44. Hedberg, A. & Lengquist, J. *Hantering av sewage inom handelssjöfarten. En fördjupning i dess problematik.* B.Sc. thesis, Chalmers University of Technology 2013.
45. HATENBOERWATER. *Sewage treatment plants. DGG number:4500171295.* 2012: Rotterdam.
46. IMO, *Revised MARPOL Annex V, Regulations for the prevention of pollution by garbage from ships*, IMO, 2011.
47. Butt, N., *The impact of cruise ship generated waste on home ports and ports of call: A study of Southampton.* Marine Policy, 2007. **31**(5): p. 591-598.
48. Polglaze, J., *Can we always ignore ship-generated food waste?* Marine Pollution Bulletin, 2003. **46**(1): p. 33-38.
49. ISO, *Ships and marine technology - Marine environment protection- Management and handling of shipboard garbage. ISO 21070:2011* 2011, International Organization for Standardization: Geneva.
50. ISO, *Ships and marine technology. Marine environment protection. Arrangement and management of port waste reception facilities. ISO 16304:2013.* 2013, International Organization for Standardization: Geneva.
51. Committee on Shipboard Pollution Control, N. S. B., *Shipboard Pollution Control:U.S. Navy Compliance with MARPOL Annex V.* 1996, Washington, D.C.: The National Academies Press.
52. IMO, *MARPOL Annex VI (including amendments). Regulations for the Prevention of Air Pollution from Ships*, in *MARPOL. Consolidated Edition 2006.* 2006, International Maritime Organization London.

53. Wey, M.-Y. *et al.*, *Pollutants in incineration flue gas.* Journal of Hazardous Materials, 2001. **82**(3): p. 247-262.
54. Kaldas, A. *et al.*, *Plasma Arc Waste Destruction System (PAWDS) A Novel Approach to Waste Elimination Aboard Ships.* Naval Engineers Journal, 2006. **118**(3): p. 139-150.
55. Seo, Y., You, H., Lee, S., Huh, C. & Chang, D., *Evaluation of CO2 liquefaction processes for ship-based carbon capture and storage (CCS) in terms of life cycle cost (LCC) considering availability.* International Journal of Greenhouse Gas Control, 2015. **35** (0): p. 1-12.
56. Abu-Khader, M. M., *Recent Progress in CO2 Capture/Sequestration: A Review.* Energy Sources, Part A: Recovery, Utilization, and Environmental Effects, 2006. **28**(14): p. 1261-1279.
57. Zhou, P. & Wang, H., *Carbon capture and storage—Solidification and storage of carbon dioxide captured on ships.* Ocean Engineering, 2014. **91**(0): p. 172-180.
58. Järvi, A. *Methane slip reduction in Wärtsilä lean burn gas engines.* in *26th CIMAC World Congress on Combustion Engines.* 2010. Bergen.
59. CIMAC. *CIMAC Position Paper WG 17 – Methane and Formaldehyde Emissions, 2014-04.* 2014, International Council of Combustion Engines Frankfurt, Germany.
60. de Jonge, E., Hugi, C. & Cooper, D. *European Commission Directorate General Environment, Service Contract on Ship Emissions: Assignment, Abatement and Market-based Instruments, Task 2b NOx Abatement.* 2005, Entec UK Limited: Nortwich, England.
61. Buhaug, Ø. *et al. Second IMO GHG Study 2009.* 2009, International Maritime Organization (IMO): London, UK.
62. IMO. *MEPC 65/4/7, Air Pollution and Energy Efficiancy, Final report of the Correspondence Group on Assessment of Technological Developments to Implement the Tier III NOx Emission Standards under MARPOL Annex VI, Submitted by the United States.* 2013, International Maritime Organization, Marine Environment Protection Committee: London.
63. Magnusson, M. *NO_X Abatement Technique for Marine Diesel Engines - Improved Marine SCR Systems* Doctor of Philosophy thesis, Chalmers University of Technology, 2014.
64. Lundström, A. *Urea Decomposition for Urea-SCR Applications* Doctor of Philosophy thesis, Chalmers University of Technology, 2010.
65. Tuenter, G., Vanleeuwen, W. F. & Snepvangers, L. J. M., *Kinetics and mechanism of the NOx reduction with NH3 on V2O5-WO3-TIO2 catalyst.* Industrial & Engineering Chemistry Product Research and Development, 1986. **25**(4): p. 633-636.
66. Koebel, M., Elsener, M. & Madia, G., *Reaction pathways in the selective catalytic reduction process with NO and NO2 at low temperatures.* Ind. Eng. Chem. Res., 2001. **40**(1): p. 52-59.
67. Koebel, M., Madia, G. & Elsener, M., *Selective catalytic reduction of NO and NO2 at low temperatures.* Catal. Today, 2002. **73**(3-4): p. 239-247.
68. Koebel, M., Madia, G., Raimondi, F. & Wokaun, A., *Enhanced reoxidation of vanadia by NO2 in the fast SCR reaction.* J. Catal., 2002. **209**(1): p. 159-165.
69. Madia, G., Koebel, M., Elsener, M. & Wokaun, A., *Side reactions in the selective catalytic reduction of NOx with various NO2 fractions.* Ind. Eng. Chem. Res., 2002. **41**(16): p. 4008-4015.
70. Koebel, M., Elsener, M. & Kleemann, M., *Urea-SCR: a promising technique to reduce NOx emissions from automotive diesel engines.* Catal. Today, 2000. **59**(3-4): p. 335.
71. Brandenberger, S., Kroecher, O., Tissler, A. & Althoff, R., *The State of the Art in Selective Catalytic Reduction of NOx by Ammonia Using Metal-Exchanged Zeolite Catalysts.* Catalysis Reviews-Science and Engineering, 2008. **50**(4): p. 492-531.
72. Cavataio, G. *et al.*, *Laboratory Testing of Urea-SCR Formulations to Meet Tier 2 Bin 5 Emissions.* 2007: p. SAE 2007-2001-1575.

73. Cavataio, G. *et al.*, *Impact and Prevention of Ultra-Low Contamination of Platinum Group Metals on SCR Catalysts Due to DOC Design.* SAE International Journal of Fuels and Lubricants, 2009. **2**(1): p. SAE 2009-2001-0627.
74. Moden, B., Donohue, J. M., Cormier, W. E. & Li, H.-X., *The Uses and Challenges of Zeolites in Automotive Applications.* Topics in Catalysis, 2010. **53**(19-20): p. 1367-1373.
75. Kieger, S., Delahay, G., Coq, B. & Neveu, B., *Selective catalytic reduction of nitric oxide by ammonia over Cu-FAU catalysts in oxygen-rich atmosphere.* J. Catal., 1999. **183**(2): p. 267-280.
76. Sjovall, H., Olsson, L., Fridell, E. & Blint, R. J., *Selective catalytic reduction of NOx with NH3 over Cu-ZSM-5 - The effect of changing the gas composition.* Applied Catalysis B: Environmental, 2006. **64**(3-4): p. 180-188.
77. Grossale, A., Nova, I. & Tronconi, E., *Study of a Fe-zeolite-based system as NH3-SCR catalyst for diesel exhaust aftertreatment.* Catal. Today, 2008. **136**(1-2): p. 18-27.
78. Colombo, M., Nova, I. & Tronconi, E., *A comparative study of the NH3-SCR reactions over a Cu-zeolite and a Fe-zeolite catalyst.* Catal. Today, 2010. **151**(3-4): p. 223-230.
79. Kamasamudram, K., Currier, N. W., Chen, X. & Yezerets, A., *Overview of the practically important behaviors of zeolite-based urea-SCR catalysts, using compact experimental protocol.* Catal. Today, 2010. **151**(3-4): p. 212-222.
80. Park, J. H. *et al.*, *Hydrothermal stability of CuZSM5 catalyst in reducing NO by NH3 for the urea selective catalytic reduction process.* J. Catal., 2006. **240**(1): p. 47-57.
81. Shwan, S. *et al.*, *Hydrothermal Stability of Fe-BEA as an NH3-SCR Catalyst.* Ind. Eng. Chem. Res., 2012. **51**(39): p. 12762-12772.
82. Wilken, N. *et al.*, *Mechanistic investigation of hydrothermal aging of Cu-Beta for ammonia SCR.* Applied Catalysis B-Environmental, 2012. **111**: p. 58-66.
83. Li, J. *et al.*, *Characterization of Active Species in Cu-Beta Zeolite by Temperature-Programmed Reduction Mass Spectrometry (TPR-MS).* Topics in Catalysis, 2013. **56**(1-8): p. 201-204.
84. Wilken, N. *et al.*, *Investigation of the Effect of Accelerated Hydrothermal Aging on the Cu Sites in a Cu-BEA Catalyst for NH3-SCR Applications.* Topics in Catalysis, 2013. **56**(1-8): p. 317-322.
85. Cheng, Y., Montreuil, C., Cavataio, G. & Lambert, C., *Sulfur Tolerance and DEeSOx Studies on Diesel SCR Catalyst.* SAE International Journal of Fuels and Lubricants, 2008. **1**(1): p. SAE 2008-2001-1023.
86. Theis, J. R., *The Poisoning and Desulfation Characteristics of Iron and Copper SCR Catalysts.* SAE International Journal of Fuels and Lubricants, 2009. **2**(1): p. SAE 2009-2001-0900.
87. Cheng, Y., Montreuil, C., Cavataio, G. & Lambert, C., *The Effects of SO2 and SO3 Poisoning on Cu/Zeolite SCR Catalyst.* 2009: p. SAE 2009-2001-0898.
88. Girard, J., Cavataio, G., Snow, R. & Lambert, C., *Combined Fe-Cu SCR Systems with Optimized Ammonia to NOx Ratio for Diesel NOx Control.* SAE International Journal of Fuels and Lubricants, 2008. **1**(1): p. SAE 2008-2001-1185.
89. Metkar, P. S., Harold, M. P. & Balakotaiah, V., *Selective catalytic reduction of NOx on combined Fe- and Cu-zeolite monolithic catalysts: Sequential and dual layer configurations.* Applied Catalysis B-Environmental, 2012. **111**: p. 67-80.
90. Heck, R. M. & Farrauto, R. J., *Catalytic Air Pollution Control - Commercial Technology.* 1995, New York, USA: Van Nostrand Reinhold.
91. Schmieg, S. J. & Lee, J.-H., *Evaluation of Supplier Catalyst Formulations for the Selective Catalytic Reduction of NOx With Ammonia.* 2005: p. SAE 2005-2001-3881.
92. Parvulescu, V. I., Grange, P. & Delmon, B., *Catalytic removal of NO.* Catal. Today, 1998. **46** (4): p. 233.
93. Bartholomew, C. H., *Mechanisms of catalyst deactivation.* Applied Catalysis A: General, 2001. **212**(1-2): p. 17-60.

94. Ham, S.-W., Soh, B.-W. & Nam, I.-S., *Sulfur Poisoning and Tolerance of SCR Catalyst to Remove NO by NH3.* Journal of the Korean Industrial and Engineering Chemistry, 2004. **15** (4): p. 373-385.
95. Chen, J. P. & Yang, R. T., *Role of WO3 in mixed V2O5-WO3/TIO2 catalysts for selective catalytic reduction of nitric-oxide with ammonia.* Appl. Catal. A-Gen., 1992. **80**(1): p. 135-148.
96. Kijlstra, W. S., Komen, N. J., Andreini, A., Poels, E. K. & Bliek, A., *Promotion and deactivation of V2O5/TiO2 SCR catalysts by SO2 at low temperature.* 11th International Congress on Catalysis - 40th Anniversary, Pts a and B, 1996. **101**: p. 951-960.
97. Zhu, Z. P., Liu, Z. Y., Liu, S. J. & Niu, H. X., *Catalytic NO reduction with ammonia at low temperatures on V2O5/AC catalysts: effect of metal oxides addition and SO2.* Applied Catalysis B: Environmental, 2001. **30**(3-4): p. 267-276.
98. Huang, Z. G., Zhu, Z. P. & Liu, Z. Y., *Combined effect of H2O and SO2 on V2O5/AC catalysts for NO reduction with ammonia at lower temperatures.* Applied Catalysis B: Environmental, 2002. **39**(4): p. 361-368.
99. Huang, Z. G., Zhu, Z. P., Liu, Z. Y. & Liu, Q. Y., *Formation and reaction of ammonium sulfate salts on V2O5/AC catalyst during selective catalytic reduction of nitric oxide by ammonia at low temperatures.* J. Catal., 2003. **214**(2): p. 213-219.
100. Zhu, Z. P., Liu, Z. Y., Liu, S. J. & Niu, H. X., *A novel carbon-supported vanadium oxide catalyst for NO reduction with NH3 at low temperatures.* Applied Catalysis B: Environmental, 1999. **23**(4): p. L229-L233.
101. Svachula, J. *et al.*, *Oxidation of SO2 to SO3 over Honeycomb DeNoxing Catalyst.* Ind. Eng. Chem. Res., 1993. **32**(5): p. 826-834.
102. Orsenigo, C., Lietti, L., Tronconi, E., Forzatti, P. & Bregani, F., *Dynamic investigation of the role of the surface sulfates in NOx reduction and SO2 oxidation over V2O5-WO3/TiO2 catalysts.* Ind. Eng. Chem. Res., 1998. **37**(6): p. 2350-2359.
103. Klimczak, M., Kern, P., Heinzelmann, T., Lucas, M. & Claus, P., *High-throughput study of the effects of inorganic additives and poisons on NH3-SCR catalysts Part I: V2O5-WO3/TiO2 catalysts.* Applied Catalysis B: Environmental, 2010. **95**(1-2): p. 39-47.
104. Ciardelli, C. *et al.*, *Reactivity of NO/NO2-NH3 SCR system for diesel exhaust aftertreatment: Identification of the reaction network as a function of temperature and NO2 feed content.* Applied Catalysis B: Environmental, 2007. **70**(1-4): p. 80-90.
105. D.E.C. Marine AB. *D.E.C. Marine AB - Diesel Emission Control, [Personal communication, April 2014].* (2 April 2014).
106. Kolle, L., Hennie, E. & Nielsen, J. B. *NOx Fund supported NOx abatement from 2008 to 2011 - User Experience.* 2012, Marintek, Norwegian Marine Technology Research Institute: Trondheim, Norway.
107. Brynolf, S., Magnusson, M., Fridell, E. & Andersson, K., *Compliance possibilities for the future ECA regulations through the use of abatement technologies or change of fuels.* Transportation Research Part D: Transport and Environment, 2014. **28**: p. 6-18.
108. Woodyard, D., *Pounder's Marine Diesel Engines* Ninth edition ed. 2009, Oxford, UK: Elsevier Ltd.
109. Æsøy, V., Einang, P. M., Stenersen, D., Hennie, E. & Valberg, I., *LNG-Fuelled Engines and Fuel Systems for Medium-Speed Engines in Maritime Applications.* SAE Technical Paper 2011-01-1998, 2011.
110. Valberg, I., Nerheim, L. M. & Hennie, E. *NOx abatement in marine sector - review of new technologies and their potential.* 2012, Marintek, Norwegian Marine Technology Research Institute: Trondheim, Norway.
111. Petzold, A. *et al.*, *Operation of marine diesel engines on biogenic fuels: modification of emissions and resulting climate effects.* Environ Sci Technol, 2011. **45**(24): p. 10394-10400.
112. Kuiken, K., *Diesel Engines Part II for ship propulsion and power plants, from 0 to 100,000 kW.* Vol. Part II. 2008, Onnen, The Netherlands: Target Global Energy Training.

113. Haraldsson, L. *General Manager Alternative NOx Solutions, Environmental Solutions - Retrofits, Wärtsilä Corporation - Ship Power, [Personal communication, June 2013].*
114. Heywood, J. B., *Internal combustion engine fundamentals.* 1988, USA: McGraw-Hill, Inc.
115. Henningsen, S., *Chapter 14 - Air Pollution from Large Two-Stroke Diesel Engines and Technologies to Control It,* in *Handbook of Air Pollution From Internal Combustion Engines*, Sher Eran, ed. 1998, Academic Press: San Diego. p. 477-534.
116. Lövblad, G. & Fridell, E. *Experiences from use of some techniques to reduce emissions from ships.* 2006: Gothenburg, Sweden.
117. Sarvi, A., Kilpinen, P. & Zevenhoven, R., *Emissions from large-scale medium-speed diesel engines: 3. Influence of direct water injection and common rail.* Fuel Processing Technology, 2009. **90**(2): p. 222-231.
118. Stone, R., *Introduction to Internal Combustion Engines.* Third Edition ed. Houndmills, Basingstoke, Hampshire and London UK. 1999: Palgrave Macmillan.
119. Haraldsson, L. *General Manager Alternative NOx Solutions, Environmental Solutions - Retrofits, Wärtsilä Corporation - Ship Power, [Personal communication, February 2014].*
120. Endresen, Ø. *et al.*, *Emission from international sea transportation and environmental impact.* Journal of Geophysical Research: Atmospheres, 2003. **108**(D17): p. n/a-n/a.
121. CONCAWE Refinery Technology Support Group (RTSG). *Impact of marine fuels quality legislation on EU refinerries at the 2020 horizon.* 2009: Brussels.
122. Kalli J. Karvonen T. Makkonen T. *Sulphur content in ships bunker fuel in 2015. A study on the impacts of the new IMO regulations on transportations costs.* 2009, Publications of the ministry of transport and communications 31/2009: Helsinki.
123. Lemper B. Hader A. Hübscher A. Maatsch S. Tasto M. *Reducing the sulphur content of shipping fuels further to 0.1 % in the North Sea and Baltic Sea in 2015: Consequences for shipping in this shipping area.* 2010, Institute of Shipping Economics and Logistics: Bremen.
124. Lloyd´s Register Marine. *Using hybrid fuels for ECA-SOx compliance. operational guidance for shipowners and operators.* 2014.
125. ISO, *Petroleum products - Fuels (class F) - Specifications of marine fuels, ISO 8217:2012.* 2012, International Organization for Standardization: Geneva.
126. Entec. *European Commission Directorate General Environment, Service Contract on Ship Emissions: Assignment, Abatement and Market-based Instruments, Task 2c-SO2 Abatement Final Report.* 2005, Entec UK Limited.
127. Lloyd´s Register. *Understanding exhaust gas treatment systems. Guidance for shipowners and operators* 2012.
128. Wärtsilä. *Exhaust Gas Scrubber Installed Onboard MT "Suula". Public test report.* 2010.
129. Hansen J. P. . *Exhaust Gas Scrubber Installed Onboard MV Ficaria Seaways. Public Test Report. Environmental Project No. 1429.* 2012, Danish Ministry of the Environment.
130. Dickson, A. G., *An exact definition of total alkalinity and a procedure for the estimation of alkalinity and total inorganic carbon from titration data.* Deep Sea Research Part A. Oceanographic Research Papers, 1981. **28**(6): p. 609-623.
131. Smith W.S., *Atmospheric Emissions From Fuel Oil Combustion . An Inventory Guide.* Environmental Health Series. Air Pollution. 1962, Cincinnati, Ohio: U.S. Department of Health, Education and Welfare.
132. Kjølholt Jesper, Aakre Stian, Jürgensen Carsten & Lauridsen Jørn. *Assesment of possible impacts of scrubber water discharges on the marine environment.* 2012: Copenhagen.
133. Karle, I.-M. & Turner, D. *Seawater Scrubbing- reduction of SOx emissions from ship exhausts.* 2007, The Alliance For Global Sustainability Gothenburg.
134. Reynolds K. J. *Exhaust Gas Cleaning Systems. Selection Guide. Prepared for Ship Operations Cooperative Program (SOCP).* 2011, U.S. Department of Transportation. The Glosten Associates.
135. IMO, *Resolution MEPC.259(68), 2015 Guidelines for exhaust gas cleaning systems. Adopted on 15 May 2015.* 2015, International Maritime Organization, Marine Environment Protection Committee: London.

136. den Boer, E. & ‘t Hoen, M. *Scrubbers - an economic and environmental assessment*. 2015, CE Delft: Delft.
137. Martens, O., M, Oldervik, O. & Terje, S., *Control of VOC emissions from crude oil tankers*. Marine Technology, 2001. **38**(3): p. 208-217.
138. Rudd, J., Howard & Hill, A., Nicholas. *Measures to Reduce Emissions of VOCs during Loading and Unloading of Ships in the EU, European Commission - Directorate General Environment*. 2001.
139. Stone, R., *Introduction to Internal Combustion Engines*. 4th edition ed. 2012, England: Palgrave Macmillan.
140. Anderson, M., Salo, K., Hallquist, Å. M. & Fridell, E., *Characterization of particles from a marine engine operating at low loads*. Atmospheric Environment, 2015. **101**: p. 65-71.
141. Anderson, M., Salo, K. & Fridell, E., *Particle- and gaseous emissions from a LNG powered ship*. Environmental Science & Technology, 2015. **49**: p.12568–12575.
142. Fridell, E. & Salo, K., *Measurements of abatement of particles and exhaust gases in a marine gas scrubber*. Proceedings of the Institution of Mechanical Engineers, Part M: Journal of Engineering for the Maritime Environment, 2014.
143. Lack, D. A. & Corbett, J. J., *Black carbon from ships: a review of the effects of ship speed, fuel quality and exhaust gas scrubbing*. Atmospheric Chemistry and Physics, 2012. **12**(9): p. 3985-4000.
144. Corbett, J., Winebrake, J. J. & Green, E. H., *An assessment of technologies for reducing regional short-lived climate forcers emitted by ships with implications for Artic shipping*. Carbon Management, 2010. **1**(2): p. 207-225.
145. Lieke, K. I. *et al.*, *Micro- and Nanostructural Characteristics of Particles Before and After an Exhaust Gas Recirculation System Scrubber*. Aerosol Science and Technology, 2013. **47** (9): p. 1038-1046.
146. Lack, D. A., Thuesen, J. & Elliot, R. *Annex to BLG 17/INF. 7: Investigation of appropriate control measures (abatement technologies) to reduce Black Carbon emissions from international shipping - Study Report*. 2012, Litehauz.
147. Lappi, M., Borkowski, T. & Myskow, J. *Ship emissions and abatement technology assessment*. 2012, VTT Technical Research Centre of Finland, Maritime University of Szczecin and Clean Shipping Currents.
148. Jayaram, V., Nigam, A., Welch, W. A., Miller, J. W. & Cocker, D. R., *Effectiveness of Emission Control Technologies for Auxiliary Engines on Ocean-Going Vessels*. J. Air Waste Manage. Assoc., 2011. **61**(1): p. 14-21.
149. Liu, Z. G., Berg, D. R., Swor, T. A. & Schauer, J. J., *Comparative Analysis on the Effect of Diesel Particulate Filter and Selective Catalytic Reduction Systems on a Wide Spectrum of Chemical Species Emissions* Environ. Sci. Technol., 2008. **42**: p. 6080-6085.
150. Hallquist, A. M., Fridell, E., Westerlund, J. & Hallquist, M., *Onboard measurements of nanoparticles from a SCR-equipped marine diesel engine*. Environ Sci Technol, 2013. **47**(2): p. 773-780.
151. Memet, F. & Stan, L., *The impact of HCFC phase-out on marine refrigeration and air conditioning*. Constanta Maritime university Analele, 2010. **11**(13): p. 107-110.
152. McMullan, J. T., *Refrigeration and the environment — issues and strategies for the future*. International Journal of Refrigeration, 2002. **25**(1): p. 89-99.
153. Calm, J. M., *The next generation of refrigerants – Historical review, considerations, and outlook*. International Journal of Refrigeration, 2008. **31**(7): p. 1123-1133.
154. Bolaji, B. O. & Huan, Z., *Ozone depletion and global warming: Case for the use of natural refrigerant – a review*. Renewable and Sustainable Energy Reviews, 2013. **18**: p. 49-54.
155. N/A, *2010 ASHRAE Handbook - Refrigeration (I-P Edition)*. American Society of Heating, Refrigerating and Air-Conditioning Engineers, Inc.
156. Weilgart, L. S., *The impacts of anthropogenic ocean noise on cetaceans and implications for management*. Canadian Journal of Zoology 2007(85): p. 1091-1116.

157. The Consortium of the SONIC project. *SONIC. Suppression of underwater noise induced by cavitation.* . [Website], http://sonic-project.eu/, last accessed on 28 August 2015.
158. The Consortium of the AQUO project. *AQUO. Achieve Quiter Oceans by Shipping noise Footprint Reduction* [website], http://www.aquo.eu/, last accessed on 28 August 2015.
159. Kalli J., Alhosalo M., Erkkilä A., Åkerström J. & Sundberg P. *Ship originated air emissions, solid waste and wastewaters- a Feasibility Study of the New Hansa Project.* 2005, Centre for maritime studies, University of Turku: Turku.
160. Khoo I-H. & Nguyen T-H. *Study of the noise pollution at container terminals and the surroundings. Final Report.* 2011, California State University.
161. Braathen, N. A., *Environmental Impacts of International Shipping: The Role of Ports.* 2011: OECD Publishing.
162. PIANC, *Considerations to reduce environmental impact of vessels: Navigation, ports, waterways.* 2008.
163. Erftemeijer, P. L. A. & Lewis, R. R. R., *Environmental impacts of dredging on seagrasses: A review.* Marine Pollution Bulletin, 2006. **52**(12): p. 1553-1572.
164. Hossain, M. & Islam, M. *Ship breaking activities and its impact on the coastal zone of Chittagong, Bangladesh.* 2006: Chittagong, Bangladesh.
165. Díez, S. *et al.*, *The Prestige oil spill. I. Biodegradation of a heavy fuel oil under simulated conditions.* Environmental Toxicology and Chemistry, 2005. **24**(9): p. 2203-2217.
166. Hassellöv, I. M., *Förstudie om vraksanering.* 2007: Alliance for Global Sustainability (AGS).
167. Michel, J. *et al. Potentially polluting wrecks in marine waters.* in *International oil spill conference.* 2005.
168. Ehler, C. *A guide to evaluate marine spatial plans.* 2014: Paris.

Part IV
Outlook

12 Improving Environmental Performance in Shipping

Selma Brynolf, J. Fredrik Lindgren, Karin Andersson, Magda Wilewska-Bien, Francesco Baldi, Lena Granhag, Hannes Johnson, Philip Linné, Erik Svensson and Maria Zetterdahl

Abstract

This book addresses the environmental issues related to shipping and the natural environment, including descriptions of and proposed solutions to the issues. Currently, challenges exist that must be addressed if shipping is to become sustainable and fulfil the zero vision of no harmful emissions to the environment. In this chapter, we evaluate the steps that have been taken (if any) to limit the various environmental issues and discuss possible steps to be taken to improve environmental performance. Furthermore, future challenges must also be addressed, e.g., the current trend of increasing ship operations in the Arctic. In general, three factors could be addressed in order to reach environmentally sustainable shipping: regulations, technical solutions, and increased environmental awareness.

S. Brynolf (✉) · J.F. Lindgren · K. Andersson · M. Wilewska-Bien · F. Baldi · L. Granhag · H. Johnson · M. Zetterdahl
Shipping and Marine Technology, Chalmers University of Technology, Gothenburg, Sweden
e-mail: selma.brynolf@chalmers.se

J.F. Lindgren
e-mail: f.lindgren@chalmers.se

K. Andersson
e-mail: karin.andersson@chalmers.se

M. Wilewska-Bien
e-mail: magda.w.bien@chalmers.se

F. Baldi
e-mail: francesco.baldi@chalmers.se

L. Granhag
e-mail: lena.granhag@chalmers.se

H. Johnson
e-mail: hannes.johnson@chalmers.se

K. Andersson et al. (eds.), *Shipping and the Environment*,
DOI 10.1007/978-3-662-49045-7_12

Keywords
Sustainability · Future challenges · Regulation · Technical solutions · Awareness · Mindset

This book addresses the environmental issues related to shipping and the natural environment, including descriptions of the issues and a discussion of proposed solutions. The chapters in the book have shown that challenges exist that must be handled if shipping is to fulfil the zero vision of no harmful emissions to the environment. Of course, shipping is not the only actor in the global transport system, and the challenges discussed are also at least partially related to other competing means of transport, such as road, rail, and air. These modes of transportation are also struggling to fulfil the goal of sustainability. Sustainability and sustainable development not only include environmental aspects but also social and economic components, which was discussed in Chap. 1. This book focuses mainly on one step (or one dimension or prerequisite, depending on one's perspective of sustainability) on the path towards sustainable development: an environmentally sustainable shipping.

The international economy is strongly intertwined with the possibility of trade between States and continents. Furthermore, States are increasingly dependent on trade with other States to supply themselves with various necessaries, such as oil, grain, or ore. As shipping activity increases with the size of the global economy, so will its impact on the environment and on human health. The transport of large amounts of goods over long distances has a well-grounded reputation as being an energy-efficient approach that minimises emissions. The congestion on land in many areas makes local and regional sea transport an interesting alternative and provides low cost infrastructure, e.g., for short-sea shipping via inland waterways or the local transport of passengers and goods in urban areas.

What are the short- and long-term challenges for shipping? What are the possible directions in which to proceed and what are the prerequisites for success in reaching the vision of environmentally sustainable shipping?

M. Zetterdahl
e-mail: maria.zetterdahl@chalmers.se

P. Linné
Department of Law, University of Gothenburg, Gothenburg, Sweden
e-mail: philip.linne@law.gu.se

E. Svensson
Researcher, Gothenburg, Sweden
e-mail: e-svensson@outlook.com

12.1 Policy Goals and Consumer Demands

The Secretary General of the International Maritime Organization (IMO) between 2004 and 2011, Mr. Efthimios E. Mitropoulos, acknowledged the importance of sustainable development at the World Maritime Day in 2009:

> Mankind is on the horns of a dilemma. For, whether we like it or not, our collective way of life has become unsustainable and we need to do something about it – and soon. The choices we have made about the way we lead our lives have been slowly eating away at the very support system that enables us to live and breathe. This cannot, and should not, go on. We need to make some tough decisions, we need to make them now and we need to act on them as one, with total and undivided commitment – today and in the future. Faced with facts we cannot argue against, we need to consider our priorities and accept that we have to make certain sacrifices; we need to start putting 'life' ahead of 'lifestyle'. [1]

Although this statement is a strong one, it has not been followed by corresponding actions. However, IMO has developed the concept of a Sustainable Maritime Transportation System (described in Sect. 1.5), which includes a set of goals and actions that highlight the importance of maritime transportation by focusing on, e.g., safety culture, environmental stewardship, energy efficiency and ocean governance. Thus, interest in future sustainable shipping has been shown by IMO's member States. Another example of governments' efforts toward environmentally sustainable shipping is found at the EU level in the form of targets on carbon dioxide (CO_2) emissions from ships set in a "white paper" in 2011:

> The environmental record of shipping can and must be improved by both technology and better fuels and operations: overall, the EU CO_2 emissions from maritime transport should be cut by 40 % (if feasible 50 %) by 2050 compared to 2005 levels. [2]

The white paper also stresses a shift from road to rail and waterborne transport. A third example is the proposal for setting a greenhouse gas emission reduction target for ship operations by the Marshall Islands submitted for the 68th session of Marine Environmental Protection Committee (MEPC) in May 2015:

> In this submission, the Marshall Islands provide the justification for and request the Committee to undertake the work necessary to establish a GHG emission reduction target for international shipping consistent with keeping global warming below 1.5°C and to agree the measures necessary to reach that target (p. 1). [3]

The proposal was discussed, although decisions and further discussions was postponed to a future MEPC session. In addition to international and national policies, there has been an increasing consumer demand for products produced in an ecological and fair trade manner. Inclusion of such values in the transport chain of a product has also shown growing interest. Different indices or eco-labels are available for sea transport, although these are often regionally based and are not applicable to other components of the transport chain. Standardised and generally accepted labels for sustainable transport chains could facilitate the development of sustainable sea transport.

12.2 The Current Situation and Future Challenges

This book has described the current environmental impacts associated with shipping. This section summarises the current situation and considers gaps between the current and future targets and goals set by governments at international or regional levels for shipping to become environmentally sustainable.

12.2.1 Discharges to the Sea

Specific discharges to the sea received attention many years ago (e.g., oil pollution) and have been regulated for a long time. Antifouling paints were recognised as a threat to the marine environment in the 1970s, although the International Convention on the Control of Harmful Anti-fouling Systems on Ships (AFS convention), banning the use of tributyltin (TBT), was not adopted until 2001 and did not enter into force until 2008 (last date for the application of organotin paints on ships was 1 January 2003). Ballast water discharges are subject to regulations that are awaiting entry into force, established in the International Convention for the Control and Management of Ships' Ballast Water and Sediments (BWM Convention). However, there are currently no rules that must be followed; there are only recommendations. Discharges from antifouling paints and ballast water are only associated with the shipping industry; anthropogenic oil spills are primarily related to the shipping industry, while other discharges occur for which shipping is only one of the contributing sectors. This situation also applies to discharges of garbage and sewage.

The discharge of *oil* to the ocean is perhaps the environmental issue that has received the most stringent regulations because this issue was the first to be recognised as an environmental problem. However, although accidents with oil tankers resulting in spills of enormous volumes of oil have decreased, large amounts of anthropogenic oil still enter the ocean every year. Operational spills from the shipping industry could be further regulated. For example, the oils used in propeller shaft bearings that are a common source of operational oil spills could be regulated such that ship owners are forced to use non-toxic and more readily biodegradable lubricating oils.

Ballast water is a major source of invasive species, which lead to immense annual monetary and environmental costs, although the BWM Convention (International Convention for the Control and Management of Ships Ballast Water and Sediments) regulating ballast water discharges has still not entered into force. The BWM Convention needs additional ratifications that correspond to 35 % of the world merchant shipping tonnage. However, the target for sufficient ratifications might soon be reached because 32.86 % (44 states) of the world merchant shipping tonnage has already accepted the BWM Convention according to IMO in September 2015. When the ballast water convention enters into force, is implemented in national law, and sufficient enforcement measures for ensuring compliance are in place, it has the potential to largely resolve problems related to invasive species in ballast water. Another important vector for transfer of invasive species are ship hull fouling. There are currently guidelines for the control and management of ships' biofouling [4]. However, to minimize the transfer of invasive aquatic species there is a further need to cover biofouling as vector in extended international regulations.

The discharge of *solid waste* from ships into the sea has been consistently limited by MARPOL 73/78 Annex V. The majority of the litter found in the sea is believed to have land-based sources. As a result of MARPOL 73/78 Annex V, discharges of solid waste from ships into the sea decreased to a large extent, although the regulations must be properly upheld by the individual flag states. Currently, the revised regulations in MARPOL 73/78 Annex V allow for the discharge of food waste, cargo residues (classified as non-harmful to the marine environment), cleaning agents and additives (classified as non-harmful to the environment) contained in wash water, and carcasses of animals. Particular restrictions are set for certain types of waste in special areas designated by IMO. Nevertheless, this material might cause local impacts on the environment in areas with dense ship traffic.

Commendable initiatives have been issued from individual States that provide equipment to fishermen for retrieving garbage, e.g., damaged or lost fishing equipment that is caught in their netting instead of throwing it back into the ocean [5]. The amounts of generated garbage should be estimated and entered (in units of cubic metres) in a garbage record book. In certain cases, obtaining precise records of amounts can be a challenge. The problem of inadequate reception facilities in ports has been recognised, and global and regional efforts have been put forth to improve the situation.

The *antifouling* paint industry was regulated via The International Convention on the Control of Harmful Anti-fouling Systems on Ships in 2001 and its subsequent implementation. This convention placed a global ban on one of the most widespread man-made toxic chemicals on our planet (TBT) and was a large step towards sustainable shipping in this area. However, the industry is currently using paint with copper and booster biocides; hence, the situation has not been fully resolved. Copper is a less harmful alternative because it is only toxic to biota in its ionic form. Discovering or developing a completely non-toxic alternative would be the most advantageous solution, one that both scientists and industry are desperately trying to find.

The contribution to *sewage* in the sea by the ship operations is small compared with that of land-based sources. However, large passenger ships represent point sources and can affect the local aquatic environment in sensitive areas, and these areas often coincide with regions that are attractive for the cruise tourism. The discharge of sewage from passenger ships has been recognised as an issue, and the Baltic Sea became the first special area under MARPOL 73/78 Annex IV: an area where the discharge of untreated sewage from passenger ships is prohibited. Originally, the ban was intended to take effect in 2016 for new passenger ships and 2018 for existing passenger ships. However, because the required conditions were not fulfilled in time (i.e., the Baltic ports should have adequate capacity to receive sewage from the ships that would be covered by the planned ban), the original dates have been postponed. Currently, 2019 (new ships) and 2021 (existing ships) are under discussion. These regulations might require expansion to other areas that are sensitive to additional eutrophication effects. Sewage can be treated on ships in sewage treatment plants, although there is no requirement to test the effluent (water discharged to the sea) after the equipment is installed. Publications show that the quality of effluent often does not meet the requirements set by MARPOL 73/78 Annex IV.

Another important issue is the lack of global regulations that cover the discharge of *grey water* to the sea (local bans on the discharge of grey water have been set by individual States). Although it is considered to be less contaminated than sewage, grey water contains components of concern.

12.2.2 Emissions to the Air

Some emissions to the air are directly regulated (e.g., sulphur oxides (SO_x), nitrogen oxide (NO_x), ozone-depleting substances, and volatile organic compounds), whereas others are indirectly regulated through regulations of other emissions (e.g., CO_2 and particulate matter). Additionally, other emissions are not regulated at all (e.g., non-CO_2 greenhouse gases, such as methane). However, the regulation of a substance does not guarantee that the problem is solved. The substance can be completely reduced, thus solving the problem. However, it can also be only marginally reduced or the regulation can have the effect of switching the emission to another type that is as hazardous to the environment as the previous one.

As emphasised in Chap. 5, SO_x emissions from ships cause considerable environmental and health problems. Figure 5.4 in Sect. 5.3 showed a steady increase of SO_x emissions from international shipping. As has been speculated, a successful implementation of the revised MARPOL 73/78 Annex VI will slow the increase and ultimately lead to declining emissions [6]. However, it has been emphasised that a global limit of 3–3.5 % sulphur content will not lead to any notable reduction in SO_x emissions [7, 8]. Any global reductions would not be observed until the global sulphur limit is below the global average sulphur content of the fuels used in ships [7], i.e., 2020 or 2025, depending on the review in 2018. The regional approach of Regulation 14 in MARPOL 73/78 Annex VI has thus resulted in SO_x Emission Control Areas (ECAs) being the current main control on SO_x emissions from ships, and this could be the situation up to 2025 [9].

The regulation of SO_x emissions from ships is still weaker than the regulation of land-based sources in primarily Europe and North America. This is especially true in comparison with transport modes on land. An example is the European directive that limits the sulphur content in road diesel to 10 ppm sulphur [10]. At the same time, the European and North American focus on this issue could shift towards a more global concern with increased attention to health impacts of air pollutants e.g.

in Asian States. A SO_x ECA in for example Hong Kong is a possible new chapter in the long debated sulphur issue for international shipping. To establish more SO_x ECAs is one possible approach to deal with health impacts from SO_x emissions from ships around the world. However, this needs to be followed by future reductions down to levels comparable with land-based emissions. Another approach could be to push for new talks in IMO for introducing a stepwise reduction of the global limit down to the current SO_x ECA limit, and to remove the SO_x ECA limit when levelled. As with the regional SO_x ECA approach, the global approach would require future reductions down to levels comparable with land-based emissions. Nonetheless, future stricter sulphur limits should be expected, although it is recommended that shipping companies take on a proactive approach and go beyond existing regulations.

The question of how ship owners and ship operators should best address these regulations remains unanswered, even though technologies are available (see Sect. 11.8.2). However, follow-up issues arise. For example, will the solutions chosen by ship owners also reduce other emissions, such as NO_x and particles? This would be the case if changing from heavy fuel oil to liquefied natural gas. Furthermore, sulphur regulations could also cause a secondary problem if many ships in a small geographical area use open-loop scrubbers. The potential environmental impacts related to the use of scrubbers, especially open-loop scrubbers, require further evaluation because the lowered pH of the released treated sea water and the release of contaminants from the fuel directly into the ocean might negatively affect ocean life.

The problem with *ozone-depleting substances* is considered solved because emissions of CFC gases were significantly halted after the Montreal Protocol. The ozone holes over Antarctica and the Arctic are decreasing. However, the emission of some ozone-depleting substances can be further reduced. Another related problem is that the substances used to replace a subset of CFC gases are also strong greenhouse gases.

Measurement campaigns suggest that *non-methane volatile organic compounds (NMVOCs)* are underestimated in the current emission inventories [11], and this could also be the case for NMVOCs from marine applications. A lack of measurements makes it difficult to estimate the potential impact of NMVOCs from marine operations and to determine if further measures are needed.

NO_x emissions contribute to various environmental issues, including acidification, eutrophication, particulate matter formation and photochemical ozone formation, and the contribution from shipping is significant. NO_x emissions have also been shown to be the dominant constituent of fuel life-cycle emissions [12]. Although NO_x emissions are regulated, this regulation can be considered weak because it will have limited impacts. The Tier III regulation will only apply in existing NO_x ECAs beginning in 2016. Could additional NO_x ECAs be a solution for further reducing NO_x emissions? Arguments exist for solving these problems with a global approach instead of creating a patchwork of regulations and regulated regional areas. However, there are also good arguments for regional regulation. In the case of ECAs, those States (or members of the EU) that are willing to move forward early with their environmental protection might have to apply for ECAs (as one of few officially accepted alternatives) instead of waiting for international action (see the arguments for international versus regional regulation in Box 3.2 in Chap. 3). In addition, another problem is that the Tier III regulations only apply to new buildings and major modifications; the date for Tier III compliance in future NO_x ECAs can be selected when applying for a NO_x ECA.

For *greenhouse gases* (GHGs) (mainly CO_2 and methane), the problem has not yet been resolved. Some argue that the contribution from the shipping industry is small, approximately 2.8 % of global emissions in 2012 [13]. However, the problem of GHGs from the shipping industry lies in the fact that the *projected* emissions are estimated to increase. Projections made for IMO have shown that existing regulation (the SEEMP and the EEDI) will be far from sufficient in curbing emission growth [14] and that it is also very difficult to find scenarios in which total emissions from shipping are reduced at all by 2050 [13]. The demand for transport by sea is expected to continue being tied to the size of the world economy—which is expected to continue growing exponentially. In other words, what the projections say is that it appears to be very difficult to increase *GHG emission efficiency* (i.e., GHG emissions per transport work) to keep pace with the rapid growth of the world's economy. Attempts to produce scenarios that lead to reduced emissions involve significant slow-steaming, increased scrapping of inefficient vessels, and changing fuels (including sails and kites), all enabled by changes in business models in the industry and foremost a price on CO_2 emissions [15].

In policy discussions, only CO_2 is discussed in the climate change context. However, *methane* could potentially pose a problem if it is used as an energy source for large-scale ship operations. If methane is used as an energy carrier in marine engines, it will be important to regulate methane emissions from marine engines and along the supply chain to limit greenhouse gas emissions.

The impacts of *particles* on human health and climate are well known, although no regulations that directly limit these emissions from ship operations have been adopted. Currently, particles are included under the sulphur requirements for marine fuels in MARPOL Annex VI, and one driving force behind the revised Regulation 14 was to reduce particle emissions. However, even if the sulphur content in marine fuels is reduced, other compounds in the fuel will still form particles during the combustion process. This regulation will mainly reduce particles that contain sulphate, which are mainly secondary particles formed after the combustion when the exhaust gases are diluted and cooled in the atmosphere. Furthermore, the regulation has resulted in that new types of low-sulphur heavy fuel oils (so-called hybrid fuels) have entered the market. How these fuels influence particle emissions is a question that has not yet been investigated, and further knowledge is required on this topic before it is possible to state that this regulation reduces particle emissions significantly and no further regulation is needed. To further regulate particle emissions, it will likely be necessary to limit the use of hybrid fuels or other low-quality fuels with a low sulphur content and instead encourage the use of fuel types with the same quality as distillates. This limitation has already been included in the local regulations of California in the United States [16].

Another aspect of particle emission regulations is the question of what *characteristics* are to be regulated. Until 2014, only particle mass was regulated for road traffic; however, with the recent version of the so-called Euro standards (i.e., Euro VI, which came into effect in 2014), even the number of particles is now regulated. Moreover, it is also important to consider the size of the emitted particles (i.e., the sizes that dominate the emissions and how particle emissions will develop/change in the atmosphere) because smaller particles (<100 nm) can reach further into the respiratory system and have more severe impacts on human health. The ideal case for the future is that the regulation of particle emissions from ship operations should be developed by considering both the particle mass and number of particles and be applied globally (i.e., not only a regional level). The other component of particle emissions, i.e., volatile particles, is difficult to measure and varies with measurement conditions. These particles should therefore not be included but still considered in evaluation and measurements.

12.2.3 Noise

Although underwater noise is recognised as an environmental stressor in some international instruments that aim to protect marine ecosystems, there is currently no globally applicable international agreement on methods for protection against underwater noise [17]. Nevertheless, IMO has adopted (voluntary) guidelines that recognise the potential short-term and long-term impacts that shipping might have on marine life and has called for guidelines to reduce noise by up to 10 dB [18]. As shipping activities intensify and additional scientific data show detrimental effects on marine biota caused by underwater anthropogenic noise, the need arises to regulate and decrease noise in the oceans. The increased concentration of shipping on specific routes in sensitive areas also requires an assessment of the impact of noise. In addition, noise from oil and gas prospecting and naval vessels that use active sonar might result in large-scale and long-range effects.

The IMO guidelines include practical measures that are suggested to reduce shipping noise. The guidelines encourage the proper maintenance of propellers, which are the main source of underwater noise, to avoid cavitation. Wake flow designed to pass through the water without turbulence around the stern of the hull is also encouraged to reduce this source of noise. These measures are easier to design into newly constructed ships, although they can also be implemented in existing ships. Furthermore, slow steaming is another way to reduce underwater noise from ships.

Noise in port areas is a nuisance that causes problems if residential areas are positioned nearby and is more common as cities become increasingly dense and areas close to waterfronts become more attractive. To avoid this problem, guidelines for reducing noise in port areas can be developed to providing port authorities with assistance in managing noise.

12.2.4 The Arctic

Because one of the effects of anthropogenic-induced climate change is an increased global temperature, the ice caps around the poles are experiencing increased melting during the warmer months of the year. This melting is expected to lead to increased shipping activity in polar regions. The decreased ice cap the Arctic has led to the possibility of shorter shipping transport routes from Europe to Asia and vice versa, which can lead to time and fuel consumption savings. The distance can be shortened by 40 % for a ship that takes the Northern Sea route from the Netherlands to Japan instead of passing through the Suez Canal. However, the total price for transport through this route will be equivalent to the former due to higher charges/fee for using the Northern Sea route, as well as the need for additional insurance. Furthermore, icebreaker assistance is necessary to use the Northern Sea route. However, the transport of goods through the Northern Sea route is not the factor that represents the largest number of ships. Fisheries represent a major portion of Arctic traffic because some areas, such as the Barents and Bering Seas, are highly productive sources of fish and crustaceans. The traffic from fishing vessels is significant and is expected to increase. Passenger traffic from cruise ships is also expected to increase as the area becomes more accessible. The last major segment contributing to increased maritime activity in the Arctic is exploration of natural resources, primarily oil and gas, which are natural resources that can be found in several locations.

Many migrating animals live in the Arctic, including mammals, fish and birds. Increased shipping traffic will increase the risk of collisions, especially in the narrow sounds through which both ships and mammals pass. The risk of oil pollution will increase due to increased traffic and oil exploration, which is related to great difficulties and dangers. The risk of transferring alien species between geographical areas via ship ballast water is also expected to increase due to the opening of trans-Arctic routes. The Arctic climate, with low temperatures and ice formation in combination with a lack of infrastructure, requires safety measures for work on ships and from an environmental perspective. For example, oil spills are highly difficult to remediate in Arctic climates and ice-covered waters because oil is mixed into or below the ice, which significantly reduces the capability of oil remediation methods, e.g., skimmers. Hence, the detrimental effects of environmental impacts in the Arctic are higher, and the remediation methods are less applicable.

The Polar Code was approved by IMO in May 2015 and is expected to enter into force 2017. The Polar Code will require that ships trading in polar regions comply with strict safety and environmental provisions. These provisions range from ship design rules, such as requirements for barriers that separate fuel tanks from the ship's outer hulls, to navigational advice on how to avoid marine mammals.

Furthermore, the Polar Code includes prohibitions on the release of oil or oily mixtures from any ship into the sea and the prevention of pollution from garbage and noxious liquid substances.

12.3 Pathways to Obtain Environmentally Sustainable Shipping

Although regulations and technical improvements are making the environmental impact from shipping less detrimental to the environment, geographical areas and/or toxic chemicals are at risk of becoming new or increased problems in the future. A summary of the current situation and future challenges shows that these different issues are related, and the need exists for a system perspective that addresses possible paths forward and the selection of solutions. The tools described in Chap. 9 could be useful in such assessments.

What pathways must be used or further expanded to achieve the goal of environmentally sustainable shipping? Next, we discuss three different tracks: increased environmental awareness, regulations and enforcement, and technical solutions. The tracks are not intended as alternatives; instead, they should be used in collaboration to increase the rate of progress towards environmentally sustainable shipping. These suggestions are not the only possible tracks; they are presented for inspiration and because we, based on our individual knowledge within the different fields covered in this book, view them as important.

12.3.1 Environmental Awareness

Environmental awareness is essential for changing the mindset of people. Increased awareness, that is, knowledge of the environmental issues with which we are confronted, and the possible outcomes of these issues are necessary for all actors in all steps of the transportation chain. This awareness should also include public and governmental bodies that are involved in the business and the consumers of the goods that are transported. One key to this process is educating the students who will soon begin their careers in the business and further education of personnel who already work within the industry.

Education plays a role in raising environmental awareness and can contribute in shaping the attitudes and behaviours needed to change our society. Preferably, environmental awareness should be incorporated into education from the beginning [19]. Furthermore, to adapt to the emerging challenges, a new set of skills must be acquired by workers within the shipping industry. In 2010, a major revision of the 1978 International Convention on Standards of Training, Certification and Watchkeeping for Seafarers (STCW) and the STCW Code took place in Manila in the Philippines to "address issues that are anticipated to emerge in the foreseeable future". The new requirements for marine environment awareness training were listed as part of these issues [20]. According to the revised STCW Code, the officers should understand the "importance of proactive measures to protect the marine environment" [21]. Although it is brief, the mandatory section on environmental issues introduced in the STCW Code is a step in the right direction from IMO. In addition, increased environmental awareness by education in shipping-related areas that are not covered by the STCW Code, as well in shipping and logistics companies, can lead to increased competitiveness of companies.

When trying to achieve environmental sustainability the shipping industry could benefit from being proactive and anticipate upcoming events. This anticipation is of special significance when accounting for the long life expectancy of vessels. Therefore, it is important to prepare in the long run for a future without fossil oil and natural gas in the shipping industry; over the short run, it is prudent to prepare for even stricter environmental regulations. The regulations on emissions to air imposed on the shipping industry are not nearly as strict as those imposed on other transportation modes and land-based industries in primarily Europe and North America. Thus, it is realistic to assume that the regulations will become even tighter in the future.

Increased awareness, a changed mindset in the industry, and a public with a will to introduce, invest in and pay for sustainable solutions could lead to a viable feedback loop of actions that increase the rate of progress towards an environmentally sustainable shipping. If the shipping industry is proactive in finding and adapting to solutions that minimise the impact on the environment, despite a lack of regulations, it could enhance the progress towards environmental sustainability. Furthermore, governmental agencies can provide support via economic incentives, lowered port dues, or financial support for exchanging or installing technologically superior alternatives to the minimally required option according to the regulations, e.g., bilge water equipment or scrubbers. Existing tools, such as the clean shipping index, are another approach that can aid customers in choosing companies with more sustainable alternatives.

In order for change to occur, increased environmental awareness should also include awareness of how society is vitally dependent on nature. For that, we need to abandon certain things we take for granted and open up ourselves for new ways of thinking. To learn more about the concepts of planetary boundaries and resilience thinking (Chap. 1) can be helpful in this regard.

12.3.2 Regulations and Enforcement

For many issues raised in this book, there is a need for stricter regulations and stricter enforcement of existing regulations in a near future in order to steer ship operations on a sustainable path. The following are merely examples of issues that should be regulated or require stricter regulations:

- The current ECA regulations do not include any quantitative emission standards on particles, although such requirements have been included in the European standards for trucks, non-road machinery and inland waterway vessels
- CO_2 and other greenhouse gases must be decreased if shipping is to fulfil the goal of matching the ambitions of other sectors.
- From an environmental point of view, there is a need to make sewage and grey water discharge regulations more strict. Currently, only local regulations forbid the discharge of sewage and greywater.
- There is still a lack of knowledge regarding noise and its environmental impacts. Regulations governing noise from shipping is still in an early phase of development, partly due to the preliminary nature of the research regarding impacts on biota in several areas. Effects on various animals have been examined; however, data on the effects of mitigating the noise and the subsequent responses of the animals are needed. In addition, information regarding the levels of noise reduction needed to reduce any effects needs to be provided to policy-makers.

One important lesson from Chap. 3 with regard to the crafting of new regulations on pollution from ships is that the way forward is through lengthy negotiations in IMO with the end result being a compromise reflecting the different interest and priorities of the member States. Several restrictions must be met for the MEPC of IMO to address new measures, e.g., a compelling need for regulation must be demonstrated. Nonetheless, Chap. 3 also underlined that the time-consuming preparatory work conducted by IMO plays a crucial role in the crafting of conventions regulating pollution from ships, and once a convention has entered into force, the MEPC has the ability to adopt amendments.

Among the several approaches that States can take to regulate pollution from ships, steps related to the enforcement of regulations (which were only mentioned briefly in this book) are certainly essential for controlling pollution from ships. As stated in Chap. 3, related to the primary responsibility of flag States to create and enforce regulation for ships, the main issue *has not been* that laws based on international standards were not adopted, but whether flag States "have taken sufficiently effective measures to ensure that they are fully implemented and complied with in practice" [22]. Given the differing interests of the world's States in enforcing regulations and that a majority of the world merchant fleet is foreign flagged, the problem of effective enforcement of international pollution standards for ships is not an easy problem to solve. One way to compensate for this weakness is to strengthen cooperation between states with equal levels of ambition, for example, in the form of port State control. In addition, another way to support regulation has been to experiment with various other methods of controlling behaviour, such as via economic incentives or different voluntary measures (see Chap. 9).

In the future, the significance of shipping is not likely to decrease, and ship operations will continue to be regulated. Preferably, issues of regulating pollution from ships are solved in fora in which States can cooperate to find constructive solutions to environmental problems. Such cooperation should continue in globally inclusive fora, particularly IMO, although cooperation at other levels should not be ruled out as unimportant. However, it is important to ensure that future prerequisites exist for reaching progress in IMO. Factors causing large conflicts between environmental and economic interests in IMO should be identified and evaluated in order to avoid postponed or strongly compromised decisions.

12.3.3 Technical Solutions

In this book, we have focused on understanding the impact of shipping on the natural environment and on addressing how to reduce this impact using various techniques in accordance with the international regulations that are in place today or will enter into force in the foreseeable future. However, what if we take a longer time perspective? Looking 50–100 years in the future, it is hoped that humanity will have reached a point at which environmental awareness will be such that only "zero-emission vessels" will be considered a sustainable solution for shipping.

12.3.3.1 Zero-Emission Vessels

In this scenario, the solutions for decreasing ship fuel consumption described in Chap. 10 will be necessary but not sufficient to completely eliminate ship emissions to the environment. Several classification societies and shipowners have begun to look into this challenge, which has resulted in a number of conceptual designs (e.g., ReVolt [23], M/S Orcelle [24], and Super Eco Ship 2030 [25]). Technical solutions already exist for a hypothetical zero-emission ship.

Despite the large variety of available alternatives for achieving the "zero-emission" goal, certain technologies and concepts are expected to become a common trend in the design of future ships, with virtually no impact on the natural environment:

- *Ultra-slow steaming*: To completely eliminate harmful emissions, future ships must require much less power for propulsion than they do today. Although improving hull and propeller designs can have profound effects, there is little doubt that future ships will need to sail at slower speeds. As discussed in Chap. 10, reducing speed by a factor of two can reduce the required propulsion power by a factor of eight (in principle). According to the ReVolt ship concept for a feeder containership estimates, an average power of only 120 kW is required to transport 100 TEU at 6 knots, which is less power needed than for a sports car. However, it should be noted that requirements for manoeuvrability will impose a minimum installed engine power. Providing sufficient installed power to be able to manoeuvre in adverse conditions while maintaining a high-energy conversion efficiency under standard operational conditions is expected to become an important challenge.
- *Renewable energy*: Wind has been used as a source of propulsion power since the dawn of humanity, although it was abandoned with the growth of

combustion engines. New technologies are available today that combine improved availability and thrust and reduced requirements for manpower and maintenance. Wind power will be a key component of future zero-emission ships, although the technique is in its infancy and suffering from a slow start at the moment. Solar panels, although they contribute to a low portion of the total energy demand, are also a factor in zero-emission ship concepts.

- *On-board energy conversion*: Although wind will play a major role in the propulsion of future ships, a power source will be necessary to ensure manoeuvrability under any conditions and the ability to propel a vessel, even if the wind is low or blows in the wrong direction. If the "zero-emission" concept is to be interpreted literally, a major shift must occur in the type of prime mover that is installed on vessels. Sulphur emissions from diesel engines can be avoided with the use of sulphur-free fuels, and CO_2 emissions can be compensated through the use of biofuels or other carbon neutral fuels. However, the possibility of completely eliminating NO_x and PM emissions is not feasible unless a different technology is used.

  - *Fuel cells*: Fuel cells produce virtually no harmful emissions and deliver notably high efficiency. Current fuel cells suffer from low energy density, although rapid development is proceeding to increase their energy density. A methanol-powered fuel cell has been tested on board a car carrier and has proven capable of enduring the harsh marine environment while delivering power at high efficiency [26].
  - *Hydrogen, biofuels and other carbon neutral fuels*: Starting from the assumption that future "zero-emission" vessels will be powered by fuel cells, shipping fuels should fulfil the requirements of suitability for use in these energy converters. The most common type of fuel cell, the proton exchange membrane (PEM), operates on hydrogen. Certain fuel cell types, typically those based on solid oxides (SOFC) and molten carbonates (MCFC), can be operated using light hydrocarbons, such as methane and methanol. A switch to these fuels as compared to hydrogen poses a much smaller barrier to the implementation of fuel cells in future ships because they are already under discussion for use in internal combustion engines. However, to maintain the "zero-emission" principle over the entire life cycle, methane and methanol should be generated on land from biomass or renewable electricity and CO_2 (see the following section on fuels).
  - *Batteries*: For short-sea shipping, batteries can play a major role. Battery technology is currently developing at a rapid pace, pushed by the development of the automotive market; batteries are expected to improve in performance and decrease in cost in the near future. Higher power, energy densities, and durability are required for the implementation of batteries on vessels, although they have only experienced limited applications. The ReVolt concept ship features a fully battery-powered propulsion system with an expected range of 100 nautical miles.

- *Unmanned vessels* might offer a significant reduction in energy requirement and also improve transport capacity. Currently, all ships require energy (for heating and power) and space (for cabins and common areas) for crew accommodations. Unmanned vessels will not have any food waste or sewage and will require no storage or holding tanks for waste. Although not crucial for the development of future zero-emission vessels, future unmanned ships will reduce costs and emissions and allow additional space to be dedicated to cargo.

- *Ballast-free, composite hull*: Most zero-emission ship design concepts focus on emissions to the air. However, these are not the only emissions from a ship to the environment; other types of impacts can be just as or more dangerous. For this reason, ballast-free designs are based on multi-hull concepts to ensure stability even with no loaded cargo on-board and will be a crucial component of future "zero-emission" ships. The use of low-weight materials, such as composite thermoplastics and aluminium, reduces a ship's lightweight, thereby increasing the payload for the same total displacement. Reductions in weight related to new materials, lighter machinery and the absence of ballast water could account for an estimated 20 % reduction in a ship's weight, contributing to a 9 % reduction in ship emissions [25]. However, it is important to consider these materials from cradle to grave, e.g., they should be intended for recycling (see Box 7.1 in Chap. 7).
- *Antifouling solutions without emissions*: To mimic surfaces that are naturally free from fouling, such as blue mussel shells [27], is a possible future antifouling solution that does not have by-products or emissions. In addition, natural substances produced by marine organisms to deter fouling are of interest for future use in antifouling paints [28]. Alternative antifouling technologies (except for paints) are designed to create an oxygen-free layer close to the hull, thereby making it hostile to many fouling organisms [29], or to perform frequent mechanical cleaning of the hull via the use of brushes or water jets. For emission-free mechanical cleaning, hard paints that are not affected by cleaning are a prerequisite in conjunction with a plan for the capture of possible invasive species.

12.3.3.2 Future Marine Fuels

What will future propulsion look like, and what types of fuels will be used? These questions are impossible to answer today, although there are many possible solutions with different advantages and drawbacks. Many fuels could contribute to more environmentally sustainable shipping in the future. However, questions remain that will require answers before we know which fuels will be used in the future. The possibility also exists that future shipping will switch radically from the use of stored energy in the form of fuels into a reliance on intermittent energy sources, such as the wind and the sun.

Historically, criteria on reliability, efficiency and cost have dominated the changes in marine fuels. These criteria might also act as important characteristics when selecting future fuels. It could also be that new characteristics will be important, such as various environmental criteria that include climate change and local and regional environmental impacts. During the twentieth century, we observed two fuel changes in shipping: one from coal to diesel and one from diesel to heavy fuel oil (HFO) (see Box 1.3 in Chap. 1). It is possible that further changes will be necessary during the twenty-first century. One possible future scenario for shipping is a transition from HFO to marine gas oil (MGO) in 2020, followed by a shift to natural-gas-based fuels. These natural gas fuels could consist of liquefied natural gas (LNG), methanol or dimethyl ether (DME).

Life cycle assessments suggest that LNG is preferable to natural gas based methanol in terms of minimising emissions [30], although this assessment might change in the future. Currently, large energy requirements are associated with methanol production from synthesis gas. However, methods used to convert methane directly to methanol without a synthesis gas step might render the routes from methane to methanol more efficient and increase the attractiveness of methanol as a fuel. Another possibility for shipping is the use of methanol with water content of approximately 10 % (crude methanol), thereby omitting the distillation step after methanol production and thus potentially saving energy and lowering costs. This omission could pose an interesting method for differentiating the methanol used in shipping from other uses but sets requirements on more corrosion resistant materials in fuel tanks and engines.

The type of natural gas used to produce such fuels as LNG, methanol and DME is also in question. The current major development in shale gas makes it a potential source in the future, although before such a development is initiated, it is also important to evaluate whether it is desirable, as shale gas extraction is associated with several environmental issues (see Box 2.6 in Chap. 2).

Even if the shipping industry shifts to natural gas fuels, another fuel shift will still be necessary for the shipping industry to reduce its climate impact. The third shift during the twenty-first century would likely be a shift to a low-carbon-emitting fuel, such as biofuels, hydrogen or electrofuels.

Glycerol is a potentially interesting biofuel. Because glycerol is produced as a by-product in the production of first-generation biofuels or fatty acid methyl ester (FAME) fuels, it is important to consider whether it could be a viable fuel in the long run, even if FAME fuels are no longer produced. Tests and evaluation of use of glycerol in marine engines are still lacking.

The biofuels produced from synthesis gas are perhaps the most promising when considering overall environmental performance [31]. The commercialisation of biomass gasification is important for the further development of the use of these fuels. It has been suggested that biomass will play an important role in the global effort to reduce greenhouse gas emissions. For example, EU Directive 2009/28/EC sets a mandatory target for all member states to use a fuel mix with a minimum of 10 % biofuels in the transport sector and an overall target of deriving 20 % of energy from renewable sources in the EU by 2020 [10]. Could this mandate also have implications for the use of biofuels in shipping? Although biofuels have been shown to have potential, their availability is limited. Additionally, biofuels are only viewed as being cost-effective in the shipping industry with a combination of tough atmospheric CO_2 concentration targets and if the yearly bioenergy supply exceeds 200 EJ [32] (see Sect. 2.6.2 for additional information on global biomass availability).

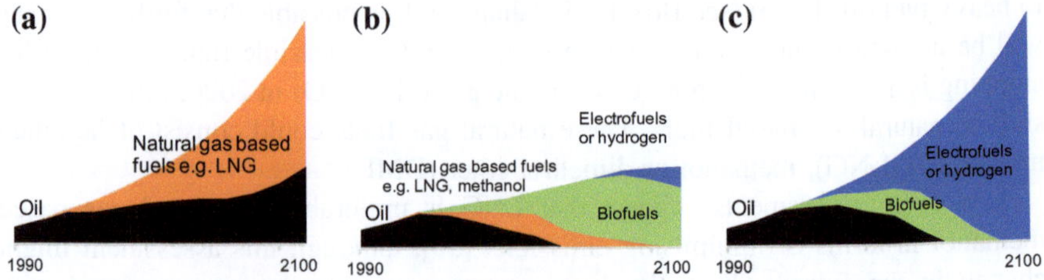

Fig. 12.1 Possible future marine fuel shifts. The **a** scenario depicts business as usual, **b** shows low growth, and **c** indicates a high-growth scenario

For a long time, hydrogen has been discussed as the future fuel for transport, although many issues remain that are related to its production, infrastructure and storage (as a few examples). The use of hydrogen could be a potential long-term method for reducing the impact on climate change if hydrogen is produced from renewable resources or from non-renewable resources in combination with CCS. Hydrogen could also be used in shipping. A design concept for a zero-emission container feeder vessel has been developed that uses liquid hydrogen as fuel to generate power in a combined fuel cell and battery system [33].

A potential future fuel category that exhibits low carbon emissions is electrofuels, and these options could be of interest in the future if large amounts of electricity are available at low costs. Electrofuels also represent an alternative method for producing methane and methanol. This process could serve as a way to store intermittent energy and simultaneously supply renewable fuels to the shipping industry and other sectors that do not easily use electricity directly.

The fuel shifts during the twenty-first century must be much more rapid than the fuel shifts that occurred during the twentieth century (which spanned approximately a half century) if shipping is to reduce its climate impact significantly. The possibility also exists for skipping the shift from HFO to MGO and instead shifting directly to natural-gas-based fuels, a shift that is already occurring. An illustration of possible future marine fuel shifts is shown in Fig. 12.1. Fuel shifts are only one component of the solution. It is also necessary to continuously increase the energy efficiency in shipping during this century.

12.4 Final Remarks

Currently, challenges exist that must be addressed if shipping is to become sustainable and fulfil the zero vision of no harmful emissions to the environment. In this chapter, we have evaluated the steps that have been taken today to limit the various environmental issues, and we have also discussed steps that could be taken towards environmental sustainability. The outcome and possibilities for shipping to contribute to sustainability are not isolated from the rest of the world; on the contrary, the decisions taken and strategies selected are strongly dependent on external conditions. The development of world economy and trade, energy supply and prices are examples of external factors that have a strong impact on the decisions taken. However, if relevant actions are taken, shipping has a large potential to develop towards becoming a strong actor as a sustainable mean of transport.

References

1. IMO. *World Maritime Day 2009. Climate change: a challenge for IMO too! A message from the Secretary-General of the International Maritime Organization, Mr. Efthimios E. Mitropoulos*. 2009.
2. European Commission, *WHITE PAPER Roadmap to a Single European Transport Area – Towards a competitive and resource efficient transport system*. 2011: Brussels.
3. IMO. *Reduction of GHG emission from ships - Setting a reduction target and agreeing associated measures for international shipping - Submitted by the Marshall Islands. Doc. MEPC 58/5/1*. 2015, International Maritime Organization: London.
4. IMO. *Resolution MEPC.207(62). 2011 Guidelines for the control and management of ships biofouling to minimize the transfer of invasive aquatic species*. 2012, International Maritime Organization: London.
5. KIMO. *Fishing for Litter Scotland, Final Report 2005 - 2008*. 2008, KIMO (Local Authorities International Environmental Organisation).
6. Klimont, Z., Smith, S. J. & Cofala, J., *The last decade of global anthropogenic sulfur dioxide: 2000–2011 emissions*. Environmental Research Letters, 2013. **8**(1): p. 014003.
7. Mestl, T., Løvoll, G., Stensrud, E. & Le Breton, A., *The Doubtful Environmental Benefit of Reduced Maximum Sulfur Limit in International Shipping Fuel*. Environmental Science & Technology, 2013. **47**(12): p. 6098-6101.
8. IMO. *A regional approach of air pollution from ships. Submitted by the Netherlands, 22 June 1993. Sub-Committee on Bulk Chemicals, 23rd session, agenda item 7. BCH 23/7/4*. 1993, International Maritime Organization: London.
9. Svensson, E. *Sulphur Regulations for Shipping—Why a Regional approach?: Scientific and Economic Arguments in IMO Documents 1988–1997*. Doctor of Philosophy thesis, Chalmers University of Technology, 2014.
10. European Commission, *Directive 2009/28/EC of the European Parliament and of the Council of 23 April 2009 on the promotion of the use of energy from renewable sources and amending and subsequently repealing Directives 2001/77/EC and 2003/30/EC (Text with EEA relevance)*. 2009: Journal of the European Union.
11. Johansson, J. K. E. *et al.*, *Emission measurements of alkenes, alkanes, SO2, and NO2 from stationary sources in Southeast Texas over a 5 year period using SOF and mobile DOAS*. Journal of Geophysical Research: Atmospheres, 2014. **119**(4): p. 1973-1991.
12. Brynolf, S. *Environmental assessment of present and future marine fuels* Doctor of Philosophy thesis, Chalmers University of Technology, 2014.
13. Smith, T. W. P. *et al. Third IMO GHG Study 2014*. 2014, International Maritime Organization (IMO): London, UK.
14. Bazari, Z. & Longva, T. *Assessment of IMO mandated energy efficiency measures for international shipping*. 2011, International Maritime Organization: London, UK.
15. Anderson, K. & Bows, A., *Executing a Scharnow turn: reconciling shipping emissions with international commitments on climate change*. Carbon Management, 2012. **3**(6): p. 615-628.
16. California Environmental Protection Agency Air Resources Board, *Fuel Sulfur and other Operational Requirements for Ocean-Going Vessels within California Waters and 24 Nautical Miles of the California Basline, 17 CCR, section 93118.2. Airborne Toxic Control Measure for Fuel Sulfur and Other Operational Requirements for Ocean-Going Vessels Within*

California Waters and 24 Nautical Miles of the California Baseline. 2011, California Environmental Protection Agency Air Resources Board: Sacramento, CA.
17. Erbe, C., *International regulation of underwater noise*. Acoustics Australia, 2013. **41**(1).
18. IMO. *Resolution MSC.337(91). Adoption of the code on noise levels on board ships*. 2012, International Maritime Organization: London.
19. OECD. *Trends Shaping Education 2014 Spotlight*. 2014, The Organisation for Economic Co-operation and Development. http://www.oecd.org/edu/ceri/Spotlight4-ThinkGreen.pdf, last accessed on 12 June 2015.
20. International Maritime Organisation. *International Convention on Standards of Training, Certification and Watchkeeping for Seafarers 1978* 2015, http://www.imo.org/OurWork/HumanElement/TrainingCertification/Pages/STCW-Convention.aspx, last accessed on 12 June 2015.
21. International Maritime Organization, *STCW. Including 2010 Manilla Amendments. STCW Convention and STCW Code*. Third consolidated edition 2011 ed. 2011, London: International Maritime Organisation.
22. De la Rue & Anderson, B. A., *Shipping and the Environment*. Second Edition ed. 2009: Informa Law.
23. Tvete, H. A. *The ReVolt - a new, innovative ship concept*. In focus - The future is hybrid, 2015.
24. WW. *Green flagship*. Wallenius Wilhelmsen. http://www.2wglobal.com/globalassets/environment/orcelle-green-flagship.pdf.
25. NYK, *NYK Super Eco Ship 2030 - our concept ship in the future*. 2010.
26. Fontell, E., *Wärtsilä marine SOFC for Wallenius car carrier*. Fuel Cells Bullettin, 2010. **2010**(6).
27. Bers, A. V. *et al.*, *Relevance of mytilid shell microtopographies for fouling defence – a global comparison*. Biofouling, 2010. **26**(3): p. 367-377.
28. Dobretsov, S., Dahms, H.-U. & Qian, P.-Y., *Inhibition of biofouling by marine microorganisms and their metabolites*. Biofouling, 2006. **22**(1): p. 43-54.
29. Lindgren, J. F., M, H., C.T, E. & Jonsson, P. R., *Oxygen-depleated surfaces: a new antifouling technology*. Biofouling, 2009. **25**(5): p. 455-461.
30. Brynolf, S., Fridell, E. & Andersson, K., *Environmental assessment of marine fuels: liquefied natural gas, liquefied biogas, methanol and bio-methanol*. Journal of cleaner production, 2014. **74**: p. 86-95.
31. Bengtsson, S., Fridell, E. & Andersson, K., *Environmental assessment of two pathways towards the use of biofuels in shipping*. Energy Policy, 2012. **44**(0): p. 451-463.
32. Taljegard, M., Brynolf, S., Grahn, M., Andersson, K. & Johnson, H., *Cost-Effective Choices of Marine Fuels in a Carbon-Constrained World: Results from a Global Energy Model*. Environmental Science & Technology, 2014. **48**(21): p. 12986-12993.
33. Sames, P. *A vision for a zero emission container feeder vessel*. in *Green Ship Technology Conference 2012*. 2012. Copenhagen: Germanisher Lloyds.

Index

K. Andersson et al. (eds.), *Shipping and the Environment*,
DOI 10.1007/978-3-662-49045-7

MIX
Papier aus verantwortungsvollen Quellen
Paper from responsible sources
FSC® C105338

If you have any concerns about our products,
you can contact us on
ProductSafety@springernature.com

In case Publisher is established outside the EU,
the EU authorized representative is:
Springer Nature Customer Service Center GmbH
Europaplatz 3, 69115 Heidelberg, Germany

Printed by Libri Plureos GmbH
in Hamburg, Germany